CRC HANDBOOK OF CHROMATOGRAPHY

Gunter Zweig and Joseph Sherma
Editors-in-Chief

GENERAL DATA AND PRINCIPLES

Editors
Gunter Zweig, Ph.D.
   U.S. Environmental Protection Agency
   Washington, D.C.
Joseph Sherma, Ph.D.
   Lafayette College
   Easton, Pennsylvania

CARBOHYDRATES

Editor
Shirley C. Churms, Ph.D.
   Research Associate
   C.S.I.R. Carbohydate Chemistry Research Unit
   Department of Organic Chemistry
   University of Cape Town, South Africa

DRUGS

Editor
Ram Gupta, Ph.D.
      Department of Laboratory Medicine
   St. Joseph's Hospital
   Hamilton, Ontario, Canada

CATECHOLAMINES
TERPENOIDS

Editor
Carmine J. Cosica, Ph.D.
   Professor of Biochemistry
   St. Louis University School of Medicine
   St. Louis, Missouri

STEROIDS

Editor
Joseph C. Touchstone, B.S., M.S., Ph.D.
   Professor
   School of Medicine
   University of Pennsylvania
   Philadelphia, Pennsylvania

PESTICIDES

Editors
Joseph Sherma, Ph.D.
   Lafayette College
   Easton, Pennsylvania
Joanne M. Follweiler, Ph.D.
   Lafayette College
   Easton, Pennsylvania

LIPIDS AND TECHNICAL LIPID DERIVATIVES

Editor
H. K. Mangold
   Executive Director and Professor
   Federal Center for Lipid Research
   4400 Munster, Germany

HYDROCARBONS

Editors
Willie E. May, Ph.D.
   Center for Analytical Chemistry
   U.S. Department of Commerce
   Washington, D.C.
Walter L. Zielinski, Jr., Ph.D.
   Air Program Manager
   National Bureau of Standards
   Washington, D.C.

INORGANICS

Editor
M. Qureshi, Ph. D.
   Professor, Chemistry Section
   Zakir Husain College of Engineering and Technology
   Aligarh Muslim University
   India

PHENOLS AND ORGANIC ACIDS

Editor
Toshihiko Hanai, Ph.D.
   University of Montreal
   Quebec, Canada

AMINO ACIDS AND AMINES

Editor
Dr. S. Blackburn
   Leeds, England

POLYMERS

Editor
Charles G. Smith
   Dow Chemical, USA
   Midland, Michigan

PLANT PIGMENTS

Editor
Dr. Hand-Peter Kost
   Botanisches Institut der Universitat Munchen
   Munchen, West Germany

CRC Handbook
of
Chromatography

Carbohydrates
Volume I

Editor

Shirley C. Churms, Ph.D.
Research Associate
C.S.I.R. Carbohydrate
Chemistry Research Unit
Department of Organic Chemistry
University of Cape Town
Republic of South Africa

Editors-in-Chief

Gunter Zweig, Ph.D.
School of Public Health
University of California
Berkeley, California

Joseph Sherma, Ph.D.
Professor of Chemistry
Chemistry Department
Lafayette College
Easton, Pennsylvania

CRC Press
Taylor & Francis Group
Boca Raton London New York

CRC Press is an imprint of the
Taylor & Francis Group, an **informa** business

CRC HANDBOOK OF CHROMATOGRAPHY

SERIES PREFACE

This *Handbook of Chromatography, Carbohydrates* by Shirley C. Churms, is one in a series of separate volumes devoted to a single class of chemical compounds or to compounds with a similar use pattern, like the prospective volumes on pesticides and drugs. When Volumes I and II of the *Handbook of Chromatography* were first published in 1972, the editors made an attempt to select the data that would cover most organic and inorganic compounds in a volume of about one thousand pages. However, during the ensuing 10 years, the literature of chromatography, especially high-performance liquid chromatography (HPLC), has grown to such an extent that, after an initial intent to update Volumes I and II, it was decided to publish separate volumes devoted to specific subjects. The present volume on the chromatography of carbohydrates is an example of the expanded handbook series. In selecting Volume Editors, the Editors-in-Chief endeavored to select scientists with extensive knowledge and expertise in the chromatography of specific compounds. The Editor of this volume, Dr. Shirley C. Churms, is renowned in the field of chromatography of carbohydrates, which is evident from the comprehensive and authoritative treatment of the subject found in this volume. We have given each Volume Editor wide latitude in designing a format that would be most useful to the reader and do justice to the particular subject being covered. Subsequent volumes of this series will include the chromatography of drugs, steroids, lipids and fatty acids, terpenoids, plant pigments, hydrocarbons, amino acids, inorganic compounds, polymers, and nucleic acids and associated compounds.

We invite readers to communicate with the Volume Editor for comments and corrections and to the Editors-in-Chief for suggestions for future volumes. The Editors-in-Chief want to thank Dr. Shirley C. Churms for her outstanding effort and the cooperation of her associates.

Gunter Zweig, Ph.D.
Joseph Sherma, Ph.D.
Spring 1981

PREFACE

This handbook is intended to serve as a working manual and reference book for carbohydrate chemists and biochemists using the chromatographic methods that are indispensable in this field. Emphasis is on newer methods, such as high-performance liquid chromatography (HPLC) and other automated liquid chromatography systems; and the material included was compiled mainly from literature published during the years 1970 to 1978. Data appearing in Volumes I and II of the *Handbook of Chromatography* are not repeated here, but references to relevant tables in Volumes I and II are given at the start of corresponding sections of this handbook. In some cases material published before 1970 that was omitted from Volumes I and II of the series is included here: this applies particularly to the sections dealing with paper chromatography and electrophoresis.

THE EDITORS-IN-CHIEF

Gunter Zweig, Ph.D., received his undergraduate and graduate training at the University of Maryland, College Park, where he was awarded the Ph.D. in biochemistry in 1952. Two years following his graduation, Dr. Zweig was affiliated with the late R. J. Block, pioneer in paper chromatography of amino acids. Zweig, Block, and Le Strange wrote one of the first books on paper chromatography which was published in 1952 by Academic Press and went into three editions, the last one authored by Gunter Zweig and Dr. Joe Sherma, the co-Editor-in-Chief of this series. *Paper Chromatography* (1952) was also translated into Russian.

From 1953 till 1957, Dr. Zweig was research biochemist at the C. F. Kettering Foundation, Antioch College, Yellow Springs, Ohio, where he pursued research on the path of carbon and sulfur in plants, using the then newly developed techniques of autoradiography and paper chromatography. From 1957 till 1965, Dr. Zweig served as lecturer and chemist, University of California, Davis and worked on analytical methods for pesticide residues, mainly by chromatographic techniques. In 1965, Dr. Zweig became Director of Life Sciences, Syracuse University Research Corporation, New York (research on environmental pollution), and in 1973 he became Chief, Environmental Fate Branch, Environmental Protection Agency (EPA) in Washington, D.C.

During his government career, Dr. Zweig continued his scientific writing and editing. Among his works are (many in collaboration with Dr. Sherma) the now 11-volume series on *Analytical Methods for Pesticides and Plant Growth Regulators* (published by Academic Press); the pesticide book series for CRC Press; co-editor of *Journal of Toxicology and Environmental Health;* co-author of basic review on paper and thin-layer chromatography for *Analytical Chemistry* from 1968 to 1980; co-author of applied chromatography review on pesticide analysis for *Analytical Chemistry*, beginning in 1981.

Among the scientific honors awarded to Dr. Zweig during his distinguished career are the Wiley Award in 1977, Rothschild Fellowship to the Weizmann Institute in 1963/64; the Bronze Medal by the EPA in 1980.

Dr. Zweig has authored or co-authored over 75 scientific papers on diverse subjects in chromatography and biochemistry, besides being the holder of three U.S. patents.

At the present time (1980/81), Dr. Zweig is Visiting Scholar in the School of Public Health, University of California, Berkeley, where he is doing research on farmworker safety as related to pesticide exposure.

Joseph Sherma, Ph.D., received a B.S. in chemistry from Upsala College, East Orange, N.J. in 1955 and a Ph.D. in analytical chemistry from Rutgers University in 1958. His thesis research in ion exchange chromatography was under the direction of the late William Rieman III. Dr. Sherma joined the faculty of Lafayette College in September 1958, and is presently full professor there in charge of two courses in analytical chemistry. At Lafayette he has continued research in chromatography and has additionally worked a total of 12 summers in the field with Harold Strain at the Argonne National Laboratory, Illinois, James Fritz at Iowa State University, Ames, Gunter Zweig at Syracuse University Research Corporation, New York, Joseph Touchstone at the Hospital of the University of Pennsylvania, Philadelphia, Brian Bidlingmeyer at Waters Associates, Framingham, Mass., and Thomas Beesley at Whatman Inc., Clifton, N.J.

Dr. Sherma and Dr. Zweig [who is now with U.S. Environmental Protection Agency (EPA)] co-authored Volumes I and II of the *CRC Handbook of Chromatography*, a book on paper chromatography, and 6 volumes of the series *Analytical Methods for Pesticides and Plant Growth Regulators*. Other books in the pesticide series and further volumes of the *CRC Handbook of Chromatography* are being edited with Dr. Zweig, and Dr. Sherma will co-author the handbook on pesticide chromatography. A book on quantitative TLC (published by Wiley-Interscience, New York) was edited jointly with Dr. Touchstone. Dr. Sherma has been co-author of 7 biennial reviews of liquid chromatography (1968 to 1980) and the 1981 review of pesticide analysis for the journal *Analytical Chemistry*. Dr. Sherma has authored major invited chapters and review papers on chromatography and pesticides in *Chromatographic Reviews* (analysis of fungicides), *Advances in Chromatography* (analysis of fungicides), *Advances in Chromatography* (analysis of nonpesticide pollutants), Heftmann's *Chromatography* (chromatography of pesticides), Race's *Laboratory Medicine* (chromatography in clinical analysis), *Food Analysis: Principles and Techniques* (TLC for food analysis), *Treatise on Analytical Chemistry* (paper and thin layer chromatography), and *CRC Critical Reviews in Analytical Chemistry* (pesticide residue analysis). A general book on thin layer chromatography co-authored by Dr. Sherma is now in press at Marcel Dekker.

Dr. Sherma spent 6 months in 1972 on sabbatical leave at the EPA Perrine Primate Laboratory, Perrine, Fla., with Dr. T. M. Shafik, and two additional summers (1975, 1976) at the U.S. Department of Agriculture (USDA) in Beltsville, Md., with Melvin Getz doing research on pesticide residue analysis methods development. He spent 3 months in 1979 on sabbatical leave with Dr. Touchstone developing clinical analytical methods. A total of more than 200 papers, books, book chapters, and oral presentations concerned with column, paper, and thin layer chromatography of metal ions, plant pigments, and other organic and biological compounds; the chromatographic analysis of pesticides; and the history of chromatography have been authored by Dr. Sherma, many in collaboration with various co-workers and students. His major research area at Lafayette is currently quantitative TLC (densitometry), applied mainly to clinical analysis and pesticide residue determinations.

Dr. Sherma has written an analytical quality control manual for pesticide analysis under contract with the U.S. EPA and has revised this and the EPA Pesticide Analytical Methods Manual under a 4-year contract (EPA) jointly with Dr. M. Beroza of the Association of Official Analytical Chemists (AOAC). Dr. Sherma has also written an instrumental analysis quality assurance manual and other analytical reports for the U.S. Consumer Product Safety Commission, and is currently preparing a manual on the analysis of food additives for the U.S. Food and Drug Administration, both of these projects also in collaboration with Dr. Beroza of the AOAC.

Dr. Sherma taught the first prototype short course on pesticide analysis with Henry Enos of the EPA for the Center for Professional Advancement. He is editor of the Kontes TLC quarterly newsletter and also teaches short courses on TLC for Kontes and the Center for Professional Advancement. He is a consultant for several industrial companies and federal agencies on chemical analysis and chromatography and regularly referees papers for analytical journals and research proposals for government agencies.

Dr. Sherma has received two awards for superior teaching at Lafayette College and the 1979 Distinguished Alumnus Award from Upsala College for outstanding achievements as an educator, researcher, author, and editor. He is a member of the American Chemical Society, Sigma Xi, Phi Lambda Upsilon, Society for Applied Spectroscopy, and the American Institute of Chemists.

THE EDITOR

Shirley C. Churms, Ph.D., is Research Associate in the Carbohydrate Chemistry Research Unit (sponsored by the South African Council for Scientific and Industrial Research) of the Department of Organic Chemistry at the University of Cape Town, Republic of South Africa.

Dr. Churms (born Macintosh) was educated at Rustenburg School, Rondebosch, Cape Town and the University of Cape Town, where she obtained the degrees of B.Sc. (with distinction in chemistry) in 1957 and B.Sc. (Hons) in 1958. She then undertook research, at the same University, on cation-exchange processes in aqueous monoethanolamine, for which the degree of Ph.D. was awarded in 1962. During this period she served as a Junior Lecturer, and in 1961 as Lecturer in a temporary capacity, in the Department of Chemistry. In 1962 she was awarded the Ohio State Fellowship by the International Federation of University Women, which enabled her to spend her postdoctoral year at the Imperial College of Science and Technology in London, England. While there she carried out an extensive survey of the properties of inorganic ion-exchangers in general and commenced an investigation of the ion-exchange properties of hydrated alumina in particular. This work was continued after her return to the University of Cape Town in 1964, and during the period 1964 to 1968 Dr. Churms was also involved in the supervision of research students who were extending her own earlier work on ion-exchange in nonaqueous solvents. Another research project, concerned with fundamental aspects of cation exchange on the high-capacity carboxylic acid resins, to which she made a major contribution resulted in the publication, in 1967, of a paper for which she was awarded, with two co-authors, the African Explosives and Chemical Industry medal conferred annually by the South African Chemical Institute.

In 1965 and 1967 Dr. Churms held temporary lecturing posts in the Department of Chemistry at the University of Cape Town, while in 1964 and 1966 her work was supported by grants administered by the South African Council for Scientific and Industrial Research. In 1968, shortly after the latter body had commenced support of a Carbohydrate Chemistry Research Unit, under the direction of Professor Alistair M. Stephen, at the University of Cape Town, she assumed her present position in this Unit.

Her involvement in carbohydrate studies resulted in a change of research interest from ion exchange *per se* to the application of chromatographic methods, and in particular gel-permeation chromatography, to the examination of polysaccharides and of the complex mixtures of products obtained in the degradation processes used for investigation of the molecular structures of polysaccharides. Dr. Churms has contributed chapters on this topic to two books already published, and a further two at present in the course of publication. She was also the author of a series of five review articles on inorganic ion-exchangers, published in 1965, and the research papers of which she has been author or co-author number approximately 40.

Dr. Churms is a Member of the Royal Society of Chemistry (London) and is thus a Chartered Chemist. She is also a member of the South African Chemical Institute and of the interdisciplinary Experimental Biology Group of the Cape, and has served as Secretary of the latter association.

CONTRIBUTORS

John E. Brewer, Ph.D.
Product Group Manager
Chromatographic Media
Pharmacia Fine Chemicals
Uppsala, Sweden

Michele Ghebregzabher, Ph.D.
Dietitian Assistant
Istituto di Clinica Pediatrica
Universitá di Perugia
Perugia, Italy

Kirsti Granath, D.Sc.
Department of Polymer Physical
  Chemistry
Pharmacia Fine Chemicals
Uppsala, Sweden

Leslie Hough, D.Sc., F.R.I.C.
Department of Chemistry
Queen Elizabeth College
London, England

Michele Lato, Ph.D.
Director
Istituto di Clinica Pediatrica
Universitá di Perugia
Perugia, Italy

Edwin H. Merrifield, Ph.D.
Research Assistant
C.S.I.R. Carbohydrate Chemistry
  Research Unit
Department of Organic Chemistry
University of Cape Town
Republic of South Africa

Frederic M. Rabel, Ph.D.
Staff Scientist
Whatman Inc.
Clifton, New Jersey

Stefano Rufini, Ph.D.
Assistant Lecturer
Istituto di Clinica Pediatrica
Universitá di Perugia
Perugia, Italy

Ramon L. Sidebotham, Ph.D.
Department of Chemistry
Queen Elizabeth College
London, England

Helmut Weigel, Ph.D.
Department of Chemistry
Royal Holloway College
Surrey, England

ACKNOWLEDGMENTS

I would like to thank the members of the Advisory Board and all other contributors for their invaluable help and advice during the compilation of this handbook. The kindness of Dr. D. Anderle (Institute of Chemistry, Slovak Academy of Sciences, Bratislava, Czechoslavakia), Professor N. K. Kochetkov (N. D. Zelinsky Institute of Organic Chemistry, USSR Academy of Sciences, Moscow) and Professor M. Torii (Research Institute for Microbial Diseases, Osaka, Japan) in making available copies of certain material from their publications not otherwise accessible to me, is appreciated.

Special thanks are due to Miss Margaret Dreyer for her sterling efforts in typing most of the manuscript of this handbook. The assistance of Ms. Eleanor Stenton and Mrs. Jean Goode in the typing is also gratefully acknowledged, as is the cooperation of the editorial staff of CRC Press.

DEDICATION

My interest in carbohydrate chemistry in general, and in carbohydrate chromatography in particular, has developed under the inspiring guidance of Professor Alistair M. Stephen, to whom this volume is dedicated.

Shirley C. Churms
April 1979

TABLE OF CONTENTS

CARBOHYDRATES

VOLUME I

SECTION I — CHROMATOGRAPHIC DATA

SECTION II — DETECTION TECHNIQUES

SECTION III — SAMPLE PREPARATION AND DERIVATIZATION

SECTION IV — PRODUCTS AND SOURCES OF CHROMATOGRAPHIC MATERIALS

SECTION V — LITERATURE REFERENCES

Section I

Chromatographic Data

Section 1

Section I.I

GAS CHROMATOGRAPHY TABLES

These tables are grouped according to the type of volatile derivative prepared for GC of the carbohydrates, with the derivatives arranged in alphabetical order. Some earlier GC data for alditol acetates, isopropylidene acetals, trifluoroacetates, and tri-methylsilyl ethers of carbohydrates will be found in Section A, Volume I, Section II.I, Tables GC 13-15, 76, and 124.

The co-operation of Dr. E. H. Merrifield (University of Cape Town) in the compilation of Tables GC1, GC2, and GC22 is gratefully acknowledged.

Table GC 1
PERACETYLATED ALDITOLS, AMINODEOXYALDITOLS, AND REDUCED OLIGOSACCHARIDES

Packing	P1	P2	P3	P4	P5
Temperature (°C)	220	T1	190	240	260
Gas; flow rate (ml/min)	N₂, 40	N₂, 34	N₂, 40	He, 100	N₂, 20
Column					
Length, cm	152	150	170	152	183
Diameter (I.D.), cm	0.6	0.3	0.6(O.D.)	0.6	0.3
Form	na	na	coiled	U-tube	na
Material	glass	glass	glass	glass	glass
Detector	FI	FI	FI	FI	FI
Reference	1	2	3	4	5
Parent sugar	r[a]	r[a]	r[a]	r[b]	r[c]
Rhamnose	—	—	0.72	—	—
Fucose	—	0.76	0.80	—	—
Ribose	0.88	—	—	—	—
Arabinose	1.00	1.00	1.00	—	—
Xylose	1.23	1.24	1.32	—	—
Mannose	2.32	2.11	2.76	—	—
Galactose	2.75	2.27	3.04	—	—
Glucose	2.68	2.43	3.24	1.00	—
2-Amino-2-deoxy-D-glucose	—	4.7	—	2.87	—
2-Amino-2-deoxy-D-galactose	—	5.1	—	3.32	—
2-Amino-2-deoxy-D-mannose	—	—	—	3.54	—
2-Amino-2-deoxy-D-gulose	—	—	—	3.59	—
2-Amino-2-deoxy-D-talose	—	—	—	2.69	—
Sucrose[d]	—	—	—	—	0.83
Trehalose[d]	—	—	—	—	1.00
Maltose	—	—	—	—	1.34
Lactose	—	—	—	—	1.65

Note: This table supplements Table GC 14 in Section A, Volume I, Section II.I.

[a] t, relative to arabinitol pentaacetate (absolute retention 6.4 min for Reference 1, 9.1 min for 2, 5 min for 3).
[b] t, relative to glucitol hexaacetate (8 min).
[c] t, relative to trehalose derivative (79 min).
[d] Peracetylated derivatives of sugars.

Packing	P1	=	5% poly(ethylene glycol adipate) on acid-washed Diatomite C (85 to 100 mesh BSS).
	P2	=	1% OV-225 on Chromosorb G-HP (80 to 100 mesh).
	P3	=	3% OV-225 on Chromosorb W, acid-washed, DMCS (80 to 100 mesh).
	P4	=	10% neopentyl glycol sebacate polyester on acid-washed Chromosorb W (80 to 100 mesh).
	P5	=	3% OV-225 on Gas-Chrom Q (100 to 120 mesh).
Temperature	T1	=	temperature programmed, 170 → 230°C, 1°/min to end temperature, 15 min isothermal.

REFERENCES

1. **Holligan, P. M. and Drew, E. A.,** Routine analysis by gas-liquid chromatography of soluble carbohydrates in extracts of plant tissues. II. Quantitative analysis of standard carbohydrates, and the separation and estimation of soluble sugars and polyols from a variety of plant tissues, *New Phytol.*, 70, 271, 1971.

Table GC 1 (continued)
PERACETYLATED ALDITOLS, AMINODEOXYALDITOLS, AND REDUCED OLIGOSACCHARIDES

2. **Metz, J., Ebert, W., and Weicker, H.,** Gas chromatographic analysis of protein and carbohydrate portions of erythrocyte membrane glycoproteins, *Clin. Chim. Acta,* 34, 31, 1971.
3. **Merrifield, E. H.,** unpublished laboratory data, 1973.
4. **Perry, M. B. and Webb, A. C.,** Analysis of 2-amino-2-deoxyhexoses by gas-liquid partition chromatography, *Can. J. Biochem.,* 46, 1163, 1968.
5. **Schwind, H., Scharbert, F., Schmidt, R., and Katterman, R.,** Gas chromatographic determination of di- and trisaccharides, *J. Clin. Chem. Clin. Biochem.,* 16, 145, 1978.

Table GC 2
ALDITOL ACETATES DERIVED FROM PARTIALLY METHYLATED SUGARS

	P1	P2	P3	P4	P5	P4	P6	P7	P8	P4	P4	P9	P10
Phase	P1	P2	P3	P4	P5	P4	P6	P7	P8	P4	P4	P9	P10
Temperature (°C)	200	190	245	170	170	120	120	—	150	180	175	180	170
Gas; flow rate (ml/min)	Ar, 20	Ar, 100	He, 100	N$_2$, 20—25	N$_2$, 15—20	He, 60	He, 60	He, 6	N$_2$, 50	na	na	N$_2$, 70—80	N$_2$, 40
Column (code)						T1	T1	T2	T3	T4	T4		
Length, cm	120	120	152	180	180	120	120	1500	150	180	180	220	170
Diameter (I.D.), cm	0.5	0.5	0.3	0.2	0.15	0.3 (O.D.)	0.3 (O.D.)	0.051	0.4	0.3	0.3	0.22	0.6 (O.D.)
Form	straight	straight	U-tube	na	na	na	na	SCOT	na	na	na	na	coiled
Material	glass	glass	glass	glass	glass	copper	copper	SS	na	glass	glass	glass	glass
Detector	D1	D1	FI	FI	FI	FI	FI	FI	FI	FI	FI	FI	FI
Reference	1, 2	2, 3	2, 4	5—7	7, 8	9	9	9	10	11	11	12	13
Parent sugar	r^{c}	r^{b}	r^{c}	r^{d}	r^{d}	r^{e}	r^{e}	r^{e}	r^{e}	r^{d}	r^{d}	r^{f}	r^{d}
L-Arabinose													
2-O-methyl	4.70	—	—	—	1.39	1.56	1.47	1.38	1.69	—	—	—	—
3-O-methyl	4.95	—	—	—	1.48	1.60	1.51	1.44	1.65	—	—	—	—
4-O-methyl	4.90	—	—	—	—	—	—	—	—	—	—	—	—
5-O-methyl	3.23	—	—	—	—	—	—	—	—	—	—	—	—
2,3-di-O-methyl	2.56	—	—	—	1.07	1.14	1.10	1.03	1.19	—	—	0.74	—
2,4-di-O-methyl	—	—	—	1.40	1.10	1.00	1.00	0.91	1.06	—	—	0.61	—
2,5-di-O-methyl	—	—	—	1.10	0.84	—	—	—	—	—	—	0.77	—
3,4-di-O-methyl	2.76	—	—	1.38	—	—	—	—	—	—	—	—	—
3,5-di-O-methyl	1.95	—	—	0.91	0.80	0.91	0.92	0.85	0.96	—	—	0.63	—
2,3,4-tri-O-methyl	1.32	—	—	0.73	0.54	0.68	0.73	0.65	0.74	—	—	0.57	—
2,3,5-tri-O-methyl	1.00	—	—	0.48	0.41	0.52	0.56	0.53	0.60	—	—	0.46	—
D-Xylose													
2-O-methyl	—	—	—	2.92	2.15	1.65	—	1.50	1.52	—	—	—	—
3-O-methyl	—	—	—	2.92	2.15	1.65	1.55	1.50	1.84	—	—	—	—
4-O-methyl	—	—	—	2.92	2.15	1.65	—	1.50	1.52	—	—	—	—
2,3-di-O-methyl	—	—	—	1.54	1.19	1.22	1.16	1.09	1.31	—	—	0.74	—
2,4-di-O-methyl	—	—	—	1.34	1.06	1.16	—	1.02	1.12	—	—	—	—
3,4-di-O-methyl	—	—	—	1.54	1.19	1.22	1.16	1.09	1.31	—	—	—	—
3,5-di-O-methyl	—	—	—	1.08	—	—	—	—	—	—	—	—	—
2,3,4-tri-O-methyl	—	—	—	0.68	0.54	0.68	0.71	0.66	0.76	—	—	0.56	—

	1	2	3	4	5	6	7	8	9	10	11	12	13
D-Ribose													
3,5-di-O-methyl	—	—	—	0.77	—	—	—	—	—	—	—	—	—
2,3,5-tri-O-methyl	—	—	—	0.40	—	—	—	—	—	—	—	—	—
D-Glucose													
2-O-methyl	—	1.60	—	7.9	6.6	—	—	—	—	—	—	—	—
3-O-methyl	—	1.81	—	9.6	7.6	—	—	—	—	—	—	—	—
4-O-methyl	—	2.00	—	11.5	8.4	—	—	—	—	—	—	—	—
6-O-methyl	—	1.14	—	5.62	5.0	—	—	—	2.45	—	—	—	3.24
2,3-di-O-methyl	—	1.00	—	5.39	4.50	2.14	1.90	2.01	2.09	—	—	—	—
2,4-di-O-methyl	—	1.00	—	5.10	4.21	2.10	1.87	1.98	1.98	—	—	—	—
2,6-di-O-methyl	—	0.75	—	3.83	3.38	1.88	1.73	1.81	1.86	—	—	—	—
3,4-di-O-methyl	—	0.97	—	5.27	4.26	—	—	—	—	—	—	—	—
3,6-di-O-methyl	—	0.82	—	4.40	3.73	1.96	1.76	1.88	2.22	—	—	—	—
4,6-di-O-methyl	—	0.80	—	4.02	3.49	1.94	1.47	1.84	—	—	—	—	—
2,3,4-tri-O-methyl	—	0.47	—	2.49	2.22	1.58	1.50	1.52	1.54	—	—	—	—
2,3,6-tri-O-methyl	—	0.48	—	2.50	2.32	1.63	1.38	1.56	1.65	—	—	—	—
2,4,6-tri-O-methyl	—	0.41	—	1.95	1.82	1.44	—	1.39	1.18	—	—	—	1.72
3,4,6-tri-O-methyl	—	0.39	—	1.98	1.83	—	—	—	1.91	—	—	—	—
2,3,4,6-tetra-O-methyl	—	0.24	—	1.00	1.00	0.98	0.99	1.00	1.00	—	—	—	1.00
D-Galactose													
2-O-methyl	—	—	—	8.1	—	—	—	—	—	—	—	2.22	—
3-O-methyl	—	—	—	11.1	—	—	—	—	—	—	—	2.52	—
4-O-methyl	—	—	—	11.1	—	—	—	—	—	—	—	2.52	—
6-O-methyl	—	—	—	5.10	—	—	—	—	—	—	—	1.75	—
2,3-di-O-methyl	—	—	—	5.68	4.7	2.19	1.95	2.04	2.12	—	—	1.83	—
2,4-di-O-methyl	—	—	—	6.35	5.1	2.27	2.03	2.09	2.28	—	—	2.19	—
2,5-di-O-methyl	—	—	—	5.8	4.65	—	—	—	—	—	—	—	—
2,6-di-O-methyl	—	—	—	3.65	3.14	1.88	1.72	1.77	2.34	—	—	1.44	—
3,4-di-O-methyl	—	—	—	6.95	5.5	2.32	2.05	2.14	2.52	—	—	—	—
3,5-di-O-methyl	—	—	—	6.35	5.1	—	—	—	—	—	—	—	—
3,6-di-O-methyl	—	—	—	4.35	—	1.99	1.81	1.87	2.56	—	—	1.57	—
4,6-di-O-methyl	—	—	—	3.64	—	—	1.67	1.70	1.57	—	—	1.54	—
2,3,4,-tri-O-methyl	—	—	—	3.44	2.89	1.82	1.48	1.52	—	—	—	1.61	—
2,3,5-tri-O-methyl	—	—	—	3.25	2.76	—	1.47	1.47	1.59	—	—	—	—
2,3,6-tri-O-methyl	—	—	—	2.42	2.22	1.59	—	—	1.56	—	—	—	—
2,4,6-tri-O-methyl	—	—	—	2.28	2.03	1.55	—	1.52	—	—	—	1.17	1.98
2,5,6-tri-O-methyl	—	—	—	2.25	1.95	—	—	—	1.82	—	—	1.35	—
3,4,6-tri-O-methyl	—	—	—	2.50	2.15	1.61	1.49	1.52	—	—	—	—	—

Table GC 2 (continued)
ALDITOL ACETATES DERIVED FROM PARTIALLY METHYLATED SUGARS

Phase	P1	P2	P3	P4	P5	P4	P6	P7	P8	P4	P4	P9	P10
Temperature (°C)	200	190	245	170	170	T1	T1	T2	T3	T4	T4	180	170
Gas; flow rate (ml/min)	Ar, 20	Ar, 100	He, 100	N_2, 20—25	N_2, 15—20	He, 60	He, 60	He, 6	N_2, 50	na	na	N_2, 70—80	N_2, 40
Column													
Length, cm	120	120	152	180	180	120	120	1500	150	180	180	220	170
Diameter (I.D.), cm	0.5	0.5	0.3	0.2	0.15	0.3 (O.D.)	0.3 (O.D.)	0.051	0.4	0.3	0.3	0.22	0.6 (O.D.)
Form	straight	straight	U-tube	na	na	na	na	SCOT	na	na	na	na	coiled glass
Material	glass	glass	glass	glass	glass	copper	copper	SS	na	glass	glass	glass	glass
Detector	DI	DI	FI	FI	FI	FI	FI	FI	FI	FI	FI	FI	FI
Reference	1, 2	2, 3	2, 4	5—7	7, 8	9	9	9	10	11	11	12	13
Parent sugar	r^a	r^b	r^c	r^d	r^e	r^f	r^g	r^h	r^i	r^j	r^k	r^l	r^m
2,3,4,6-tetra-O-methyl	—	—	—	1.25	1.19	1.14	1.13	1.10	1.16	—	—	1.00	1.14
2,3,5,6-tetra-O-methyl	—	—	—	1.15	1.10	—	—	—	—	—	—	—	—
D-Mannose													
2-O-methyl	—	—	—	7.0	5.64	—	—	—	—	3.44	7.26	—	—
3-O-methyl	—	—	—	8.8	6.8	—	—	—	—	4.25	9.58	—	—
4-O-methyl	—	—	—	8.8	6.8	—	—	—	—	3.44	7.26	—	—
5-O-methyl	—	—	—	—	—	—	—	—	—	—	—	—	—
6-O-methyl	—	—	—	4.48	4.1	—	—	—	—	2.39	4.45	—	3.66
2,3-di-O-methyl	—	—	—	4.83	3.69	2.06	—	—	—	2.52	4.80	—	—
2,4-di-O-methyl	—	—	—	5.44	4.51	—	—	—	—	—	—	—	—
2,5-di-O-methyl	—	—	—	—	—	—	—	—	—	2.09	3.66	—	—
2,6-di-O-methyl	—	—	—	3.35	—	—	1.67	—	2.08	1.96	3.35	—	—
3,4-di-O-methyl	—	—	—	5.37	4.36	—	1.87	—	2.09	—	—	—	—
3,5-di-O-methyl	—	—	—	5.44	4.51	—	—	—	—	2.71	5.30	—	—
3,6-di-O-methyl	—	—	—	4.15	3.67	—	—	—	—	2.25	4.12	—	—
4,6-di-O-methyl	—	—	—	3.29	2.92	—	—	—	—	—	—	—	—
2,3,4-tri-O-methyl	—	—	—	2.48	2.19	1.59	1.49	—	1.62	—	—	—	2.69
2,3,5-tri-O-methyl	—	—	—	—	—	—	—	—	—	1.74	2.78	—	—
2,3,6-tri-O-methyl	—	—	—	2.20	2.03	1.53	1.44	—	1.40	1.50	2.17	—	—
2,4,6-tri-O-methyl	—	—	—	2.09	1.90	1.47	1.40	—	1.45	—	—	—	1.89
2,5,6-tri-O-methyl	—	—	—	—	—	—	—	—	—	1.35	1.79	—	—

Compound	1	2	3	4	5	6	7	8	9	10	11
3,4,6-tri-O-methyl	1.64	—	—	—	—	—	1.37	1.44	1.82	1.95	—
2,3,4,6-tetra-O-methyl	0.98	—	—	—	1.00	1.00	1.00	1.00	0.99	1.00	—
2,3,5,6-tetra-O-methyl	—	—	1.02	1.03	—	—	—	—	—	—	—
L-Fucose	—	—	—	—	—	—	—	—	—	—	—
2-O-methyl	—	—	—	—	—	—	—	1.33	1.43	1.67	—
3-O-methyl	—	—	—	—	—	—	—	—	—	2.05	—
4-O-methyl	—	—	—	—	—	—	—	—	1.71	2.08	—
5-O-methyl	—	—	—	—	—	—	—	—	1.33	—	—
2,3-di-O-methyl	—	—	—	—	—	—	—	—	—	1.18	—
2,4-di-O-methyl	—	—	—	—	—	—	—	1.06	1.02	1.12	—
3,5-di-O-methyl	—	—	—	—	—	—	—	—	0.96	—	—
2,3,4-tri-O-methyl	—	—	—	—	—	0.69	0.71	0.69	0.58	0.65	—
2,3,5-tri-O-methyl	—	—	—	—	—	—	—	—	0.47	0.62	—
L-Rhamnose	—	—	—	—	—	—	—	—	—	—	—
2-O-methyl	—	0.84	—	—	—	1.21	1.18	1.26	1.37	1.52	—
3-O-methyl	—	0.96	—	—	—	1.33	1.33	1.38	1.67	1.94	—
4-O-methyl	1.33	—	—	—	—	—	—	—	1.57	1.72	—
2,3-di-O-methyl	0.83	0.78	—	—	—	0.95	0.97	0.97	0.92	0.98	—
2,4-di-O-methyl	0.89	0.72	—	—	—	0.90	0.92	0.92	0.94	0.99	—
3,4-di-O-methyl	0.77	0.55	—	—	—	0.55	0.55	0.50	0.87	0.92	—
2,3,4-tri-O-methyl	0.46	0.55	—	—	—	—	—	—	0.35	0.46	—
D-*xylo*-Hexose,3,6-dideoxy	—	—	—	—	—	—	—	—	—	—	—
2,4-di-O-methyl	—	—	—	—	—	—	—	—	—	0.32	—
D-*ribo*-Hexose,3,6-dideoxy	—	—	—	—	—	—	—	—	—	—	—
2,4-di-O-methyl	—	—	—	—	—	—	—	—	—	—	—
D-*arabino*-Hexose,3,6-dideoxy	—	—	—	—	—	—	—	—	—	—	—
2,4-di-O-methyl	—	—	—	—	—	—	—	—	—	0.34	—
2-Amino-2-deoxy-D-glucose	—	—	—	—	—	—	—	—	—	0.29	—
3-O-methyl	—	—	—	—	—	—	—	—	—	—	0.72
4-O-methyl	—	—	—	—	—	—	—	—	—	—	0.67
6-O-methyl	—	—	—	—	—	—	—	—	—	—	0.59
3,4-di-O-methyl	—	—	—	—	—	—	—	—	—	—	0.54
3,6-di-O-methyl	—	—	—	—	—	—	—	—	—	—	0.33
4,6-di-O-methyl	—	—	—	—	—	—	—	—	—	—	0.51
3,4,6-di-O-methyl	—	—	—	—	—	—	—	—	—	—	0.29

Table GC 2 (continued)
ALDITOL ACETATES DERIVED FROM PARTIALLY METHYLATED SUGARS

* t_r relative to 1,4-di-O-acetyl-2,3,5-tri-O-methyl-L-arabinitol (4.4 min).

* t_r relative to 1,4,5,6-tetra-O-acetyl-2,3-di-O-methyl-D-glucitol (29 min).

* t_r relative to 2-acetamido-1,3,4,5,6-penta-O-acetyl-2-deoxy-D-glucitol.

* t_r relative to 1,5-di-O-acetyl-2,3,4,6-tetra-O-methyl-D-glucitol (about 11 min in Reference 5, 10 min in Reference 12).

* t_r relative to 1,5-di-O-acetyl-2,3,4,6-tetra-O-methyl-D-mannitol.

* t_r relative to 1,5-di-O-acetyl-2,3,4,6-tetra-O-methyl-D-galactitol (14.8 min).

Phase
P1 = 15% LAC-4R-886 on Chromosorb W (100 to 120 mesh).

P2 = 10% LAC-4R-886 on Chromosorb W (60 to 80 mesh).

P3 = 10% neopentyl glycol sebacate polyester on acid-washed Chromosorb W (80 to 100 mesh).

P4 = 3% ECNSS-M on Gas-Chrom Q (100 to 120 mesh).

P5 = 3% OV-225 on Gas-Chrom Q (100 to 120 mesh).

P6 = mixture of 0.2% poly(ethylene glycol adipate), 0.2% poly(ethylene glycol succinate) and 0.4% silicone XF-1150 on Gas-Chrom P (100 to 120 mesh).

P7 = OV-225 SCOT capillary column (Perkin-Elmer)

P8 = mixture of 0.3% OV-275 and 0.4% XF-1150 on Gas-Chrom Q (100 to 120 mesh).

P9 = 15% Apiezon T on Gas-Chrom Q (100 to 120 mesh).

P10 = 3% OV-17 on acid-washed Chromosorb W (80 to 100 mesh).

Temperature
T1 = temperature programmed, 110 → 180°C at 1°C/min.

T2 = temperature programmed, 150 → 190°C at 0.5°C/min.

T3 = temperature 120°C for 5 min, then 120—180°C at 1°C/min.

T4 = temperature programmed, 140 → 185°C at 1.5°C/min.

Detector
D1 = β-ionization (⁹⁰Sr).

REFERENCES

1. Williams, S. C. and Jones, J. K. N., The synthesis, separation and identification of the methyl ethers of arabinose and their derivatives, *Can. J. Chem.*, 45, 275, 1967.

2. Jones, H. G., Gas-liquid chromatography of methylated sugars, in *Methods in Carbohydrate Chemistry*, Vol. 6, Whistler, R. L. and BeMiller, J. N., Eds., Academic Press, New York, 1972, 25.

3. Jones, H. G. and Jones, J. K. N., Separation and identification of methyl ethers of D-glucose and D-glucitol by gas-liquid chromatography, *Can. J. Chem.*, 47, 3269, 1969.

4. Perry, M. B. and Webb, A. C., Chromatographic analysis of methyl ethers of 2-amino-2-deoxy-D-glucopyranose, *Can. J. Chem.*, 47, 4091, 1969.

5. Björndal, H., Lindberg, B., and Svensson, S., Gas-liquid chromatography of partially methylated alditols as their acetates, *Acta Chem. Scand.*, 21, 1801, 1967.

6. Björndal, H., Hellerqvist, C. G., Lindberg, B., and Svensson, S., Gas-liquid chromatography and mass spectrometry in methylation analysis of polysaccharides, *Angew. Chem. Int. Ed. Engl.*, 9, 610, 1970.

7. Jansson, P. E., Kenne, L., Liedgren, H., Lindberg, B., and Lönngren, J., A practical guide to the methylation analysis of carbohydrates, *Chem. Commun. Univ. Stockholm*, 8, 1976.

8. Lönngren, J. and Pilotti, Å., Gas-liquid chromatography of partially methylated alditols as their acetates. II, *Acta Chem. Scand.*, 25, 1144, 1971.

9. Talmadge, K. W., Keegstra, K., Bauer, W. D., and Albersheim, P., The structure of plant cell walls. I. The macromolecular components of the walls of suspension-cultured sycamore cells with a detailed analysis of the pectic polysaccharides, *Plant Physiol.*, 51, 158, 1973.

10. Darvill, A. G., Roberts, D. P., and Hall, M. A., Improved gas-liquid chromatographic separation of methylated acetylated alditols, *J. Chromatogr.*, 115, 319, 1975.

11. Woolard, G. R. and Rathbone, E. B., Gas-liquid chromatography of methylated α-D-mannosides and their O-methyl ethers. I. Acetylated methyl glycosides and alditols derived from methyl α-D-mannofuranoside, *S. Afr. J. Chem.*, 30, 69, 1977.

12. Parolis, H. and McGarvie, D., G.l.c. of partially methylated alditol acetates, *Carbohydr. Res.*, 62, 363, 1978.

13. Merrifield, E. H., laboratory data, 1979.

Table GC 3
ALDITOL ACETATES DERIVED FROM PARTIALLY ETHYLATED SUGARS

Phase	P1		P2		P3		P4	
Temperature	T1		T1		T2		T3	
Gas; flow rate (ml/min)	He, 60		He, 60		He, 60		He, 60	
Column								
Length, cm	120		120		120		1520	
Diameter (O.D.), cm	0.3		0.3		0.3		0.051 (I.D.)	
Form	na		na		na		open tubular	
Material	copper		copper		copper		SS	
Detector	FI		FI		FI		FI	
Parent sugar	r^a	r^b	r^a	r^b	r^a	r^b	r^a	r^b
L-Arabinose								
2-O-ethyl	0.43	1.64	0.47	1.57	0.55	1.36	0.53	1.32
3-O-ethyl	0.45	1.73	0.49	1.62	0.57	1.42	0.55	1.36
4-O-ethyl	0.41	1.56	0.47	1.54	0.56	1.39	0.54	1.34
5-O-ethyl	0.34	1.31	0.38	1.29	0.47	1.17	0.47	1.17
2,3-di-O-ethyl	0.26	1.00	0.30	1.00	0.41	1.00	0.41	1.01
2,4-di-O-ethyl	0.27	1.04	0.32	1.04	0.41	1.00	0.42	1.03
2,5-di-O-ethyl	0.23	0.88	0.27	0.89	0.37	0.92	0.38	0.94
3,4-di-O-ethyl	0.26	1.00	0.30	1.00	0.41	1.01	0.41	1.01
3,5-di-O-ethyl	0.20	0.76	0.24	0.80	0.34	0.85	0.36	0.89
2,3,4-tri-O-ethyl	0.11	0.40	0.15	0.48	0.23	0.56	0.27	0.67
2,3,5-tri-O-ethyl	0.09	0.32	0.11	0.36	0.20	0.49	0.24	0.59
D-Xylose								
2-O-ethyl	0.48	1.83	0.51	1.68	0.60	1.49	0.57	1.45
3-O-ethyl	0.48	1.82	0.51	1.68	0.60	1.49	0.57	1.45
4-O-ethyl	0.48	1.83	0.51	1.68	0.60	1.49	0.57	1.45
2,3-di-O-ethyl	0.29	1.09	0.33	1.06	0.43	1.07	0.43	1.08
2,4-di-O-ethyl	0.27	1.00	0.31	1.00	0.41	1.00	0.40	1.01
3,4-di-O-ethyl	0.29	1.09	0.33	1.06	0.43	1.07	0.43	1.08
2,3,4-tri-O-ethyl	0.11	0.43	0.15	0.48	0.24	0.59	0.25	0.65
D-Glucose								
2-O-ethyl	0.76	2.68	0.77	2.58	0.81	1.99	0.76	1.88
3-O-ethyl	0.80	2.82	0.80	2.66	0.83	2.05	0.78	1.94
4-O-ethyl	0.82	2.94	0.82	2.76	0.87	2.14	0.81	2.02
6-O-ethyl	0.70	2.47	0.71	2.39	0.75	1.85	0.71	1.76
2,3-di-O-ethyl	0.61	2.17	0.62	2.09	0.69	1.69	0.65	1.76
2,4-di-O-ethyl	0.60	2.15	0.62	2.08	0.69	1.69	0.65	1.61
2,6-di-O-ethyl	0.55	1.94	0.57	1.90	0.63	1.56	0.61	1.52
3,4-di-O-ethyl	0.61	2.15	0.62	2.09	0.68	1.69	0.66	1.62
3,6-di-O-ethyl	0.58	2.04	0.59	1.98	0.66	1.63	0.63	1.57
4,6-di-O-ethyl	0.55	1.97	0.57	1.92	0.63	1.56	0.62	1.53
2,3,4-tri-O-ethyl	0.40	1.38	0.41	1.39	0.48	1.20	0.50	1.24
2,3,6-tri-O-ethyl	0.42	1.49	0.44	1.48	0.53	1.30	0.52	1.29
2,4,6-tri-O-ethyl	0.36	1.28	0.39	1.29	0.45	1.11	0.48	1.18
3,4,6-tri-O-ethyl	0.36	1.28	0.39	1.29	0.46	1.13	0.48	1.18
2,3,4,6-tetra-O-ethyl	0.19	0.69	0.22	0.75	0.30	0.75	0.35	0.85
D-Galactose								
2-O-ethyl	0.75	2.79	0.76	2.67	0.79	1.94	0.74	1.85
3-O-ethyl	0.81	2.98	0.82	2.81	0.85	2.09	0.80	1.98
4-O-ethyl	0.82	2.82	0.83	2.82	0.86	2.13	—	—
6-O-ethyl	0.66	2.34	0.68	2.35	0.71	1.73	0.69	1.69
2,3-di-O-ethyl	0.62	2.31	0.63	2.23	0.69	1.71	0.65	1.63

Table GC 3 (continued)
ALDITOL ACETATES DERIVED FROM PARTIALLY ETHYLATED SUGARS

Phase	P1		P2		P3		P4	
Temperature	T1		T1		T2		T3	
Gas; flow rate (ml/min)	He, 60		He, 60		He, 60		He, 60	
Column								
Length, cm	120		120		120		1520	
Diameter (O.D.), cm	0.3		0.3		0.3		0.051 (I.D.)	
Form	na		na		na		open tubular	
Material	copper		copper		copper		SS	
Detector	FI		FI		FI		FI	
Parent sugar	r^a	r^b	r^a	r^b	r^a	r^b	r^a	r^b
2,4-di-O-ethyl	0.63	2.32	0.67	2.29	0.70	1.70	0.68	1.67
2,6-di-O-ethyl	0.53	1.94	0.55	1.90	0.62	1.51	0.61	1.53
3,4-di-O-ethyl	0.63	2.32	0.67	2.26	0.71	1.72	0.68	1.68
3,6-di-O-ethyl	0.57	2.10	0.59	2.03	0.65	1.66	0.63	1.55
4,6-di-O-ethyl	0.53	1.92	0.56	1.90	0.62	1.50	0.60	1.48
2,3,4-tri-O-ethyl	0.44	1.66	0.47	1.58	0.54	1.32	0.55	1.36
2,3,6-tri-O-ethyl	0.39	1.41	0.42	1.43	0.49	1.21	0.50	1.24
2,4,6-tri-O-ethyl	0.40	1.43	0.42	1.42	0.50	1.22	0.50	1.23
3,4,6-tri-O-ethyl	0.40	1.43	0.42	1.42	0.50	1.22	0.50	1.24
2,3,4,6-tetra-O-ethyl	0.21	0.78	0.25	0.84	0.33	0.81	0.36	0.90
D-Mannose								
2-O-ethyl	0.70	2.64	0.72	2.40	0.76	1.86	0.77	1.97
3-O-ethyl	0.79	2.95	0.80	2.60	0.83	2.02	0.82	2.11
4-O-ethyl	0.79	2.95	0.80	2.60	0.83	2.02	0.82	2.11
6-O-ethyl	0.62	2.31	0.65	2.15	0.69	1.70	0.69	1.79
2,3-di-O-ethyl	0.55	2.09	0.58	1.94	—	—	0.64	1.66
2,4-di-O-ethyl	0.55	2.09	0.58	1.94	—	—	0.64	1.65
2,6-di-O-ethyl	0.50	1.88	0.54	1.78	0.59	1.45	0.59	1.50
3,4-di-O-ethyl	0.60	2.24	0.62	2.03	0.67	1.65	0.68	1.74
3,6-di-O-ethyl	0.55	2.08	0.57	1.90	0.65	1.60	0.62	1.56
4,6-di-O-ethyl	0.49	1.85	0.53	1.77	0.59	1.44	0.59	1.53
2,3,4-tri-O-ethyl	0.35	1.31	0.40	1.28	0.46	1.13	0.49	1.21
2,3,6-tri-O-ethyl	0.36	1.34	0.40	1.31	0.47	1.15	0.49	1.22
2,4,6-tri-O-ethyl	0.35	1.32	0.40	1.29	0.46	1.14	0.48	1.21
3,4,6-tri-O-ethyl	0.34	1.30	0.39	1.26	0.46	1.13	0.47	1.20
2,3,4,6-tetra-O-ethyl	0.17	0.62	0.22	0.72	0.29	0.70	0.32	0.81
L-Fucose								
2-O-ethyl	0.35	1.24	0.39	1.23	0.45	1.11	0.46	1.15
3-O-ethyl	0.40	1.42	0.43	1.37	0.50	1.22	0.50	1.24
4-O-ethyl	0.38	1.35	0.43	1.37	0.50	1.22	0.50	1.24
2,3-di-O-ethyl	0.25	0.88	0.28	0.89	0.36	0.89	0.39	0.95
2,4-di-O-ethyl	0.25	0.88	0.29	0.91	0.36	0.89	0.39	0.95
3,4-di-O-ethyl	0.25	0.88	0.30	0.93	0.36	0.89	0.40	0.97
2,3,4-tri-O-ethyl	0.11	0.40	0.14	0.46	0.22	0.53	0.26	0.64
L-Rhamnose								
2-O-ethyl	0.34	1.18	0.36	1.18	0.44	1.06	0.45	1.13
3-O-ethyl	0.39	1.34	0.40	1.32	0.50	1.21	0.49	1.23
4-O-ethyl	0.36	1.25	0.38	1.24	0.43	1.06	0.45	1.14
2,3-di-O-ethyl	0.21	0.75	0.23	0.76	0.32	0.79	0.35	0.87
2,4-di-O-ethyl	0.22	0.77	0.25	0.81	0.33	0.81	0.36	0.90
2,3,4-tri-O-ethyl	0.07	0.28	0.09	0.30	0.16	0.38	0.21	0.53

[a] t_r relative to *myo*-inositol hexaacetate.
[b] t_r relative to 1,5-di-O-acetyl-2,3,4,6-tetra-O-methyl-D-mannitol.

Table GC 3 (continued)
ALDITOL ACETATES DERIVED FROM PARTIALLY ETHYLATED SUGARS

Phase   P1 = 3% ECNSS-M on Gas-Chrom Q (100 to 120 mesh).

      P2 = mixture of 0.2% poly(ethylene glycol succinate), 0.2% poly(ethylene glycol adipate) and 0.4% GE silicone XF-1150 on Gas-Chrom P (100 to 120 mesh).

      P3 = 2% Silar 10-C on Gas-Chrom Q (100 to 120 mesh).

      P4 = open tubular capillary column coated with mixture containing 1% poly(ethylene glycol adipate), 1% poly(ethylene glycol succinate) and 4% GE silicone XF-1150 (initially dissolved in chloroform).

Temperature T1 = temperature programmed: 110°C for 6 min after injection, then 110 → 180°C at 1°/min; isothermal at 180°C for 10 min.

      T2 = 115°C for 6 min after injection, then 115 → 235°C at 1°/min.

      T3 = 135°C for 6 min after injection, then 135 → 205°C at 1°/min; isothermal at 205°C for 20 min.

From Sweet, D. P., Albersheim, P., and Shapiro, R. H., *Carbohydr. Res.,* 40, 199, 1975. With permission.

Table GC 4
PERMETHYLATED ALDITOLS AND REDUCED OLIGOSACCHARIDES

Packing	P1	P2	P1	P2
Temperature (°C)	110	197	220	265
Gas; flow rate (mℓ/min)	N_2, 60	He, na	He, na	He, na
Column				
Length, cm	180	200	200	200
Diameter (I.D.), cm	0.45	0.35	0.2	0.35
Form	coiled	na	na	na
Material	aluminium	glass	na	glass
Detector	FI	FI	FI	FI
Reference	1	2	3	4
Parent sugar	r^a	r^b	r^c	r^e
Erythrose	0.26	—	—	—
Ribose	0.60	—	—	—
Arabinose	1.05	—	—	—
Xylose	1.00	—	—	—
Glucose	2.46	—	—	—
Mannose	2.84	—	—	—
Galactose	3.88	—	—	—
2-Deoxy-D-ribose	0.40	—	—	—
Rhamnose	1.25	—	—	—
Fucose	1.47	—	—	—
2-Deoxy-D-glucose	1.39	—	—	—
Laminaribiose	—	0.61	—	—
Cellobiose	—	0.55	—	—
Maltose	—	0.61	—	—
Gentiobiose	—	0.92	—	—
Lactose	—	0.70	—	—
Melibiose	—	1.00	—	—
β-D-GalpNAc-(1→4)-D-Gal	—	2.52^d	—	—
β-D-Galp-(1→4)-D-GlcpNAc	—	3.06^d	0.87	—
α-D-Galp-(1→3)-D-GalpNAc	—	—	0.66	—
β-D-Galp-(1→3)-D-GalpNAc	—	—	0.77	—
β-D-Galp-(1→3)-D-GlcpNAc	—	—	0.72	—
α-D-Galp-(1→6)-D-GlcpNAc	—	—	0.70	—
β-D-Galp-(1→6)-D-GlcpNAc	—	—	0.83	—
β-D-Glcp-(1→2)-β-D-Glcp-(1→2)-D-Glc	—	—	—	0.83
β-D-Glcp-(1→2)-β-D-Glcp-(1→4)-D-Gal	—	—	—	0.80
β-D-Glcp-(1→3)-β-D-Glcp-(1→3)-D-Glc	—	—	—	1.01
β-D-Glcp-(1→3)-β-D-Glcp-(1→4)-D-Glc	—	—	—	0.95
β-D-Glcp-(1→3)-β-D-Glcp-(1→6)-D-Glc	—	—	—	1.56
β-D-Glcp-(1→4)-β-D-Glcp-(1→3)-D-Glc	—	—	—	1.03
α-D-Glcp-(1→4)-α-D-Glcp-(1→4)-D-Glc	—	—	—	1.00
α-D-Glcp-(1→4)-α-D-Glcp-(1→6)-D-Glc	—	—	—	1.41
β-D-Glcp-(1→6)-β-D-Glcp-(1→3)-D-Glc	—	—	—	0.87
α-D-Glcp-(1→6)-α-D-Glcp-(1→4)-D-Glc	—	—	—	1.39
β-D-Glcp-(1→6)-β-D-Glcp-(1→6)-D-Glc	—	—	—	1.33
β-D-Galp-(1→6)-β-D-Galp-(1→3)-L-Ara	—	—	—	0.97
β-D-Galp-(1→3)-β-D-Galp-(1→4)-D-Glc	—	—	—	1.23
β-D-Galp-(1→4)-β-D-Galp-(1→4)-D-Glc	—	—	—	1.04
β-D-Galp-(1→6)-β-D-Galp-(1→4)-D-Glc	—	—	—	1.39
α-D-Galp-(1→6)-α-D-Galp-(1→6)-D-Glc	—	—	—	1.69
β-D-Galp-(1→6)-β-D-Galp-(1→6)-D-Gal	—	—	—	2.42
α-D-Manp-(1→2)-α-D-Manp-(1→2)-D-Man	—	—	—	0.67
β-D-Manp-(1→4)-β-D-Manp-(1→4)-D-Man	—	—	—	1.52

Table GC 4 (continued)
PERMETHYLATED ALDITOLS AND REDUCED OLIGOSACCHARIDES

Packing	P1	P2	P1	P2
Temperature (°C)	110	197	220	265
Gas; flow rate (ml/min)	N_2, 60	He, na	He, na	He, na
Column				
Length, cm	180	200	200	200
Diameter (I.D.), cm	0.45	0.35	0.2	0.35
Form	coiled	na	na	na
Material	aluminium	glass	na	glass
Detector	FI	FI	FI	FI
Reference	1	2	3	4
Parent sugar	r^a	r^b	r^c	r^c
β-DXylp-(1→3)-β-D-Xylp-(1→4)-D-Xyl	—	—	—	0.42

$$\beta\text{-D-Glc}p$$
$$1$$
$$\downarrow \qquad\qquad — \qquad — \qquad — \qquad 1.03$$
$$6$$

β-D-Glcp-(1→3)-D-Glc

[a] t, relative to permethylated xylitol (5.7 min).
[b] t, relative to permethylated melibiitol.
[c] t, relative to permethylated maltotriitol (9 min in Reference 4).
[d] Column temperature 221°C.

Packing P1 = 3% QF-1 on Gas-Chrom Q (80 to 100 mesh).
P2 = 1% OV-22 on Supelcoport (80 to 100 mesh).

REFERENCES

1. **Whyte, J. N. C.**, Gas-liquid chromatographic analysis of permethylated alditols and aldonic acids, *J. Chromatogr.*, 87, 163, 1973.
2. **Kärkkäinen, J.**, Analysis of disaccharides as permethylated disaccharide alditols by gas-liquid chromatography-mass spectrometry, *Carbohydr. Res.*, 14, 27, 1970.
3. **Mononen, I., Finne, J., and Kärkkäinen, J.**, Analysis of permethylated hexopyranosyl-2-acetamido-2-deoxyhexitols by g.l.c.-m.s., *Carbohydr. Res.*, 60, 371, 1978.
4. **Kärkkäinen, J.**, Structural analysis of trisaccharides as permethylated trisaccharide alditols by gas-liquid chromatography-mass spectrometry, *Carbohydr. Res.*, 17, 11, 1971.

Table GC 5
ACETYLATED ALDONONITRILES DERIVED FROM NEUTRAL SUGARS AND DEAMINATED AMINODEOXYHEXOSES

Packing	P1	P2	P3	P4	P5	P6	P1
Temperature (°C)	T1	T2	185	210	210	190	T3
Gas; flow rate (mℓ/min)	N$_2$, 50	N$_2$, 40	na	na	na	N$_2$, 75	N$_2$, 24
Column							
Length, cm	100	152	150	200	150	152	274
Diameter (I.D.), cm	0.3	0.4	0.4	0.4	0.4	0.3	0.3
Form	coiled	coiled	coiled	coiled	coiled	na	na
Material	glass	na	glass	glass	glass	SS	SS
Detector	FI	FI	FI	FI	FI	FI	FI
Reference	1	2	3	3	3	4—6	7, 8
Parent sugar	r^a	r^a	r^b	r^b	r^b	r^a	r^a
Erythrose	—	—	—	—	—	0.08	—
L-Arabinose	0.59	0.73	0.23	0.24	0.28	0.33	0.70
D-Xylose	0.68	—	0.28	0.26	0.31	0.42	0.76
D-Ribose	0.52	0.69	0.17	0.20	0.23	0.27	—
D-Lyxose	—	—	0.19	0.22	0.25	0.31	—
D-Glucose	1.00	1.00	0.65	0.67	0.72	1.00	1.00
D-Galactose	1.09	1.06	0.70	0.73	0.78	1.14	1.05
D-Mannose	0.91	0.93	0.49	0.54	0.61	0.78	0.93
L-Rhamnose	0.38	0.65	0.11	0.18	0.19	0.17	0.56
L-Fucose	—	—	0.15	0.19	0.24	0.25	0.64
D-*glycero*-D-*gulo*-Heptose	1.38	—	—	—	—	—	—
D-*glycero*-D-*manno*-Heptose	1.27	—	—	—	—	—	—
2,5-Anhydro-D-mannose^c	—	—	—	—	—	0.39	0.77
2,5-Anhydro-D-talose^d	—	—	—	—	—	0.67	0.89

^{a} t, relative to peracetylated D-glucononitrile (22 min in Reference 1, 58 min in Reference 2, 47.2 min in Reference 4, 70.2 min in Reference 7).
^{b} t, relative to *myo*-inositol hexaacetate.
^{c} From deamination of 2-amino-2-deoxy-D-glucose.
^{d} From deamination of 2-amino-2-deoxy-D-galactose.

Packing	P1 = 3% poly(neopentyl glycol succinate) on Chromosorb W (60 to 80 mesh).
	P2 = 10% poly(neopentyl glycol succinate) on Gas-Chrom Q (100 to 120 mesh).
	P3 = 3% ECNSS-M on Gas-Chrom Q (100 to 120 mesh).
	P4 = 5% OV-225 on Chromosorb W AW DMCS (100 to 120 mesh).
	P5 = 3% OV-17 on Supasorb AW DMCS (100 to 120 mesh).
	P6 = 10% LAC-4R-886 on acid-washed Chromosorb W (100 to 200 mesh).
Temperature	T1 = temperature programmed, 160 → 230°C at 2°/min.
	T2 = isothermal at 160°C for 10 min, then temperature programmed, 160 → 235°C at 4°/min.
	T3 = temperature programmed, 130 → 195°C at 1°/min, then isothermal at 195°C for about 15 min.

REFERENCES

1. **Dmitriev, B. A., Backinowsky, L. V., Chizhov, O. S., Zolotarev, B. M., and Kochetkov, N. K.,** Gas-liquid chromatography and mass spectrometry of aldononitrile acetates and partially methylated aldononitrile acetates, *Carbohydr. Res.,* 19, 432, 1971.
2. **Baird, J. K., Holroyde, M. J., and Ellwood, D. C.,** Analysis of the products of Smith degradation of polysaccharides by g.l.c. of the acetylated, derived aldononitriles and alditols, *Carbohydr. Res.,* 27, 464, 1973.
3. **Morrison, I. M.,** Determination of the degree of polymerisation of oligo- and polysaccharides by gas-liquid chromatography, *Carbohydr. Res.,* 108, 361, 1975.
4. **Varma, R., Varma, R. S., and Wardi, A. H.,** Separation of aldononitrile acetates of neutral sugars by gasliquid chromatography and its application to polysaccharides, *J. Chromatogr.,* 77, 222, 1973.

Table GC 5 (continued)
ACETYLATED ALDONONITRILES DERIVED FROM NEUTRAL SUGARS AND DEAMINATED AMINODEOXYHEXOSES

5. **Varma, R., Varma, R. S., Allen, W. S., and Wardi, A. H.**, Gas chromatographic determination of neutral sugars from glycoproteins and acid mucopolysaccharides as aldononitrile acetates, *J. Chromatogr.*, 86, 205, 1973.
6. **Varma, R. S., Varma, R., Allen, W. S., and Wardi, A. H.**, A gas chromatographic method for determination of hexosamines in glycoproteins and acid mucopolysaccharides, *J. Chromatogr.*, 93, 221, 1974.
7. **Varma, R. and Varma, R. S.**, Simultaneous determination of neutral sugars and hexosamines in glycoproteins and acid mucopolysaccharides (glycosaminoglycans) by gas-liquid chromatography, *J. Chromatogr.*, 128, 45, 1976.
8. **Varma, R. and Varma, R. S.**, A simple procedure for combined gas chromatographic analysis of neutral sugars, hexosamines and alditols. Determination of degree of polymerization of oligo- and polysaccharides and chain weights of glycosaminoglycans, *J. Chromatogr.*, 139, 303, 1977.

Table GC 6
ACETYLATED ALDONONITRILES DERIVED FROM METHYLATED SUGARS

Packing	P1	P2	P3	P4	P5	P6	P6
Temperature (°C)	200	180	200	T1	T2	T3	175
Gas; flow rate (ml/min)	He, 75	N₂, 18	N₂, 30	He, 35	He, 30	N₂, 40	N₂, 40
Column							
Length, cm	120	300	300	183	183	180	180
Diameter (I.D.), cm	0.38	0.216	0.216	0.318	0.318	0.3	0.3
Form	U-tube	na	na	na	na	na	na
Material	glass	SS	SS	SS	SS	glass	glass
Detector	FI	FI	FI	FI	FI	FI	FI
Reference	1	2	2	3	3, 4	5	5
Parent sugar	r^a	r^b	r^b	r^b	r^c	r^c	r^c
D-Xylose							
2-O-methyl	1.64	—	—	—	—	—	—
3-O-methyl	2.07	—	—	—	—	—	—
4-O-methyl	1.55	—	—	—	—	—	—
2,3-di-O-methyl	1.00	—	—	—	—	—	—
2,4-di-O-methyl	0.74	—	—	—	—	—	—
2,5-di-O-methyl	0.65	—	—	—	—	—	—
3,4-di-O-methyl	1.00	—	—	—	—	—	—
3,5-di-O-methyl	0.65	—	—	—	—	—	—
2,3,4-tri-O-methyl	0.39	—	—	—	—	—	—
2,3,5-tri-O-methyl	0.34	—	—	—	—	—	—
D-Glucose							
2-O-methyl	—	4.79	5.26	—	—	—	—
3-O-methyl	—	7.16	8.85	—	—	—	—
4-O-methyl	—	6.70	6.81	—	—	—	—
6-O-methyl	—	4.62	4.63	—	—	—	—
2,3-di-O-methyl	—	4.04	5.11	2.12	3.50	—	—
2,4-di-O-methyl	—	3.34	3.28	2.60	2.84	—	—
2,6-di-O-methyl	—	2.67	2.77	—	—	—	—
3,4-di-O-methyl	—	5.13	5.31	2.06	3.58	—	—
3,6-di-O-methyl	—	3.56	4.07	—	—	—	—
4,6-di-O-methyl	—	2.93	2.89	—	—	—	—
2,3,4-tri-O-methyl	—	2.23	2.27	1.67	2.00	—	—
2,3,6-tri-O-methyl	—	2.06	2.50	1.42	2.00	—	—
2,4,6-tri-O-methyl	—	1.44	1.40	1.33	1.45	—	—
3,4,6-tri-O-methyl	—	2.23	2.13	—	1.85	—	—
2,3,4,6-tetra-O-methyl	—	1.00	1.00	1.00	0.94	—	—
D-Mannose							
2-O-methyl	—	—	—	—	—	2.06	6.53
3-O-methyl	—	—	—	—	—	2.40	9.89
4-O-methyl	—	—	—	—	—	2.34	9.53
5-O-methyl	—	—	—	—	—	1.91	4.99
6-O-methyl	—	—	—	—	—	1.74	3.61
2,3-di-O-methyl	—	—	—	—	2.55	1.82	4.30
2,4-di-O-methyl	—	—	—	—	3.16	1.89	4.95
2,5-di-O-methyl	—	—	—	—	—	1.91	4.99
2,6-di-O-methyl	—	—	—	—	2.28	1.61	2.96
3,4-di-O-methyl	—	—	—	—	3.68	2.13	7.16
3,5-di-O-methyl	—	—	—	—	—	1.91	4.99
3,6-di-O-methyl	—	—	—	—	2.85	1.82	4.30
4,6-di-O-methyl	—	—	—	—	2.45	1.66	3.16
5,6-di-O-methyl	—	—	—	—	—	1.43	2.28
2,3,4-tri-O-methyl	—	—	—	—	2.03	1.53	2.58
2,3,5-tri-O-methyl	—	—	—	—	—	1.47	2.38
2,3,6-tri-O-methyl	—	—	—	—	1.65	1.37	2.11

Table GC 6 (continued)
ACETYLATED ALDONONITRILES DERIVED FROM METHYLATED SUGARS

Packing	P1	P2	P3	P4	P5	P6	P6
Temperature (°C)	200	180	200	T1	T2	T3	175
Gas; flow rate (ml/min)	He, 75	N₂, 18	N₂, 30	He, 35	He, 30	N₂, 40	N₂, 40
Column							
Length, cm	120	300	300	183	183	180	180
Diameter (I.D.), cm	0.38	0.216	0.216	0.318	0.318	0.3	0.3
Form	U-tube	na	na	na	na	na	na
Material	glass	SS	SS	SS	SS	glass	glass
Detector	FI	FI	FI	FI	FI	FI	FI
Reference	1	2	2	3	3, 4	5	5
Parent sugar	r^a	r^b	r^b	r^b	r^c	r^c	r^c
2,4,6-tri-O-methyl	—	—	—	—	1.59	1.31	1.71
2,5,6-tri-O-methyl	—	—	—	—	—	1.29	1.65
3,4,6-tri-O-methyl	—	—	—	—	1.89	1.51	2.47
3,5,6-tri-O-methyl	—	—	—	—	—	1.54	2.65
2,3,4,6-tetra-O-methyl	—	—	—	—	1.00	1.00	1.00
2,3,5,6-tetra-O-methyl	—	—	—	—	—	0.95	0.86

[a] t_r relative to derivatives from 2,3- and 3,4-di-O-methyl-D-xylose (9 min).

[b] t_r relative to 5-O-acetyl-2,3,4,6-tetra-O-methyl-D-gluconitrile (8.03 min with P2, 6.53 min with P3).

[c] t_r relative to 5-O-acetyl-2,3,4,6-tetra-O-methyl-D-mannononitrile (6.4 min in Reference 4, about 40 min with T3 in Reference 5).

Packing
  P1 = 10% LAC-4R-886 on acid-washed Chromosorb W (100 to 120 mesh).
  P2 = 3% QF-1 on Gas-Chrom Q (100 to 120 mesh).
  P3 = 3% SP-2340 on Supelcoport (100 to 120 mesh).
  P4 = 5% Apiezon L on Gas-Chrom W (60 to 80 mesh).
  P5 = 5% butanediol succinate polyester on Supelcoport (60 to 80 mesh).
  P6 = 3% ECNSS-M on Gas-Chrom Q (100 to 120 mesh).

Temperature
  T1 = temperature programmed, 150 → 210°C at 1°/min.
  T2 = temperature programmed, 185 → 210°C at 1°/min, then held at 210°C.
  T3 = temperature programmed, 110 → 180°C at 1°/min.

REFERENCES

1. Lance, D. G. and Jones, J. K. N., Gas chromatography of derivatives of the methyl ethers of D-xylose, *Can. J. Chem.*, 45, 1995, 1967.
2. Anderle, D. and Kováč, P., Separation of O-acetyl-O-methyl-D-gluconitriles by gas chromatography, *Chem. Zvesti.*, 30, 355, 1976.
3. Seymour, F. R., Slodki, M. E., Plattner, R. D., and Jeanes, A., Six unusual dextrans: methylation structural analysis by combined g.l.c.-m.s. of per-O-acetyl-aldononitriles, *Carbohydr. Res.*, 53, 153, 1977.
4. Seymour, F. R., Plattner, R. D., and Slodki, M. E., Gas-liquid chromatography-mass spectrometry of methylated and deuteriomethylated per-O-acetyl-aldononitriles from D-mannose, *Carbohydr. Res.*, 44, 181, 1975.
5. Woolard, G. R., Rathbone, E. B., and Dercksen, A. W., Gas-liquid chromatography of methyl α-D-mannosides and their O-methyl ethers. II. Acetylated aldononitriles derived from methyl α-D-mannopyranoside and methyl α-D-mannofuranoside, *S. Afr. J. Chem.*, 30, 169, 1977.

Table GC 7
BUTANEBORONIC ESTERS AND TRIMETHYSILYLATED
BUTANEBORONATES OF SOME SUGARS AND ALDITOLS

Packing	P1		P2
Temperature (°C)	200		T1
Gas; flow rate (ml/min)	na		N_2, 45
Column			
Length, cm	180		213
Diameter (I.D.), cm	0.5		0.6
Form	U-tube		coiled
Material	glass		glass
Detector	D1		FI
Reference	1		2
Parent compound	r^a	r^b	r^b (TMS)
L-Arabinose	—	1.24	1.24
D-Xylose	—	1.39	1.36
D-Ribose	—	1.40[c]	0.60; 0.90[d]
L-Rhamnose	—	1.10;[d] 1.44	0.45; 1.43[d]
L-Fucose	—	1.07	1.07
D-Galactose	0.88	[e]	1.28
D-Mannose	0.44; 0.79	[e]	1.34
D-Glucose	0.70; 0.86;[d] 1.02	[e]	1.56
D-Fructose	—	[e]	1.21
Erythritol	—	0.73	0.44;[d] 0.47;[d] 0.69;[d] 0.78
L-Arabinitol	—	2.03[f]	0.61;[d] 0.95;[d] 0.11
Xylitol	—	2.09[f]	1.06
D-Glucitol	0.79	1.97	1.93
D-Mannitol	0.70	1.76	1.22;[d] 1.27;[d] 1.74
Galactitol	0.96	2.10	1.18; 2.07[d]

[a]   t, relative to *myo-* inositol buraneboronate (5.7 min).
[b]   t, relative to methyl palmitate.
[c]   Peak with shoulder.
[d]   Minor peak.
[e]   No peak.
[f]   Broad peak.

Packing   P1   = 3% OV-17 on Gas-Chrom Q (100 to 120 mesh).
           P2   = 1% ECNSS-M on Gas-Chrom Q (80 to 100 mesh).
Temperature T1   = temperature programmed, 115 → 190°C at 5°/min.
Detector   D1   = argon ionization detector.

REFERENCES

1.   Eisenberg, F., Cyclic butaneboronic acid esters: novel derivatives for the rapid separation of carbohydrates by gas-liquid chromatography, *Carbohydr. Res.*, 19, 135, 1971.
2.   Wood, P. J., Siddiqui, I. R., and Weisz, J., The use of butaneboronic esters in the gas-liquid chromatography of some carbohydrates, *Carbohydr. Res.*, 42, 1, 1975.

Table GC 8
CHIRAL GLYCOSIDES: RESOLUTION OF ENANTIOMERS

Phase	P1	P1	P2
Temperature	T1	T2	230°C
Gas; flow rate (ml/min)	N₂, 1 ml/min	N₂, 1 ml/min	N₂, 0.4 ml/min
Column			
Length, m	25	25	25
Diameter (I. D.), mm	0.31	0.31	0.25
Form	capillary	capillary	WCOT
Material	glass	glass	glass
Detector	Fl	Fl	Fl
Reference	1	1	2

Glycoside	TMS(−)-2-butyl		Acetylated(+)-2-octyl
Parent sugar	r^a	r^a	r^b
D-Arabinose	0.55;^{c} 0.56;^{d} 0.58;^{c} 0.67^d	0.70;^{c} 0.71;^{d} 0.73;^{c} 0.80^d	2.82; 3.21; 3.31
L-Arabinose	0.54;^{d} 0.55;^{c} 0.58;^{c} 0.70^d	0.69;^{d} 0.70;^{c} 0.73;^{c} 0.82^d	2.75; 2.93; 3.41
D-Xylose	0.80;^{c} 1.04^c	0.88;^{c} 1.02^c	2.41; 2.63; 3.11; 3.28
L-Xylose	0.79;^{c} 1.06^c	0.87;^{c} 1.03^c	2.48; 2.97; 3.02; 3.28
D-Ribose	0.62; 0.64^c	0.75; 0.76^c	2.83; 3.00; 3.26; 3.47
L-Ribose	0.65;^{c} 0.66	0.77;^{c} 0.78	3.02; 3.13; 3.36
D-Rhamnose	0.59;^{c} 0.71^c	0.72;^{c} 0.82^c	1.69; 2.21; 2.39
L-Rhamnose	0.57;^{c} 0.70^c	0.71;^{c} 0.81^c	1.69; 1.96; 2.21; 2.44
D-Fucose	0.59;^{d} 0.69;^{c} 0.72;^{c} 0.82^d	0.72;^{d} 0.79;^{c} 0.82;^{c} 0.89^d	2.05; 2.25; 2.40; 2.96
L-Fucose	0.62;^{d} 0.67;^{c} 0.78;^{c} 0.80^d	0.74;^{d} 0.78;^{c} 0.87;^{c} 0.88^d	2.15;^{e} 2.61^e
D-Glucose	1.66;^{c} 2.10^c	1.36;^{c} 1.53^c	7.24; 8.03; 10.30
L-Glucose	1.58;^{c} 2.10^c	1.32;^{c} 1.53^c	7.24; 8.22; 8.46; 8.94
D-Galactose	1.27;^{d} 1.47;^{c} 1.61^c	1.18;^{d} 1.28;^{c} 1.34^c	6.56; 8.08; 10.47
L-Galactose	1.30;^{d} 1.41;^{c} 1.70^c	1.21;^{d} 1.25;^{c} 1.38^c	7.29; 7.78; 8.81; 9.01
D-Mannose	1.24;^{c} 1.64^c	1.15;^{c} 1.34^c	6.22; 7.88; 8.40; 8.50
L-Mannose	1.18;^{c} 1.61^c	1.12;^{c} 1.32^c	6.86; 7.13; 7.91; 8.90

[a] t_r relative to TMS methyl α-D-galactopyranoside (∼ 26 min at 150°C, ∼ 10 min at 175°C, ∼ 31 min at program T2).

[b] t_r relative to 1,5-di-O-acetyl-2,3,4,6-tetra-O-methyl-D-glucitol (∼ 7 min).

[c] Pyranose (ring sizes from mass spectrometry).

[d] Furanose.

[e] Peak shape indicated more than one component.

Phase   P1 = SE-30, wall-coated.
           P2 = SP-1000 WCOT column (Varian).
Temperature T1 = isothermal; pentoses and 6-deoxyhexoses at 150°C, hexoses at 175°C.
                 T2 = temperature programmed, 135 → 200°C at 1°/min.

REFERENCES

1. Gerwig, G. J., Kamerling, J. P., and Vliegenthart, J. F. G., Determination of the D and L configuration of neutral monosaccharides by high-resolution capillary g.l.c., *Carbohydr. Res.*, 62, 349, 1978.
2. Leontein, K., Lindberg, B., and Lönngren, J., Assignment of absolute configuration of sugars by g.l.c. of their acetylated glycosides formed from chiral alcohols, *Carbohydr. Res.*, 62, 359, 1978.

Table GC 9
O-ISOPROPYLIDENE DERIVATIVES OF COMMON ALDOSES

		P1	P2
Packing		P1	P2
Temperature (°C)		T1	T2
Gas; flow rate (ml/min)		N_2, 20	N_2, 20
Column			
Length, cm		200	183
Diameter (I.D.), cm		0.22	0.15
Form		na	na
Material		SS	glass
Detector		FI	FI

Parent sugar	Main acetal	r[a]	r[a]
L-Arabinose	1,2:3,4 (β)	0.24	0.38
D-Xylose	1,2:3,5 (α)	0.39	0.49
D-Ribose	2,3	0.33	0.45
	1,5-anhydro-2,3-*O*-isopropylidene-β-D-ribofuranose	0.097	0.17
L-Rhamnose	2,3	0.72	0.78;[b] 0.82
L-Fucose	1,2:3,4 (α)	0.20	0.32
D-Glucose	1,2:5,6 (α)	0.59;[c] 0.82;[b] 0.90	0.60;[c] 0.87;[b] 0.94
D-Galactose	1,2:3,4 (α)	0.86	0.90
D-Mannose	2,3:5,6	1.00	1.00

Note: This table supplements Table GC 13 in Section A, Volume I, Section II.1.

[a]   t, relative to 2,3:5,6-di-*O*-isopropylidene-D-mannose.
[b]   Minor peak.
[c]   Trace.

Packing   P1 = 3% OV-225 on Gas-Chrom Q (80 to 100 mesh).
              P2 = 3% XE-60 on Chromosorb G AW DMCS (80 to 100 mesh).
Temperature T1 = isothermal at 110°C for 12 min, then programmed at 1.5°/min for 30 min.
              T2 = 120°C for 5 min, then programmed at 2°/min for 30 min.

From Morgenlie, S., *Carbohydr. Res.,* 41, 285, 1975. With permission.

Table GC 10
ACETATES OF SOME PARTIALLY METHYLATED SUGARS

Packing	P1	P1	P2
Temperature (°C)	170	185	T1
Gas; flow rate (ml/min)	He, 86	He, 86	He, 86
Column			
Length, cm	244	244	122
Diameter (I.D.), cm	0.48	0.48	0.64
Form	na	na	na
Material	glass	glass	glass
Detector	Fl	Fl	Fl

Methylated sugar	r[a]	r[a]	r[a]
D-Glucose			
2,3-di-O-methyl	—	6.69; 7.63	—
2,3,6-tri-O-methyl	—	2.85; 3.27	—
2,3,4,6-tetra-O-methyl	0.82; 1.00	0.92; 1.00	0.92; 1.00
D-Galactose			
2,3,4,6-tetra-O-methyl	1.16; 1.76	—	—
2,3,5,6-tetra-O-methyl	1.31	—	—
D-Mannose			
2,3,6-tri-O-methyl	—	6.0	—
2,3,4,6-tetra-O-methyl	1.56; 2.02	—	—
L-Rhamnose			
2,3-di-O-methyl	—	—	1.22; 1.35

[a] t_r relative to 1,5-di-O-acetyl-2,3,4,6-tetra-O-methyl-D-glucose, second peak (11.5 min at 170°C, 5.2 min at 185°C with P1, 30.2 min with P2 and programme T1).

Packing   P1 = 3% ECNSS-M on Gas-Chrom Q (100 to 120 mesh).
                P2 = 5% butanediol succinate polyester on Diatoport S (80 to 100 mesh).
Temperature T1 = temperature programmed, 110 → 210°C at 2°C/min.

REFERENCE

1. Bebault, G. M., Dutton, G. G. S., and Walker, R. H., Separation by gas-liquid chromatography of tetra-O-methylaldohexoses and other sugars as acetates, *Carbohydr. Res.*, 23, 430, 1972.

Table GC 11
METHYL GLYCOSIDES OF METHYLATED SUGARS

Packing	P1	P2	P3	P4	P5	P6	P7	P8	P9	P10
Temperature (°C)	175	200	155	167	200	190	112	160	175	195
Gas; flow rate (ml/min)	Ar, 80—100	Ar, 80—100	He, 60	Ar, 62	He, 25	He, 55	Ar, na	N₂, 39	N₂, 14.1	N₂, 20
Column										
Length, cm	120	120	91	120	200	365	120	183	120	300
Diameter (I.D.), cm	0.5	0.5	0.6 (O.D.)	0.5	0.4	0.6	0.5	0.6 (O.D.)	0.3 (O.D.)	0.3
Form	na	na	na	na	na	na	na	na	na	na
Material	glass	glass	glass	glass	glass	glass	glass	glass	brass	glass
Detector	D1	D1	FI	D1	FI	D2	D1	FI	FI	FI
Reference	1	1	2	3	4	5	6	7	8	9
Methyl glycoside	r^t	r^t	r^t	r^t	r^t	r^t	r^t	r^t	r^f	r^e
L-Arabinoside										
2-O-methyl	6.10	1.10; 1.45	—	—	—	—	—	—	—	—
3-O-methyl	4.47; 6.95	1.26; 1.45; 1.58	4.55; 6.57; 11.1; 14.6	—	—	—	—	—	—	—
2,3-di-O-methyl	1.56; 1.93	0.63; 0.95	1.96; 2.29; 2.54	—	—	—	—	—	—	—
2,4-di-O-methyl	2.26;[a] 2.37	1.09;[a] 1.13	2.94;[a] 3.16	—	—	—	—	—	—	—
2,5-di-O-methyl	1.89; 3.47	0.70; 1.03	2.33; 4.63	—	—	—	—	—	—	—
3,4-di-O-methyl	2.15	0.99; 1.52	2.82	—	—	—	—	—	—	—
3,5-di-O-methyl	1.08; 2.55	0.60; 0.84	1.28; 3.21	—	—	—	—	—	—	—
2,3,4-tri-O-methyl	1.04	0.83	1.26	—	—	—	—	—	—	—
2,3,5-tri-O-methyl	0.56; 0.72	0.47; 0.59	0.59; 0.79	—	—	—	—	—	—	—
D-Xyloside										
2-O-methyl	4.11; 6.23	1.01; 1.34	5.88; 9.35	—	—	—	—	3.72;[i] 8.21[i]	—	—
3-O-methyl	3.55; 5.57	0.94; 1.15	5.00; 8.12	—	—	—	—	3.06;[i] 9.43[i]	—	—
4-O-methyl	4.35	1.09	—	—	—	—	—	—	—	—
5-O-methyl	—	—	—	—	—	—	—	3.75;[i] 8.16[i]	—	—
2,3-di-O-methyl	1.50; 1.79	0.64; 0.75	1.75; 1.95; 2.20	—	—	—	—	2.31;[i] 2.64[i]	—	—
2,4-di-O-methyl	1.49; 1.97	0.73; 0.93	—	—	—	—	—	—	—	—
2,5-di-O-methyl	—	—	—	—	—	—	—	2.05;[i] 2.95[i]	—	—

3,4-di-O-methyl	1.36; 1.63	0.71; 0.76	—	—	—	—	—	1.53; 4.59	—	—
3,5-di-O-methyl	—	—	—	—	—	—	—	1.00	—	—
2,3,4-tri-O-methyl	0.46; 0.57	0.45; 0.54	0.44; 0.59	—	—	—	—	1.31; 1.45	—	—
2,3,5-tri-O-methyl	—	—	—	—	—	—	—	—	—	—
L-Rhamnoside										
3-O-methyl	3.66	1.01	4.85	—	—	—	—	—	—	—
3,4-di-O-methyl	0.73; 1.01	0.61	1.10	—	—	—	—	—	—	—
2,3,4-tri-O-methyl	0.46	0.46	0.44	—	—	—	—	—	—	—
L-Fucoside										
2-O-methyl	4.25	0.99; 1.13; 1.15; 1.54	—	—	—	—	—	—	—	—
2,3,4-tri-O-methyl	0.72	—	—	—	—	—	—	—	—	—
D-Glucoside										
3-O-methyl	—	—	—	—	16.5; 23.6	—	—	—	—	—
2,3-di-O-methyl	—	2.46; 3.22	—	—	6.86	—	—	—	—	—
2,4-di-O-methyl	—	2.30; 3.22	—	—	—	—	—	—	—	—
3,4-di-O-methyl	—	—	—	—	4.26	—	—	—	—	—
4,6-di-O-methyl	—	—	—	—	4.50	—	—	—	—	—
2,3,4-tri-O-methyl	2.59; 3.70	1.35; 1.83	—	—	2.38	—	—	—	—	—
2,3,6-tri-O-methyl	3.52; 4.78	1.71; 2.18	—	—	2.20; 2.92	—	2.36; 3.55; 4.68	—	—	—
2,4,6-tri-O-methyl	3.31; 4.88	1.64; 2.24	—	—	2.15	—	—	—	—	—
3,4,6-tri-O-methyl	3.12; 3.73	—	—	—	2.00	—	—	—	—	—
2,3,4,6-tetra-O-methyl	1.00; 1.43	1.00; 1.32	1.00; 1.52	—	0.74; 1.00	—	1.00; 1.36	—	—	—
D-Galactoside										
2-O-methyl	—	—	—	—	—	—	—	—	—	2.70
3-O-methyl	—	—	—	—	—	—	—	—	—	2.70
4-O-methyl	—	—	—	—	—	—	—	—	—	2.84
6-O-methyl	—	—	—	—	—	—	—	—	—	2.55
2,3-di-O-methyl	2.46; 3.65; 4.20	—	14.6; 18.3; 19.7; 25.3	6.51; 7.72	—	—	—	—	—	2.24
2,4-di-O-methyl	3.72; 4.40	—	25.5; 30.0	—	—	—	—	—	—	2.30
2,6-di-O-methyl	2.51; 3.21; 3.77	—	14.1; 17.3; 20.1; 26.9	4.89; 5.63	—	—	—	—	—	2.15

Table GC 11 (continued)
METHYL GLYCOSIDES OF METHYLATED SUGARS

	P1	P2	P3	P4	P5	P6	P7	P8	P9	P10
Packing										
Temperature (°C)	175	200	155	167	200	190	112	160	175	195
Gas; flow rate (mℓ/min)	Ar, 80—100	Ar, 80—100	He, 60	Ar, 62	He, 25	He, 55	Ar, na	N₂, 39	N₂, 14.1	N₂, 20
Column										
Length, cm	120	120	91	120	200	365	120	183	120	300
Diameter (I.D.), cm	0.5	0.5	0.6 (O.D.)	0.5	0.4	0.6	0.5	0.6 (O.D.)	0.3 (O.D.)	0.3
Form	na	na	na	na	na	na	na	na	na	na
Material	glass	glass	glass	glass	glass	glass	glass	glass	brass	glass
Detector	D1	D1	F1	D1	F1	D2	D1	F1	F1	F1
Reference	1	1	2	3	4	5	6	7	8	9
Methyl glycoside	r^t	r^t	r^t	r^t	r^t	r^t	r^t	r^t	r^t	r^t
3,4-di-O-methyl	—	—	—	10.9; 11.1	—	—	—	—	—	2.28
3,6-di-O-methyl	—	—	—	—	—	—	—	—	—	2.02
4,6-di-O-methyl	—	—	—	—	—	—	—	—	—	1.96
2,3,4-tri-O-methyl	7.50	2.62; 2.89	9.3*; 9.8	3.16; 3.64	—	—	—	—	—	1.84
2,3,5-tri-O-methyl		1.61; 2.23; 2.49	3.90; 4.77; 5.48; 6.20	2.17; 2.23	—	—	—	—	—	—
2,3,6-tri-O-methyl	3.21; 4.30; 4.70		5.12; 6.08	1.70; 1.80; 2.18; 2.30	—	—	—	—	—	1.51
2,4,6-tri-O-methyl	4.17; 4.70	2.08; 2.38		2.19; 2.60	—	—	—	—	—	1.51
3,4,6-tri-O-methyl	—	—	—	1.54; 1.62	—	—	—	—	—	1.45
2,3,4,6-tetra-O-methyl	1.80	1.52*; 1.60	2.00	1.00	—	—	—	—	—	1.00
2,3,5,6-tetra-O-methyl				0.82						
D-Mannoside										
2,3-di-O-methyl	—	3.37	—	—	—	2.28	—	—	—	—
2,4-di-O-methyl	—	—	—	—	—	1.60	—	—	—	—
2,5-di-O-methyl	—	—	—	—	—	1.46	—	—	—	—
2,6-di-O-methyl	—	—	—	—	—	1.47; 1.93	—	—	—	—
3,4-di-O-methyl	—	2.26	—	—	—	1.45	—	—	—	—
3,5-di-O-methyl	—	—	—	—	—	1.74	—	—	—	—
3,6-di-O-methyl	—	—	—	—	—	1.32; 1.82	—	—	—	—
4,6-di-O-methyl	—	—	16.4	—	—	2.10	—	—	—	—

	Col 1	Col 2	Col 3	Col 4	Col 5
2,3,4-tri-O-methyl	3.11	1.65	—	0.73	—
2,3,6-tri-O-methyl	5.08	2.35	—	1.00	—
2,4,6-tri-O-methyl	—	—	—	0.82	—
2,5,6-tri-O-methyl	—	—	—	0.67	—
3,4,6-tri-O-methyl	3.08	1.71	3.74	0.74	—
3,5,6-tri-O-methyl	—	—	—	0.67	—
2,3,4,6-tetra-O-methyl	1.42	1.29	1.54	—	—
D-Fructoside					
3,4-di-O-methyl	—	1.72; 2.23; 3.02	—	—	—
1,3,4-tri-O-methyl	1.89; 2.49; 3.94; 4.43	1.10; 1.31; 1.71; 2.39	—	—	0.9; 3.9; 5.2
1,3,6-tri-O-methyl	—	—	—	—	1.70; 2.72; 3.21; 3.53; 4.00
1,4,6-tri-O-methyl	—	—	—	—	0.9; 2.5; 3.72; 4.6
3,4,6-tri-O-methyl	2.74; 4.12	1.52; 1.79	—	—	—
1,3,4,6-tetra-O-methyl	1.04; 1.26	1.01; 1.16	—	—	—
Methyl(methyla-D-glucopyranosid)uronate					
2-O-methyl	—	—	—	—	5.72
3-O-methyl	—	—	—	—	3.67
4-O-methyl	—	—	—	—	4.21
2,3-di-O-methyl	8.4; 9.3	2.47; 3.12	11.7; 15.8	—	2.25
2,4-di-O-methyl	—	—	—	—	2.37
3,4-di-O-methyl	—	—	—	—	1.81
2,3,4-tri-O-methyl	2.53; 3.24	1.77; 2.21	2.78; 3.72	—	1.00
Methyl(methyla-D-galactopyranosid)uronate					
2,3,4-tri-O-methyl	7.18	3.90; 4.26	—	—	—

Table GC 11 (continued)
METHYL GLYCOSIDES OF METHYLATED SUGARS

" t, relative to methyl 2,3,4,6-tetra-*O*-methyl-β-D-glucopyranoside.

ᵇ t, relative to methyl 2,3,4,6-tetra-*O*-methyl-β-D-galactopyranoside (6.2 min).

ᶜ t, relative to methyl 2,3,4,6-tetra-*O*-methyl-α-D-glucopyranoside.

ᵈ t, relative to methyl 2,3,6-tri-*O*-methyl-α-D-mannopyranoside (10.6 min).

ᵉ t, relative to methyl 2,3,4-tri-*O*-methyl-β-D-xylopyranoside (1.55 min).

ᶠ t, relative to methyl (methyl 2,3,4-tri-*O*-methyl-α-D-glucopyranosid) uronate (3.07 min).

ᵍ t, relative to methyl 2,3,4,6-tetra-*O*-methyl-α-D-galactopyranoside.

ʰ Shoulder (incomplete resolution).

ⁱ Furanoside.

Packing		
P1	=	15% butane 1,4-diol succinate polyester on acid-washed Celite (80 to 100 mesh).
P2	=	10% polyphenyl ether [*m*-bis(*m*-phenoxyphenoxy)benzene] on acid-washed Celite.
P3	=	14% ethylene glycol succinate polyester on Chromosorb W (80 to 100 mesh).
P4	=	5% neopentyl glycol succinate polyester on acid-washed Chromosorb W (60 to 80 mesh).
P5	=	5% ethylene glycol succinate polyester on Chromosorb W.
P6	=	2% neopentyl glycol succinate polyester on Chromosorb W.
P7	=	10% poly(ethylene glycol adipate) on HMDS-treated Celite (80 to 100 mesh).
P8	=	5% XE-60 on Embacel AW (60 to 70 mesh).
P9	=	4% GE-XE-60 on Gas-Chrom Z (80 to 100 mesh).
P10	=	3% Carbowax 6000 on HMDS-treated Chromosorb W (100 to 120 mesh).

Detector		
D1	=	β-ionization.
D2	=	thermal conductivity.

REFERENCES

1. Aspinall, G. O., Gas-liquid partition chromatography of methylated and partially methylated methyl glycosides, *J. Chem. Soc.*, 1676, 1963.

2. Stephen, A. M., Kaplan, M., Taylor, G. L., and Leisegang, E. C., Application of gas-liquid chromatography to the structural investigation of polysaccharides. I. The structures of the gums of *Virgilia oroboides* and of *Agave americana*, *Tetrahedron*, Suppl. 7, 233, 1966.

3. Ovodov, Yu. S. and Pavlenko, A. F., Gas-liquid chromatography of methylated D-galactose derivatives, *J. Chromatogr.*, 36, 531, 1968.

4. Heyns, K., Sperling, K. R., and Grützmacher, H. F., Combination of gas chromatography and mass spectrometry for analysis of partially methylated sugar derivatives. The mass spectra of partially methylated methyl glucosides, *Carbohydr. Res.*, 9, 79, 1969.

5. Bhattacharjee, S. S. and Gorin, P. A. J., Identification of di-*O*-, tri-*O*-, and tetra-*O*-methylmannoses by gas-liquid chromatography, *Can. J. Chem.*, 47, 1207, 1969.

6. Percival, E. and Young, M., Characterisation of sucrose lactate and other oligosaccharides found in the Cladophorales, *Carbohydr. Res.*, 20, 217, 1971.

7. Anderle, D., Kováč, P., and Anderlová, H., Separation of some methyl O-methyl-D-xylofuranosides by gas chromatography. II, *J. Chromatogr.*, 64, 368, 1972.

8. Anderle, D. and Kováč, P., Separation of methyl(methyl-O-methyl-α-D-glucopyranosid) uronates by gas chromatography, *J. Chromatogr.*, 91, 463, 1974.

9. Fournet, B., Dhalluin, J.-M., Leroy, Y., Montreuil, J., and Mayer, H., Analytical and preparative gas-liquid chromatography of the fifteen methyl ethers of methyl-α-D-galactopyranoside, *J. Chromatogr.*, 153, 91, 1978.

Table GC 12
TRIFLUOROACETATES OF COMMON MONO- AND OLIGOSACCHARIDES

	P1	P2	P3	P4
Phase	P1	P2	P3	P4
Temperature	T1	T2	T3	T4
Gas; flow rate (ml/min)	N_2, 7.5	N_2, 63	N_2, 62	N_2, 0.79
Column				
Length, cm	200	183	183	5000
Diameter (I.D.), cm	0.2	0.39	0.39	0.034
Form	U-tube	na	na	capillary
Material	glass	glass	glass	glass
Detector	FI	FI	FI	DI
Reference	1	2	2	3
Parent sugar	r^a	r^a	r^a	r^a
L-Arabinose	0.47; 0.55; 0.70[b]	—	0.64; 0.87[b]	0.47;[b] 0.54; 0.69;[b] 0.70
D-Xylose	0.39; 0.58; 0.75[b]	0.68; 0.77	0.61; 0.76; 0.81[b]	0.43; 0.61
D-Ribose	0.67; 0.78	—	—	0.64; 0.70; 0.90;[b] 0.95[b]
D-Lyxose	—	—	—	0.37; 0.61;[b] 0.74;[b] 1.32[b]
L-Rhamnose	0.33; 0.47; 0.61[b]	—	—	0.33; 0.66
L-Fucose	0.43; 0.50	—	—	—
D-Glucose	0.92; 1.00	0.95; 1.00	0.82; 1.00	0.77; 1.00
D-Galactose	1.00; 1.05;[b] 1.17[b]	—	0.76; 1.10;[b] 1.17[b]	0.84;[b] 1.15;[b] 1.17
D-Mannose	0.94; 1.11	—	0.74	1.33
D-Fructose	—	—	0.80;[b] 0.84;[b] 0.90	0.88;[b] 0.91
2-Amino-2-deoxy-D-glucose	0.86; 1.61	—	—	—
2-Amino-2-deoxy-D-galactose	1.44; 1.61;[b] 1.67[b]	—	—	—
2-Amino-2-deoxy-D-mannose	1.47; 1.56	—	—	—
Sucrose	—	—	1.77	—
Lactose	—	—	1.93; 2.00	—
Maltose	—	1.76; 1.82	1.86	—
Maltotriose	—	—	2.52	—
Raffinose	—	2.14	2.27	—
Stachyose	—	2.48	2.70	—

[a] t, relative to trifluoroacetate of β-D-glucose (36 min in Reference 1, 8.4 min and 5.6 min for P2 and P3 in Reference 2, 2.37 min in Reference 3).

[b]   Minor peak.

Phase	P1	=	5% OV-210 on Varaport 30.
	P2	=	3% OV-210 on Chromosorb W HP (80 to 100 mesh).
	P3	=	10% OV-17 on Chromosorb W HP (80 to 100 mesh).
	P4	=	SE-54 WCOT capillary column (G. C. Labor, Jaeggi, Switzerland).
Temperature	T1	=	temperature programmed, 90 → 190°C at 1°/min.
	T2	=	75°C for 1 min, then temperature programmed 75 → 225°C at 10°/min.
	T3	=	75°C for 1 min, then temperature programmed 75 → 225°C at 10°/min; isothermal at 225°C for 5 min.
	T4	=	74°C for 4 min, then temperature programmed 74 → 130°C at 1°/min.
Detector	D1	=	^{63}Ni electron-capture detector.

Table GC 12 (continued)
TRIFLUOROACETATES OF COMMON MONO- AND OLIGOSACCHARIDES

REFERENCES

1. Zanetta, J. P., Breckenridge, W. C., and Vincendon, G., Analysis of monosaccharides by gas-liquid chromatography of the *O*-methyl glycosides as trifluoroacetate derivatives. Application to glycoproteins and glycolipids., *J. Chromatogr.*, 69, 291, 1972.
2. Sullivan, J. E. and Schewe, L. R., Preparation and gas chromatography of highly volatile trifluoroacetylated carbohydrates using *N*-methylbis (trifluoroacetamide), *J. Chromatogr. Sci.*, 15, 196, 1977.
3. Eklund, G., Josefsson, B., and Roos, C., Gas-liquid chromatography of monosaccharides at the picogram level using glass capillary columns, trifluoroacetyl derivatization and electron-capture detection, *J. Chromatogr.*, 142, 575, 1977.

Table GC 13
TRIFLUOROACETATES OF SOME HEXOPYRANOSIDES AND -FURANOSIDES

Packing	P1	P2	P3	P4	P5	P6
Temperature (°C)	115	140	165	165	155	160
Gas; flow rate (mℓ/min)	N₂, na	N₂, na	N₂, na	N₂, na	N₂, na	N₂, na
Column						
Length, cm	150	150	150	150	150	150
Diameter (I.D.), cm	0.3	0.3	0.3	0.3	0.3	0.3
Form	straight	straight	straight	straight	straight	straight
Material	SS	SS	SS	SS	SS	SS
Detector	FI	FI	FI	FI	FI	FI
Parent compound	r[a]	r[a]	r[a]	r[a]	r[a]	r[a]
D-Glucopyranoside						
methyl α-	1.00	1.05	1.09	1.11	1.10	1.03
methyl β-	1.08	1.16	1.21	1.99	2.14	1.38
ethyl β-	1.31	1.38	1.46	1.48	1.96	1.75
phenyl α-	8.55	7.34	4.64	2.26[b]	2.74[c]	[d]
phenyl β-	10.43	9.35	6.26	5.22[b]	7.09[c]	[d]
D-Glucofuranoside						
methyl α-	0.86	0.93	0.93	1.06	1.10	0.98
methyl β-	0.79	0.90	0.90	0.97	1.04	0.80
ethyl β-	0.95	0.92	1.04	0.87	0.85	0.90
phenyl β-	7.19	5.70	4.12	2.09[b]	2.65[c]	[d]
D-Galactopyranoside						
methyl α-	0.92	0.97	0.98	1.19	1.16	0.96
methyl β-	1.03	1.27	1.30	2.50	2.78	1.70
ethyl α-	1.00	1.03	1.13	1.04	1.00	1.08
ethyl β-	1.29	1.38	1.51	2.57	2.83	1.35
phenyl α-	8.60	6.76	4.37	2.83[b]	3.20[c]	[d]
phenyl β-	10.29	10.47	7.03	7.52[b]	8.91[c]	[d]
D-Galactofuranoside						
methyl α-	0.79	0.91	0.97	1.23	1.34	1.08
methyl β-	0.78	0.80	0.89	1.05	1.13	0.80
ethyl β-	0.90	0.84	0.92	0.96	1.00	0.95
phenyl β-	7.33	5.98	4.64	2.70[b]	3.03[c]	[d]
D-Mannopyranoside						
methyl α-	1.00	1.00	1.00	1.00	1.00	1.00
methyl β-	1.29	1.46	1.42	2.70	3.30	1.97
D-Mannofuranoside						
methyl α-	1.05	1.18	1.09	1.32	1.43	1.30

Table GC 13 (continued)
TRIFLUOROACETATES OF SOME HEXOPYRANOSIDES AND -FURANOSIDES

[a] t, relative to trifluoroacetate of methyl α-D-mannopyranoside (1.05, 1.07, 0.82, 1.15, 1.15, and 1.00 min for P1, P2, P3, P4, P5, and P6, respectively.

[b] Column temperature 190°C.

[c] Column temperature 180°C.

[d] Not eluted during 20 min at 200°C.

Packing P1 = 1.5% SE-30 on Chromosorb W (60 to 80 mesh).
         P2 = 3% SE-52 on Chromosorb W (60 to 80 mesh).
         P3 = 2% QF-1 on Chromosorb W (60 to 80 mesh).
         P4 = 3% neopentyl glycol succinate polyester on Chromosorb G (60 to 80 mesh).
         P5 = 3% butanediol succinate polyester on Chromosorb G (60 to 80 mesh).
         P6 = 5% Ucon oil 50 LB 550X on Chromosorb W (60 to 80 mesh).

From Yoshida, K., Honda, N., Iino, N., and Kato, K., *Carbohydr. Res.*, 10, 333, 1969. With permission.

Table GC 14 (continued)
TRIFLUOROACETATES OF *O*-METHYL GLYCOSIDES PRODUCED ON METHANOLYSIS OF POLYSACCHARIDES AND GLYCOPROTEINS

Phase	P1	P2	P3
Temperature	T1	T2	T3
Gas; flow rate (mℓ/min)	N₂, 7.5	N₂, 44	Ar[a], 1—2
Column			
Length, cm	200	183	7000
Diameter (I.D.), cm	0.2	0.2	0.05
Form	U-tube	na	capillary
Material	glass	glass	glass
Detector	FI	D1	D2
Reference	1	2	3
Parent sugar	r[b]	r[b]	r[b]
L-Arabinose	0.42; 0.48	—	0.47; 0.53
D-Xylose	0.36; 0.48	—	0.45; 0.46;[c] 0.50
D-Ribose	0.45;[c] 0.48; 0.62	—	0.50; 0.54;[c] 0.55; 0.67
D-Lyxose	—	—	0.43; 0.47; 0.54[c]
L-Rhamnose	0.32; 0.45	—	0.38; 0.50
L-Fucose	0.39; 0.52	0.38; 0.40;[c] 0.64	0.44; 0.45;[c] 0.57
D-Glucose	1.00; 1.05	1.00; 1.23	1.00; 1.07
D-Galactose	0.88; 0.91; 1.08[c]	1.02; 1.42	0.89;[c] 0.92;[c] 0.94; 1.10
D-Mannose	0.95; 1.12[c]	0.96; 1.40[c]	0.96; 1.14[c]
2-Amino-2-deoxy-D-glucose	1.44; 1.52	1.80;[c] 1.90; 2.18	—
2-Amino-2-deoxy-D-galactose	1.62; 1.68[c]	1.84; 2.03;[c] 2.14; 2.49[c]	—
2-Amino-2-deoxy-D-mannose	1.52; 1.62	—	—
2-Acetamido-2-deoxy-D-glucose	1.74; 2.18	1.79; 1.90;[c] 2.18;[c] 2.44[c]	1.42;[c] 1.66; 1.73[c]
2-Acetamido-2-deoxy-D-galactose	1.77; 1.86[c]	1.83;[c] 2.02; 2.14;[c] 2.49[c]	1.51;[c] 1.57;[c] 1.62;[c] 1.72

Table GC 14 (continued)
TRIFLUOROACETATES OF *O*-METHYL GLYCOSIDES PRODUCED ON METHANOLYSIS OF POLYSACCHARIDES AND GLYCOPROTEINS

N-Acetylneuraminic acid	2.42; 2.48	3.05;[c] 3.11;[c] 3.24	2.32
D-Glucuronic acid[d]	1.30; 1.44	—	—
D-Galacturonic acid[d]	0.76; 0.88[c]	—	—

[a]   Containing methane (5%).

[b]   *t*, relative to trifluoroacetate of methyl *α*-D-glucopyranoside (33 min in Reference 1, 16.8 min in Reference 2, 37.2 min in Reference 3).

[c]   Minor peak.

[d]   Methyl ester.

Phase	P1 =	5% OV-210 on Varaport 30.
	P2 =	2% XF-1105 on Gas-Chrom P (80 to 100 mesh).
	P3 =	capillary packed with Supelcoport (80 to 100 mesh), then 10 m*l* 1% solution of OV-210 in dichloromethane-acetone (4:1) passed through column under N₂ pressure 1.5 kg/cm²; solvent evaporated by passing N₂ through column before conditioning at 210°C overnight at N₂ flow-rate 2 m*l*/min.
Temperature	T1 =	temperature programmed, 90 → 190°C at 1°/min.
	T2 =	temperature programmed, 80 → 200°C at 2°C/min.
	T3 =	temperature programmed, 120 → 210°C at 1°/min.
Detector	D1 =	⁶³Ni linear electron capture detector.
	D2 =	Hewlett-Packard 5709 electron capture detector.

REFERENCES

1.   Zanetta, J. P., Breckenridge, W. C., and Vincendon, G., Analysis of monosaccharides by gas-liquid chromatography of the *O*-methyl glycosides as trifluoroacetate derivatives. Application to glycoproteins and glycolipids, *J. Chromatogr.*, 69, 291, 1972.

2.   Pritchard, D. G. and Niedermeier, W., Sensitive gas chromatographic determination of the monosaccharide composition of glycoproteins using electron capture detection, *J. Chromatogr.*, 152, 487, 1978.

3.   Wrann, M. M. and Todd, C. W., Sensitive determination of sugars utilizing packed capillary columns and electron capture detection, *J. Chromatogr.*, 147, 309, 1978.

Table GC 15
TRIFLUOROACETATES OF ALDITOLS AND MONO-*O*-METHYL-D-GLUCITOLS

Phase	P1	P2	P3	P4	P5
Temperature (°C)	160	T1	T2	T3	T4
Gas; flow rate (mℓ/min)	He, 4	N₂, 7.5	N₂, 44	Ar[a], 1—2	He, 25
Column					
Length, cm	1524	200	183	7000	305
Diameter (I.D.), cm	0.051	0.2	0.2	0.05	0.32 (O.D.)
Form	SCOT	U-tube	na	capillary	na
Material	glass	glass	glass	glass	SS
Detector	FI	FI	D1	D2	FI
Reference	1	2	3	4	5
Parent compound	r[b]	r[c]	r[c]	r[c]	r[c]
Erythritol	0.52	—	—	—	—
Threitol	0.62	—	—	—	—
L-Arabinitol	1.00	—	0.70	0.99	—
Ribitol	0.93	—	0.65	—	—
Xylitol	1.06	—	—	—	—
D-Glucitol	1.54	1.25	0.96	1.22	—
Galactitol	1.68	1.35	1.00	1.28	—
D-Mannitol	1.43	1.19	0.87	1.18	—
Allitol	1.77	—	—	—	—
L-Iditol	1.47	—	—	—	—
D-Talitol	1.47	—	—	—	—
Fucitol	—	—	—	0.84	—
2-Amino-2-deoxy-D-glucitol	—	—	—	1.82	—
2-Amino-2-deoxy-D-galactitol	—	—	—	1.94	—
D-Glucitol					
2-*O*-methyl	—	—	—	—	0.63
3-*O*-methyl	—	—	—	—	0.79
4-*O*-methyl	—	—	—	—	0.88
6-*O*-methyl	—	—	—	—	0.29

[a] Containing methane (5%).

[b] t, relative to trifluoroacetate of L-arabinitol (3.22 min).

[c] t, relative to trifluoroacetate of *meso*-inositol (24 min in Reference 2, 26.3 min in Reference 3, 27.5 min in Reference 4, 15.7 min in Reference 5).

Phase	P1	=	FS-1265 SCOT column (Perkin-Elmer).
	P2	=	5% OV-210 on Varaport 30.
	P3	=	2% XF-1105 on Gas-Chrom P (80 to 100 mesh).
	P4	=	OV-210 on Supelcoport (80 to 100 mesh) in packed capillary (see P3 in Table GC 14 for method of preparation).
	P5	=	1% XE-60 on Gas-Chrom Z (80 to 100 mesh).
Temperature	T1	=	temperature programmed, 90 → 190°C at 1°/min.
	T2	=	temperature programmed, 80 → 200°C at 2°/min.
	T3	=	temperature programmed, 120 → 210°C at 1°/min.
	T4	=	temperature programmed, 130 → 150°C at 1°/min.
Detector	D1	=	⁶³Ni linear electron capture detector.
	D2	=	Hewlett-Packard 5709 electron capture detector.

Table GC 15 (continued)
TRIFLUOROACETATES OF ALDITOLS AND MONO-*O*-METHYL-D-GLUCITOLS

REFERENCES

1. Shapira, J., Identification of sugars as their trifluoroacetyl polyol derivatives, *Nature (London)*, 222, 792, 1969.
2. Zanetta, J. P., Breckenridge, W. C., and Vincendon, G., Analysis of monosaccharides by gas-liquid chromatography of the *O*-methyl glycosides as trifluoroacetate derivatives. Application to glycoproteins and glycolipids, *J. Chromatogr.*, 69, 291, 1972.
3. Pritchard, D. G. and Niedermeier, W., Sensitive gas chromatographic determination of the monosaccharide composition of glycoproteins using electron capture detection, *J. Chromatogr.*, 152, 487, 1978.
4. Wrann, M. M. and Todd, C. W., Sensitive determination of sugars utilizing packed capillary columns and electron capture detection, *J. Chromatogr.*, 147, 309, 1978.
5. Anderle, D. and Kováč, P., Gas chromatography of partially methylated alditols as trifluoroacetyl derivatives. Separation of mono-*O*-methyl-per-*O*-TFA-D-glucitols, *J. Chromatogr.*, 49, 419, 1970.

TABLE GC 16
TRIMETHYLSILYL ETHERS OF NEUTRAL MONOSACCHARIDES AND COMMON OLIGOSACCHARIDES

	P1	P1	P2	P2	P3	P4	P5	P6	P7
Packing									
Temperature (°C)	T1	250	140	150	170	144	T2	T3	T4
Gas; flow rate (mℓ/hr)	Ar, 75 (monosaccharides), 150 (oligo-)				H₂, 75	N₂, 15.5	N₂, 50	N₂, 40	N₂, 20
Column									
Length, cm	183	183	244	244	200	300	250	152	183
Diameter (I.D.), cm	0.6 (O.D.)	0.6 (O.D.)	0.6	0.6	0.47	0.32	0.32	0.6	0.2
Form	coiled	coiled	U—tube	U—tube	coiled	na	na	coiled	na
Material	SS	SS	glass	glass	copper	SS	glass	glass	glass
Detector	F1	F1	D1	D1	D2	F1	F1	F1	F1
Reference	1	1	1	1	2	3	4	5	6
Parent sugar	r^t	r^t	r^r	r^r	r^r	r^r	r^r	r^r	r^r
Erythrose	0.10; 0.12; 0.14	—	0.29; 0.35	—	—	—	—	0.17; 0.18	—
L-Arabinose	0.28; 0.33; 0.38*	—	0.97; 1.10;* 1.31	—	0.32(β)	0.19; 0.22; 0.26; 0.27*	0.31	0.46; 0.51; 0.56*	0.66; 0.76
D-Xylose	0.31;* 0.43; 0.54	—	1.64; 2.11	—	0.50(α)	0.30; 0.37	0.45; 0.57	0.49;* 0.61; 0.72	0.65; 0.73
D-Ribose	0.27;* 0.32; 0.35	—	1.22; 1.33; 1.48*	—	0.38; 0.40	0.23; 0.24; 0.25; 0.28	0.36; 0.37	0.50; 0.53; 0.54	0.70
D-Lyxose	0.26; 0.33	—	0.94; 1.26; 1.42	—	0.30; 0.39	0.21; 0.26*	0.31; 0.38	—	—
D-Glucose	1.00; 1.57	—	5.48	1.00; 1.94	1.14; 1.18	0.58; 1.00	0.86; 1.13	1.00; 1.20	1.00; 1.05
D-Galactose	0.76;* 0.88; 1.08	—	—	0.91;* 1.03; 1.38	1.00(α)	0.46; 0.55; 0.67;* 0.71	0.72;* 0.80; 0.90	0.87;* 0.94; 1.03	—
D-Mannose	0.70; 1.08	—	—	0.62; 1.31	0.78; 1.18	0.42; 0.69	0.66; 0.89	0.86; 1.03	—
D-Allose	0.76; 0.81; 0.91	—	—	0.78; 0.91; 1.20*	—	—	0.72; 0.76; 0.82*	—	—
D-Altrose	0.65; 0.68; 0.94*	—	—	0.63; 0.75; 1.16*	—	—	0.64; 0.67; 0.82	—	—
D-Gulose	0.66; 0.74; 0.95*	—	—	0.84; 0.95; 2.11*	—	—	0.65; 0.71; 1.00*	—	—
D-Talose	0.86; 1.00;* 1.13	—	—	1.06;* 1.22; 1.51	—	—	0.74; 0.78; 0.91	—	—
D-glycero-D-galacto-Heptose	4.27	—	—	—	—	—	—	—	—
D-glycero-D-gulo-Heptose	2.29	—	—	—	—	—	—	—	—
2-Deoxy-D-ribose	0.16	—	0.49	—	0.18	0.20	—	0.24;* 0.27	—

	C1	C2	C3	C4	C5	C6	C7	C8	C9
2-Deoxy-D-glucose	0.16;* 0.25; 0.46; 0.64	—	—	—	—	—	—	0.65; 0.78	—
2-Deoxy-D-galactose	0.42; 0.45; 0.53	—	—	0.59	—	—	—	—	—
L-Rhamnose	0.30	—	—	0.20	0.34	0.21; 0.28*	0.33; 0.42	0.48; 0.59	—
L-Fucose	0.33; 0.38; 0.45	—	—	0.25	0.42	0.20; 0.26; 0.30;* 0.32	0.35;* 0.39	0.50;* 0.55; 0.60	—
6-Deoxy-D-glucose	—	—	—	—	0.52; 0.72	—	0.45	—	—
D-*erythro*-Pentulose	0.23;* 0.25; 0.33; 0.35	—	—	—	—	—	—	0.47; 0.54; 0.59	—
D-*threo*-Pentulose	—	—	—	—	—	—	—	0.42; 0.46; 0.54; 0.61*	—
D-Fructose	0.69	—	—	0.56	0.80	0.41; 0.42; 0.44; 0.51*	—	0.85; 0.91*	0.92
L-Sorbose	0.85	—	—	0.69	—	0.34;* 0.44; 0.51;* 0.67	—	0.81;* 0.94; 1.02*	—
D-Tagatose	—	—	—	—	—	—	—	0.88;* 0.94;* 0.99	—
D-*altro*-Heptose	1.12	—	—	—	—	1.12	—	1.02	1.02
Sucrose	10.4	1.00	—	—	—	—	—	2.40	1.59
Trehalose	9.04;* 13.5	—	—	—	—	—	—	2.55	1.66
Maltose	11.7;* 13.1; 1.16	—	—	—	—	—	—	2.52; 2.63	—
Cellobiose	11.9;* 16.6; 1.14;* 1.37	—	—	—	—	—	—	2.50; 2.71	—
Gentiobiose	22.6	—	—	—	—	—	—	2.98	—
Lactose	10.5	—	—	—	—	—	—	2.41; 2.69	—
Melibiose	15.1;* 19.0; 20.0	—	—	—	—	—	—	2.85; 2.90	—
Melezitose	120.5	7.98	—	—	—	—	—	3.86	—
Raffinose	99.0	6.40	—	—	—	—	—	3.72	1.99
Stachyose	52.4	52.4	—	—	—	—	—	2.88	—

* t, relative to trimethylsilylated α-D-glucose (20 min at 140°C, 1.3 min at 210°C on P1; 25.2 min on P2 at 150°C; 11.0 min in Reference 5; 7.25 min in Reference 6).

b t, relative to trimethylsilylated sucrose (2.3 min).

c t, relative to trimethylsilylated methyl α-L-arabinopyranoside (10.85 min).

d t, relative to trimethylsilylated α-D-galactose.

e t, relative to trimethylsilylated β-D-glucose (about 26 min).

f t, relative to trimethylsilylated D-mannitol (about 58 min).

■ Minor peak.

▲ Trace.

Table GC 16 (continued)

TRIMETHYLSILYL ETHERS OF NEUTRAL MONOSACCHARIDES AND COMMON OLIGOSACCHARIDES

Packing	P1	= 3% SE-52 on acid-washed, silanized Chromosorb W (80 to 100 mesh).
	P2	= 15% poly(ethylene glycol succinate) on Chromosorb W (80 to 100 mesh).
	P3	= 10% SE-52 on acid-washed, silanized Chromosorb W (80 to 100 mesh).
	P4	= 1.2% XE-60 on acid-washed DMCS-treated Chromosorb W (60 to 80 mesh).
	P5	= 3% SE-30 on Diatoport S (80 to 100 mesh).
	P6	= 2% SE-52 on acid-washed Diatomite C (85 to 100 mesh, BSS).
	P7	= 5% Dexsil 300 GC on Chromosorb W AW DMCS (80 to 100 mesh).
Temperature	T1	= 140°C for monosaccharides, 210°C for oligosaccharides.
	T2	= temperature programmed, 140 → 200°C at 0.5°/min.
	T3	= temperature programmed, 140 → 290°C at 4°/min.
	T4	= temperature programmed, 160 → 240°C at 10°/min, then 240 → 350°C at 30°/min; isothermal at 350°C for 11 min.
Detector	D1	= argon ionization detector.
	D2	= thermal conductivity (katharometer).

REFERENCES

1. Sweeley, C. C., Bentley, R., Makita, M., and Wells, W. W., Gas-liquid chromatography of trimethylsilyl derivatives of sugars and related substances, *J. Am. Chem. Soc.*, 85, 2497, 1963.
2. Cheminat, A. and Brini, M., Separation of various monosaccharides by gas chromatography, *Bull. Soc. Chim. Fr.*, 80, 1966.
3. Ellis, W. C., Liquid phases and solid supports for gas-liquid chromatography of trimethylsilyl derivatives of monosaccharides, *J. Chromatogr.*, 41, 335, 1969.
4. Bhatti, T., Chambers, R. E., and Clamp, J. R., The gas chromatographic properties of biologically important N-acetylglucosamine derivatives, monosaccharides, disaccharides, trisaccharides, tetrasaccharides, and pentasaccharides, *Biochim. Biophys. Acta*, 222, 339, 1970.
5. Holligan, P. M. and Drew, E. A., Routine analysis by gas-liquid chromatography of soluble carbohydrates in extracts of plant tissues. II. Quantitative analysis of standard carbohydrates, and the separation and estimation of soluble sugars and polyols from a variety of plant tissues, *New Phytol.*, 70, 271, 1971.
6. Janauer, G. A. and Englmaier, P., Multi-step time program for the rapid gas-liquid chromatography of carbohydrates, *J. Chromatogr.*, 153, 539, 1978.

Table GC 17
TRIMETHYLSILYL ETHERS OF NEUTRAL OLIGOSACCHARIDES

Packing	P1	P1	P2	P3
Temperature (°C)	210	T1	228	T2
Gas; flow rate (ml/min)	Ar, 70—80	N₂, 50	N₂, 18	He, 50
Column				
Length, cm	122	100	270	183
Diameter (I.D.), cm	0.4	0.32	0.32	0.32
Form	na	na	coiled	na
Material	glass	glass	SS	SS
Detector	D1	FI	FI	FI
Reference	1	2	3	4
Parent sugar	r^a	r^a	r^a	r^b
Disaccharides				
Sucrose	1.00	1.00	1.00	—
α, α-Trehalose	—	—	1.38	—
β-β-Trehalose	—	—	1.90	—
Kojibiose	—	—	1.40; 1.82	—
Sophorose	—	1.59; 1.86	1.66; 1.99	—
Laminaribiose	1.62	1.57; 1.77	1.64; 1.80	—
Maltose	1.40	1.17; 1.36	1.19; 1.33	0.92; 1.00
Cellobiose	1.19;^{c} 1.73	1.18; 1.77	1.22; 1.70	—
Isomaltose	—	1.96; 2.36	2.37; 2.77	—
Gentiobiose	—	2.56; 2.68	2.36; 2.76	—
Lactose	—	0.99; 1.58	1.09; 1.54	—
Neolactose	—	—	0.81; 0.87	—
Melibiose	—	—	2.16; 2.33	—
Lactulose	—	—	0.94	—
Turanose	—	—	1.28; 1.56	—
Palatinose	—	1.37;^{c} 1.49	1.39;^{c} 1.48; 1.79^c	—
α-D-Glcp-(1 → 5)-D-Glcf	—	—	2.40; 2.80	—
β-D-Galp-(1 → 3)-D-Galp	1.00; 1.97	—	—	—
β-D-Galp-(1 → 6)-D-Galp	2.5	—	2.38	—
β-D-Glcp-(1 → 2)-L-Ara	—	—	0.95	—
β-D-Galp-(1 → 3)-L-Ara	0.80; 0.96; 1.19^c	—	0.73; 0.90; 1.03	—
β-L-Arap-(1 → 3)-L-Ara	0.67; 0.74;^{c} 1.10	—	—	—
β-D-Xylp-(1 → 3)-D-Xyl	0.42;^{c} 0.56; 0.92^c	—	—	—
β-D-Xylp-(1 → 4)-D-Xyl	0.69;^{c} 0.89	—	—	—
β-D-Xylp-(1 → 6)-D-Glc	—	—	1.92; 1.96	—
β-D-Manp-(1 → 4)-D-Manp	1.03; 1.40	1.22; 1.64	1.18; 1.52	—
α-D-Manp-(1 → 6)-D-Glc	—	—	1.70; 2.08	—
β-D-Manp-(1 → 4)-D-Gul	1.31	—	—	—
α-L-Fucp-(1 → 2)-L-Fucp	0.51	—	—	—
α-L-Fucp-(1 → 3)-L-Fucp	0.46, 0.95^c	—	—	—
α-L-Fucp-(1 → 4)-L-Fucp	0.51;^{d} 0.57	—	—	—
Higher oligosaccharides				
Raffinose	—	1.00^e	—	—
Maltotriose	—	—	—	1.40
Isomaltotriose	—	1.87^e	—	—
Mannotriose	—	1.22^e	—	—
Stachyose	—	1.00^f	—	—
Maltotetraose	—	—	—	1.92
Isomaltotetraose	—	1.67^f	—	—

Note: This table supplements Table GC 124 in Section A, Volume I, Section I.II.

Table GC 17 (continued)
TRIMETHYLSILYL ETHERS OF NEUTRAL OLIGOSACCHARIDES

[a]   t, relative to trimethylsilylated sucrose (12.1 min in Reference 3).
[b]   t, relative to trimethylsilylated β-maltose (about 13 min).
[c]   Minor peak.
[d]   Shoulder.
[e]   t, relative to trimethylsilylated raffinose.
[f]   t, relative to trimethylsilylated stachyose.

Packing   P1   =   3% SE-30 on Gas-Chrom P (Reference 1) or Diatoport S (Reference 2).
               P2   =   3% OV-17 on Chromosorb W—HP (80 to 100 mesh).
               P3   =   3% OV-17 on silanized Chromosorb W (60 to 80 mesh).
Temperature   T1   =   220°C for disaccharides, 300°C for tri- and tetrasaccharides.
               T2   =   temperature programmed, 150 → 325°C at 10°/min, then isothermal at 325°C
                         for 12 min.
Detector   D1   =   β-ionization.

REFERENCES

1. Percival, E., Gas chromatography of the trimethylsilyl ethers of some less-common disaccharides, *Carbohydr. Res.*, 4, 441, 1967.
2. Bhatti, T., Chambers, R. E., and Clamp, J. R., The gas chromatographic properties of biologically important N-acetylglucosamine derivatives, monosaccharides, disaccharides, trisaccharides, tetrasaccharides and pentasaccharides, *Biochim. Biophys. Acta*, 222, 339, 1970.
3. Haverkamp, J., Kamerling, J. P., and Vliegenthart, J. F. G., Gas-liquid chromatography of trimethylsilyl disaccharides, *J. Chromatogr.*, 59, 281, 1971.
4. Brobst, K. M., Gas-liquid chromatography of trimethylsilyl derivatives. Analysis of corn syrup, in *Methods in Carbohydrate Chemistry*, Vol. 6, Whistler, R. L. and BeMiller, J. N., Eds., Academic Press, New York, 1972, 3.

Table GC 18
TRIMETHYLSILYL ETHERS OF VARIOUS GLYCOSIDES OF COMMON SUGARS

Packing	P1	P1	P2	P3	P4	P5	P6
Temperature (°C)	140	200	170	160	180	160	200
Gas; flow rate (ml/min)	na, 75	N_2, na	N_2, na	N_2, na	N_2, na	N_2, na	N_2, na
Column Length, cm	183	150	150	150	150	150	150
Diameter (I.D.), cm	0.63	0.3	0.3	0.3	0.3	0.3	0.3
Form	coiled	straight	straight	straight	straight	straight	straight
Material	SS	SS	SS	SS	SS	SS	SS
Detector	FI	FI	FI	FI	FI	FI	FI
Reference	1	2	2	2	2	2	2
Parent compound	r[a]	r[a]	r[a]	r[a]	r[a]	r[a]	r[a]
L-Arabinopyranoside							
methyl α-	0.21	—	—	—	—	—	—
methyl β-	0.20	—	—	—	—	—	—
D-Xylopyranoside							
methyl α-	0.31	—	—	—	—	—	—
methyl β-	0.34	—	—	—	—	—	—
D-Glucopyranoside							
methyl α-	0.92	1.47	1.45	1.52	1.71	1.85	1.67
methyl β-	1.07	1.54	1.56	1.65	1.81	2.02	1.85
ethyl β-	—	1.70	1.95	1.74	2.20	2.13	1.88
phenyl α-	—	5.96	6.76	6.52	11.40	11.73	8.00
phenyl β-	7.16	7.00	8.57	8.60	16.80	19.47	10.38
methyl 4,6-benzylidene-α-	4.67	—	—	—	—	—	—
D-Glucofuranoside							
methyl α-	—	1.06	1.06	1.25	1.16	1.15	1.15
methyl β-	—	1.02	1.04	1.23	1.14	1.01	1.11
ethyl β-	—	1.13	1.19	1.30	1.20	1.13	1.13
phenyl β-	—	4.87	5.81	6.10	9.50	10.00	6.66
D-Mannopyranoside							
methyl α-	0.59	1.00	1.00	1.00	1.00	1.00	1.00
methyl β-	0.67	1.14	1.13	1.39	1.26	1.29	1.22

Table GC 18 (continued)
TRIMETHYLSILYL ETHERS OF VARIOUS GLYCOSIDES OF COMMON SUGARS

Packing	P1	P1	P2	P3	P4	P5	P6
Temperature (°C)	140	200	170	160	180	160	200
Gas; flow rate (ml/min)	na, 75	N₂, na	N₂, na	N₂, na	N₂, na	N₂, na	N₂, na
Column							
Length, cm	183	150	150	150	150	150	150
Diameter (I.D.), cm	0.63	0.3	0.3	0.3	0.3	0.3	0.3
Form	coiled	straight	straight	straight	straight	straight	straight
Material	SS	SS	SS	SS	SS	SS	SS
Detector	Fl	Fl	Fl	Fl	Fl	Fl	Fl
Reference	1	2	2	2	2	2	2
Parent compound	r^a	r^a	r^a	r^a	r^a	r^a	r^a
D-Mannofuranoside							
methyl α-	—	1.28	1.31	1.39	1.41	1.51	1.37
D-Galactopyranoside							
methyl α-	0.72	1.16	1.15	1.19	1.33	1.40	1.19
methyl β-	0.82	1.30	1.32	1.57	1.61	1.90	1.50
ethyl α-	—	1.17	1.19	1.14	1.20	1.13	1.09
ethyl β-	—	1.43	1.43	1.53	2.10	1.87	1.56
phenyl α-	—	4.74	5.24	5.55	9.50	9.67	6.28
phenyl β-	—	6.70	7.62	10.70	20.60	25.73	10.50
methyl 4,6-benzylidene-α-	2.77[c]	—	—	—	—	—	—
methyl 4,6-benzylidene-β-	3.00[c]	—	—	—	—	—	—
D-Galactofuranoside							
methyl α-	—	1.20	1.17	1.43	1.35	1.50	1.32
methyl β-	—	1.00	1.03	1.13	1.08	1.23	1.09
D-Galactofuranoside							
ethyl β-	—	1.09	1.10	1.11	1.10	1.07	1.06
phenyl β-	—	4.61	5.14	6.65	9.30	10.69	6.50
L-Idopyranoside							
methyl α-	0.75	—	—	—	—	—	—
methyl β-	0.64	—	—	—	—	—	—
methyl 4,6-benzylidene-α-	2.96[c]	—	—	—	—	—	—
methyl 4,6-benzylidene-β-	2.52;[c] 3.11[c]	—	—	—	—	—	—

[a] t, relative to trimethylsilylether of α-D-glucose (20 min at 140°C, 1.3 min at 210°C).

[b] t, relative to trimethylsilylether of methyl α-D-mannopyranoside (1.15, 1.05, 0.90, 0.50, 0.75, and 1.60 min for P1, P2, P3, P4, P5, and P6, respectively).

[c] Column temperature 210°C.

Packing P1 = 3% SE-52 on Chromosorb W (80 to 100 mesh in Reference 1, 60 to 80 mesh in Reference 2).

P2 = 1.5% SE-30 on Chromosorb W (60 to 80 mesh).

P3 = 2% QF-1 on Chromosorb W (60 to 80 mesh).

P4 = 3% neopentyl glycol succinate polyester on Chromosorb G (60 to 80 mesh).

P5 = 3% butanediol succinate polyester on Chromosorb G (60 to 80 mesh).

P6 = 5% Ucon oil 50 LB 550X on Chromosorb W (60 to 80 mesh).

REFERENCES

1. **Sweeley, C. C., Bentley, R., Makita, M., and Wells, W. W.**, Gas-liquid chromatography of trimethylsilyl derivatives of sugars and related substances, *J. Am. Chem. Soc.*, 85, 2497, 1963.

2. **Yoshida, K., Honda, N., Iino, N., and Kato, K.**, Gas-liquid chromatography of hexopyranosides and hexofuranosides, *Carbohydr. Res.*, 10, 333, 1969.

Table GC 19
TRIMETHYLSILYL ETHERS OF O-METHYL GLYCOSIDES OF NEUTRAL SUGARS AND OTHER SUGAR CONSTITUENTS OF GLYCOPROTEINS AND GLYCOSAMINOGLYCANS

	P1	P2	P3	P4	P5
Packing					
Temperature (°C)	T1	T2	T3	T4	130
Gas; flow rate (ml/min)	N_2, 50	He, 50	N_2, 60	N_2, 20	N_2, 13
Column					
Length, cm	250	152	200	200	200
Diameter (I.D.), cm	0.32	0.3	0.3	0.3	0.2
Form	U—tube	na	spiral	spiral	spiral
Material	glass	SS	glass	glass	SS
Detector	FI	FI	FI	FI	FI
Reference	1, 2	3	4	4	5
Parent sugar	r^c	r^c	r^c	r^c	r^c
L-Arabinose	0.22; 0.23	0.45; 0.49	—	—	—
D-Xylose	0.34; 0.37	0.59	0.66; 0.69	0.36; 0.39	0.31
D-Ribose	0.24; 0.25	—	—	—	—
D-Lyxose	0.24; 0.28ᶜ	—	—	—	—
L-Rhamnose	0.26	—	—	—	—
L-Fucose	0.25;ᶜ 0.28; 0.31	0.50;ᶜ 0.52; 0.55	0.56; 0.59; 0.62	0.29;ᶜ 0.30; 0.34	0.19;ᶜ 0.23; 0.26
D-Glucose	0.79; 0.86	0.87	0.85; 0.87	0.67; 0.71	1.00(a)
D-Galactose	0.61;ᶜ 0.68; 0.74	0.77;ᶜ 0.79; 0.83	0.73; 0.77; 0.81	0.57;ᶜ 0.60; 0.65	0.62;ᶜ 0.74; 0.84
D-Mannose	0.59; 0.65ᶜ	0.72; 0.74ᶜ	0.74	0.54	0.51; 0.57ᶜ
D-Allose	0.53; 0.55; 0.60;ᶜ 0.63	—	—	—	—
D-Altrose	0.42; 0.49; 0.57; 0.66	—	—	—	—
D-Gulose	0.37; 0.51; 0.55; 0.66ᶜ	—	—	—	—
D-Talose	0.65; 0.70ᶜ	—	—	—	—
2-Amino-2-deoxy-D-glucose	—	0.87(a)	0.87; 0.93	0.67; 0.75	—
2-Amino-2-deoxy-D-galactose	—	—	0.76;ᶜ 0.80; 0.95	0.52;ᶜ 0.64; 0.88	—

2-Acetamido-2-deoxy-D-glucose	1.05ᵃ; 1.17; 1.31; 1.38ᶜ	1.20; 1.31	0.84; 1.04ᶜ	0.58; 0.96ᶜ	—
2-Acetamido-2-deoxy-D-galactose	1.08; 1.21	1.16; 1.26	0.78; 0.99ᶜ	0.55; 0.96ᶜ	—
N-Acetylneuraminic acid	2.22	—	1.09	1.14	—
D-Glucuronic acidᶠ	0.45; 0.82	—	0.88	0.73	—
L-Iduronic acidᶠ	0.37	—	0.90	0.74	—

ᵃ tᵣ relative to trimethylsilylated D-mannitol (about 57 min in Reference 2).
ᵇ tᵣ relative to trimethylsilylated *meso*-inositol (about 9 min).
ᶜ tᵣ relative to trimethylsilylated *myo*-inositol (about 17 min for P3, 33 min for P4).
ᵈ tᵣ relative to trimethylsilylated methyl α-D-glucopyranoside (50 min).
ᵉ Minor peak.
ᶠ Methyl ester.

Packing   P1 = 3.8% SE-30 on Diatoport S (80 to 100 mesh).
           P2 = 0.05% OV-17 on glass beads (Corning GLC-110; 120 to 140 mesh).
           P3 = 3% Apiezon M on Chromosorb W AW DMCS (100 to 120 mesh).
           P4 = 3% SE-30 on Chromosorb W AW DMCS (100 to 120 mesh).
           P5 = 3% OV-17 on Gas-Chrom Q (100 to 120 mesh).

Temperature   T1 = temperature programmed, 140 → 200°C at 0.5°/min.
              T2 = temperature programmed, 80 → 250°C at 10°/min.
              T3 = temperature programmed, 110 → 280°C at 8°/min.
              T4 = temperature programmed, 110 → 200°C at 2°/min.

REFERENCES

1. Clamp, J. R., Dawson, G., and Hough, L., The simultaneous estimation of 6-deoxy-L-galactose (L-fucose), D-mannose, D-galactose, 2-acetamido-2-deoxy-D-glucose (N-acetyl-D-glucosamine) and N-acetylneuraminic acid (sialic acid) in glycopeptides and glycoproteins, *Biochim. Biophys. Acta*, 148, 342, 1967.
2. Bhatti, T., Chambers, R. E., and Clamp, J. R., The gas chromatographic properties of biologically important N-acetylglucosamine derivatives, monosaccharides, disaccharides, trisaccharides, tetrasaccharides, and pentasaccharides, *Biochim. Biophys. Acta*, 222, 339, 1970.
3. Reinhold, V. N., Gas-liquid chromatographic analysis of constituent carbohydrates in glycoproteins, in *Methods in Enzymology*, Vol. 25 (Part B), Hirs, C. H. W. and Timasheff, S. N., Eds., Academic Press, New York, 1972, 244.
4. Yokota, M. and Mori, T., Gas-liquid chromatographic analysis of the monosaccharide composition of acid glycosaminoglycans (mucopolysaccharides) derived from animal tissues, *Carbohydr. Res.*, 59, 289, 1977.
5. Cahour, A. and Hartmann, L., Study of neutral and amino-monosaccharides by gas-liquid differential chromatography: application to three reference glycoproteins, *J. Chromatogr.*, 152, 475, 1978.

Table GC 20
TRIMETHYLSILYLATED OXIME AND O-METHYL OXIME DERIVATIVES OF MONO- AND DISACCHARIDES

Phase	P1	P2	P3	P4	P4	P5	P5	P6	P6	P7	P8	P9	P10
Temperature (°C)	170	170	170	160	160	160	160	120	120	215	215	272	140
Gas; flow rate (ml/min)	Ar, 75	N₂, 35	N₂, 35	N₂, 30—35	N₂, 30—35	N₂, 30—35	N₂, 30—35	N₂, 30—35	N₂, 30—35	N₂, 40	N₂, 40	He, 0.9	N₂, na
Column													
Length, cm	244	200	200	200	200	200	200	200	200	200	200	4500	4500
Diameter (I.D.), cm	0.6 (O.D.)	0.3	0.3	0.2	0.2	0.2	0.2	0.2	0.2	0.3	0.3	0.025	0.02
Form	U—tube	na	na	na	na	na	na	na	na	coiled	coiled	capillary	capillary
Material	glass	na	na	SS	SS	SS	SS	SS	SS	SS	SS	glass	SS
Detector	DI	FI	FI	FI	FI	FI	FI	FI	FI	FI	FI	FI	FI
Reference	1	2	2	3	3	3	3	3	3	4	4	5	6
Parent compound	Oxime	MO	MO	Oxime	MO	Oxime	MO	Oxime	MO	Oxime	Oxime	MO	Oxime
	r	r	r'	r	r	r	r	r	r	r	r	r	r'
Monosaccharides													
Glyceraldehyde	—	—	—	0.06	—	0.08; 0.09	—	0.06	—	—	—	—	0.78; 1.00
Dihydroxyacetone	—	—	—	0.05	—	0.09	—	0.05	—	—	—	—	1.00
D-Erythrose	—	0.32	0.35	0.16	—	0.20; 0.23	—	0.15	—	—	—	—	2.28; 2.78
D-Threose	—	—	—	—	—	—	—	—	—	—	—	—	2.95; 3.00
L-glycero-Tetrulose	—	—	—	0.14	—	0.20	—	0.13	—	—	—	—	2.36; 2.47
L-Arabinose	—	—	—	0.43	—	0.53	—	0.40	—	—	—	—	—
D-Xylose	—	0.97	0.99	0.42	0.25	0.55	0.39	0.40	0.25	—	—	—	—
D-Ribose	—	1.00	1.00	0.48	—	0.60	—	0.46	—	—	—	—	—
D-Glucose	2.16	3.05	2.60	1.34	0.80	1.55	1.04	1.32	0.84	—	—	—	—
D-Galactose	1.98	2.96	2.46	1.27; 1.41	—	1.43; 1.57	—	1.26; 1.54	0.83; 1.02	—	—	—	—
D-Mannose	2.00	—	1.30	1.30	0.77	1.45	0.96	1.31; 1.51	—	—	—	—	—
D-Allose	2.08	—	—	—	—	—	—	—	—	—	—	—	—
D-Altrose	1.98	—	—	—	—	—	—	—	—	—	—	—	—
D-Gulose	2.10	—	—	—	—	—	—	—	—	—	—	—	—
2-Deoxy-D-ribose	—	0.60	0.68	0.28	—	0.41; 0.44	—	0.31	—	—	—	—	—
2-Deoxy-D-glucose	—	—	—	0.78	—	1.01; 1.08	—	0.89	—	—	—	—	—
2-Deoxy-D-galactose	—	—	—	0.78	—	1.02; 1.10	—	0.90	—	—	—	—	—
L-Rhamnose	—	—	—	0.58	—	0.68	—	0.58	—	—	—	—	—
L-Fucose	—	1.36	1.22	0.58	—	0.69	—	0.57	—	—	—	—	—
D-erythro-Pentulose	—	—	—	0.41	—	0.50	—	0.37	—	—	—	—	—
D-threo-Pentulose	—	—	—	0.40	—	0.51	—	0.36	—	—	—	—	—
D-Fructose	—	2.79	2.30	1.07	0.73	1.13	0.94; 1.02	1.03	0.72	—	—	—	—
L-Sorbose	—	—	—	1.07	—	1.19	—	1.02	—	—	—	—	—
D-glycero-D-gulo-Heptose	—	8.29	5.90	—	—	—	—	—	—	—	—	—	—
Methyl ethers													
D-Ribose													
2-O-methyl	—	0.69	0.82	—	—	—	—	—	—	—	—	—	—

3-O-methyl	—	—	0.71	0.84	—	—	—	—	—	—	—
2,3-di-O-methyl	—	—	0.47	0.72	—	—	—	—	—	—	—
D-Glucose, 2,3,6-tri-O-methyl	—	—	3.15ᶜ	—	—	—	—	—	—	—	—
D-Galactose, 2,3,4,6-tetra-O-methyl	—	—	2.56ᶠ	—	—	—	—	—	—	—	—
Disaccharides											
Kojibiose	—	—	—	—	—	—	1.56; 1.75ᵃ	1.36; 1.50ᵃ	1.02; 1.10ᵃ	—	—
Nigerose	—	—	—	—	—	—	1.21;ᵃ 1.48; 1.57ᵃ	1.05;ᵃ 1.23; 1.35ᵃ	0.94; 1.00ᵃ	—	—
Maltose	—	—	—	—	—	—	1.45	1.26	1.02; 1.10ᵃ	—	—
Cellobiose	—	—	—	—	—	—	1.24; 1.35ᵃ	1.09; 1.16ᵃ	0.94; 1.00ᵃ	—	—
Gentiobiose	—	—	—	—	—	—	2.01; 2.16ᶜ	1.82	1.24; 1.31ᵃ	—	—
Lactose	—	—	—	—	—	—	1.16	0.97	0.93; 0.95ᵃ	—	—
Neolactose	—	—	—	—	—	—	1.22; 1.35ᵃ	0.99; 1.08ᵃ	—	—	—
Melibiose	—	—	—	—	—	—	2.20; 2.46	1.95; 2.12ᵃ	1.39; 1.49ᵃ	—	—
Lactulose	—	—	—	—	—	—	1.15	0.94	—	—	—
Turanose	—	—	—	—	—	—	1.47	1.23	—	—	—
Palatinose	—	—	—	—	—	—	—	—	1.31	—	—

Note:　MO = O-methyl oxime.

ᵃ　tᵣ relative to trimethylsilylated α-D-glucose (6.0 min).

ᵇ　tᵣ relative to trimethylsilylated O-methyl oxime of D-ribose (about 2 min).

ᶜ　tᵣ relative to trimethylsilylated D-glucitol (13.2 min, 6.6 min, 14.4 min for P4, P5, P6, respectively).

ᵈ　tᵣ relative to trimethylsilylated trehalose (14.5 min for P7, 17.05 min for P8 in Reference 4; 30 min for P9 in Reference 5).

ᵉ　tᵣ relative to trimethylsilylated oxime of dihydroxyacetone (about 12 min).

ᶠ　Estimated from published chromatogram.

ᵍ　Minor peak.

Packing　P1　= 15% poly(ethylene glycol succinate) on Chromosorb W (80 to 100 mesh).

　　　　　P2　= 3% SE-30 on Supelcoport (100 to 120 mesh).

　　　　　P3　= 5% OV-17 on Supelcoport (100 to 120 mesh).

　　　　　P4　= 0.5% OV-1 on Chromosorb G (100 to 120 mesh).

　　　　　P5　= 0.5% OV-17 on Chromosorb G (100 to 120 mesh).

　　　　　P6　= 3% DC QF-1 on Gas-Chrom Q (100 to 120 mesh).

　　　　　P7　= 1.5% SE-52 on Chromosorb W AW DMCS (60 to 80 mesh).

　　　　　P8　= 1.5% OV-17 on Shimalite W (80 to 100 mesh).

　　　　　P9　= OV-101 coating in open tubular capillary column.

　　　　　P10 = OV-17 coating in capillary column.

Detector　D1　= argon ionization detector.

Table GC 20 (continued)
TRIMETHYLSILYLATED OXIME AND O-METHYL OXIME DERIVATIVES OF MONO- AND DISACCHARIDES

REFERENCES

1. Sweeley, C. C., Bentley, R., Makita, M., and Wells, W. W., Gas-liquid chromatography of trimethylsilyl derivatives of sugars and related substances, *J. Am. Chem. Soc.*, 85, 2497, 1963.
2. Laine, R. A. and Sweeley, C. C., O-Methyl oximes of sugars. Analysis as O-trimethylsilyl derivatives by gas-liquid chromatography and mass spectrometry, *Carbohydr. Res.*, 27, 199, 1973.
3. Petersson, G., Gas-chromatographic analysis of sugars and related hydroxy acids as acyclic oxime and ester trimethylsilyl derivatives, *Carbohydr. Res.*, 33, 47, 1974.
4. Toba, T. and Adachi, S., Gas-liquid chromatography of trimethylsilylated disaccharide oximes, *J. Chromatogr.*, 135, 411, 1977.
5. Adam, S. and Jennings, W. G., Gas chromatographic separation of silylated derivatives of disaccharide mixtures on open tubular glass capillary columns, *J. Chromatogr.*, 115, 218, 1975.
6. Anderle, D., Konigstein, J., and Kovacik, V., Separation of trioses and tetroses as trimethylsilyl oximes by gas chromatography, *Anal. Chem.*, 49, 137, 1977.

Table GC 21
TRIMETHYSILYL ETHERS OF POLYOLS, CYCLITOLS, AND REDUCED OLIGOSACCHARIDES

Packing	P1	P2	P3	P4	P4	P5	P6	P7	P8	P9	P10
Temperature (°C)	140	140	T1	T2	T3	T4	175	200	210	263	218
Gas; flow rate (ml/min)	Ar, 75	Ar, 75	N_2, 40	He, na	He, na	He, 88	N_2, 42	N_2, 45	Ar, 70—80	He, na	He, na
Column Length, cm	244	183	152	183	183	244	183	183	122	200	200
Diameter (I.D.), cm	0.63	0.63	0.63	0.32	0.32	0.63	0.4	0.4	0.4	0.35	0.35
Form	U—tube	coiled	coiled	na	na	coiled	na	na	na	na	na
Material	glass	SS	glass	SS	SS	copper	glass	glass	glass	glass	glass
Detector	D1	FI	FI	FI	FI	D2	FI	FI	D3	FI	FI
Reference	1	1	2	3	3	4	5	5	6	7	7
Parent compound	r[a]	r[b]	r[b]	r[c]	r[c]	r[d]	r[e]	r[e]	r[f]	r[g]	r[g]
Acyclic polyols Ethylene glycol	—	—	—	0.14	0.09	0.43	—	—	—	—	—

	1	2	3	4	5	6	7	8	9	10	11	12	13
Glycerol	—	—	0.11	0.09	0.51	0.07	0.45	1.57	—	—	—	—	—
Threitol	—	—	—	—	—	—	—	2.66	—	—	—	—	—
Erythritol	0.49	0.16	0.32	0.29	0.71	0.24	0.68	2.70	—	—	—	—	—
Arabinitol	2.13	0.46	0.68	0.61	0.86	0.58	0.85	—	—	—	—	—	—
Ribitol	2.06	0.46	0.67	—	—	—	—	—	—	—	—	—	—
Xylitol	2.08	0.42	0.64	—	—	—	—	—	—	—	—	—	—
Glucitol	—	1.24	1.11	1.00	1.00	1.00	1.00	—	—	—	—	—	—
Galactitol	—	1.28	1.13	—	—	—	—	—	—	—	—	—	—
Mannitol	—	1.21	1.09	1.00	1.00	1.00	1.00	—	—	—	—	—	—
Iditol	—	—	1.11	—	—	—	—	—	—	—	—	—	—
Talitol	—	1.30	1.12	—	—	—	—	—	—	—	—	—	—
2-Deoxy-D-ribitol	—	—	—	0.39	0.77	0.36	0.76	—	—	—	—	—	—
2-Deoxy-D-glucitol	—	—	—	0.79	0.93	0.77	0.92	—	—	—	—	—	—
3-Deoxy-D-glucitol	—	—	—	0.79	0.93	0.79	0.93	—	—	—	—	—	—
3-Deoxy-D-mannitol	—	—	—	0.79	0.93	0.77	0.92	—	—	—	—	—	—
L-Rhamnitol	—	—	—	0.67	0.89	0.60	0.86	—	—	—	—	—	—
1,6-Dideoxyhexitols	—	—	—	—	0.76	0.35	0.75	—	—	—	—	—	—
2,6-Dideoxy-D-ribo-hexitol	—	—	—	0.46	0.81	0.42	0.79	—	—	—	—	—	—
Cyclitols													
myo-Inositol	—	—	1.37	1.26	1.08	1.27	1.09	—	1.00	1.00	—	—	—
cis-Inositol	—	—	—	—	—	—	—	—	1.17	0.86	—	—	—
scyllo-Inositol	—	—	—	—	—	—	—	—	0.73	0.79	—	—	—
epi-Inositol	—	—	—	—	—	—	—	—	0.88	0.78	—	—	—
chiro-Inositol	—	—	—	—	—	—	—	—	0.56	0.63	—	—	—
muco-Inositol	—	—	—	—	—	—	—	—	0.50	0.55	—	—	—
neo-Inositol	—	—	—	—	—	—	—	—	0.46	0.51	—	—	—
allo-Inositol	—	—	—	—	—	—	—	—	0.45	0.49	—	—	—
myo-Inositol, 1-*O*-methyl	—	—	—	—	—	—	—	—	0.81	0.71	—	—	—
4-*O*-methyl	—	—	—	—	—	—	—	—	0.66	0.62	—	—	—
5-*O*-methyl	—	—	—	—	—	—	—	—	0.68	0.64	—	—	—
1,3-di-*O*-methyl	—	—	—	—	—	—	—	—	0.74	0.54	—	—	—
2-*C*-methyl	—	—	—	—	—	—	—	—	1.26	0.82	—	—	—
2-*C*-hydroxymethyl	—	—	—	—	—	—	—	—	1.25	1.16	—	—	—
chiro-Inositol,2-*O*-methyl	—	—	—	—	—	—	—	—	0.45	0.45	—	—	—
4-*O*-methyl	—	—	—	—	—	—	—	—	0.39	0.40	—	—	—

Table GC 21 (continued)
TRIMETHYSILYL ETHERS OF POLYOLS, CYCLITOLS, AND REDUCED OLIGOSACCHARIDES

	P1	P2	P3	P2	P2	P4	P4	P5	P6	P7	P8	P9	P10
Packing	P1	P2	P3	P2	P2	P4	P4	P5	P6	P7	P8	P9	P10
Temperature (°C)	140	140	T1	T2	T3	T2	T3	T4	175	200	210	263	218
Gas; flow rate (ml/min)	Ar, 75	Ar, 75	N₂, 40	He, na	He, na	He, na	He, na	He, 88	N₂, 42	N₂, 45	Ar, 70—80	He, na	He, na
Column													
Length, cm	244	183	152	183	183	183	183	244	183	183	122	200	200
Diameter (I.D.), cm	0.63	0.63	0.63	0.32	0.32	0.32	0.32	0.63	0.4	0.4	0.4	0.35	0.35
Form	U—tube	coiled	coiled	na	na	na	na	coiled	na	na	na	na	na
Material	glass	SS	glass	SS	SS	SS	SS	copper	glass	glass	glass	glass	glass
Detector	D1	F1	F1	F1	F1	F1	F1	D2	F1	F1	D3	F1	F1
Reference	1	1	2	3	3	3	3	4	5	5	6	7	7
Parent compound	r^a	r^b	r^b	r^c	r^c	r^c	r^c	r^d	r^e	r^e	r^f	r^g	r^g
Reduced oligosaccharides													
Laminaribiitol	—	—	—	—	—	—	—	—	—	—	1.90	0.67	0.63
Maltitol	—	—	—	—	—	—	—	—	—	—	1.90	0.76	0.68
Cellobiitol	—	—	—	—	—	—	—	—	—	—	1.70	0.64	0.59
Gentiobiitol	—	—	—	—	—	—	—	—	—	—	—	0.85	0.89
Lacitiol	—	—	—	—	—	—	—	—	—	—	2.7	0.62	0.54
Melibiitol	—	—	—	—	—	—	—	—	—	—	2.0	1.00	1.00
D-Mannitol,4-O-β-D-mannopyranosyl	—	—	—	—	—	—	—	—	—	—	—	—	—
L-Gulitol,4-O-β-D-mannopyranosyl	—	—	—	—	—	—	—	—	—	—	2.0	—	—
L-Arabinitol,3-O-β-L-arabinopyranosyl	—	—	—	—	—	—	—	—	—	—	0.37	—	—
D-Xylitol,4-O-β-D-xylopyranosyl	—	—	—	—	—	—	—	—	—	—	0.46	—	—
L-Fucitol,2-O-α-L-fucopyranosyl	—	—	—	—	—	—	—	—	—	—	0.77	—	—
3-O-α-L-fucopyranosyl	—	—	—	—	—	—	—	—	—	—	0.75	—	—
4-O-α-L-fucopyranosyl	—	—	—	—	—	—	—	—	—	—	0.69	—	—

a t, relative to trimethylsilylated methyl α-L-arabinopyranoside (10.85 min).

b t, relative to trimethylsilylated α-D-glucose (20 min in Reference 1, 11 min in Reference 2).

c t, relative to trimethylsilylated D-glucitol (22.8 min for T1, 38.6 min for T2, with P2; 20.8 min for T2, 36.3 min for T3, with P4).

d t, relative to trimethylsilylated butane-1,4-diol (about 10.5 min).

e t, relative to hexa*kis* trimethylsilyl-*myo*-inositol (3.7 min with P6, 10.8 min with P7).

f t, relative to tr methylsilylated sucrose.

t, relative to trimethylsilylated melibiitol.

Packing P1 = 15% poly(ethylene glycol succinate) on Chromosorb W (80 to 100 mesh).
　　　　 P2 = 3% SE-52 on acid-washed, silanized Chromosorb W (80 to 100 mesh).
　　　　 P3 = 2% SE-52 on acid-washed Diatomite C (85 to 100 mesh BSS).
　　　　 P4 = 3% JXR silicone gum on Gas-Chrom Q (100 to 120 mesh).
　　　　 P5 = 20% SF-96 on Diatoport S (60 to 80 mesh).
　　　　 P6 = 3% XE-60 on Supelcoport (80 to 100 mesh).
　　　　 P7 = 3% OV-1 on Gas-Chrom Q (100 to 120 mesh).
　　　　 P8 = 3% SE-30 on Gas-Chrom P.
　　　　 P9 = 2.2% SE-30 on Gas-Chrom S (100 to 120 mesh).
　　　　 P10 = 1% OV-22 on Supelcoport (80 to 100 mesh).

Temperature T1 = temperature programmed, 140 → 290°C at 4°/min.
　　　　 T2 = temperature programmed, 120 → 180°C at 2°/min.
　　　　 T3 = temperature programmed, 60 → 70°C at 1°/min, then 70 → 210°C at 4°/min.
　　　　 T4 = isothermal at 130°C for 6 min, then temperature programmed, 130 → 220°C at 3°/min.

Detector D1 = argon ionization detector.
　　　　 D2 = thermal conductivity detector.
　　　　 D3 = β-ionization detector.

REFERENCES

1. **Sweeley, C. C., Bentley, R., Makita, M., and Wells, W. W.,** Gas-liquid chromatography of trimethylsilyl derivatives of sugars and related substances, *J. Am. Chem. Soc.,* 85, 2497, 1963.

2. Holligan, P. M. and Drew, E. A., Routine analysis by gas-liquid chromatography of soluble carbohydrates in extracts of plant tissues. II. Quantitative analysis of standard carbohydrates, and the separation and estimation of soluble sugars and polyols from a variety of plant tissues, *New Phytol,* 70, 271, 1971.

3. El-Dash, A. A. and Hodge, J. E., Determination of 3-deoxy-D-*erythro*-hexosulose (3-deoxy-D-glucosone) and other degradation products of sugars by borohydride reduction and gas-liquid chromatography, *Carbohydr. Res.,* 18, 259, 1971.

4. Dutton, G. G. S., Gibney, K. B., Jensen, G. D., and Reid, P. E., The simultaneous estimation of polyhydric alcohols and sugars by gas-liquid chromatography. Applications to periodate-oxidixed polysaccharides, *J. Chromatogr.,* 36, 152, 1968.

5. Loewus, F. and Shah, R. H., Gas-liquid chromatography of trimethylsilyl ethers of cyclitols, in *Methods in Carbohydrate Chemistry,* Vol. 6, Whistler, R. L. and BeMiller, J. N., Eds., Academic Press, New York, 1972, 14.

6. Percival, E., Gas chromatography of the trimethylsilyl ethers of some less-common disaccharides, *Carbohydr. Res.,* 4, 441, 1967.

7. Kärkkäinen, J., Determination of the structure of disaccharides as trimethylsilyl derivatives of disaccharide alditols by gas-liquid chromatography — mass spectrometry, *Carbohydr. Res.,* 11, 247, 1969.

Table GC 22
TRIMETHYLSILYL ETHERS OF ALDITOLS DERIVED FROM PARTIALLY METHYLATED SUGARS

Packing	P1	P2	P3	P4	P5	P5
Temperature (°C)	175	140	140	140	140	140
Gas; flow rate (mℓ/min)	He, 20	He, 60	He, 60	He, 60	He, 60	He, 60
Column						
Length, cm	400	180	180	180	180	180
Diameter (O.D.), cm	0.2 (I.D.)	0.63	0.63	0.63	0.63	0.63
Form	na	coiled	coiled	coiled	coiled	coiled
Material	copper	glass	glass	glass	glass	glass
Detector	Fl	Fl	Fl	Fl	Fl	Fl
Reference	1	2	2	2	2	3
Parent sugar	r^a	r^b	r^b	r^b	r^b	r^b
L-Arabinose	—	0.68	1.55	1.54	1.38	—
2,3-di-O-methyl	—	0.50	0.73	0.73	0.69	—
2,5-di-O-methyl	—	0.41	0.62	0.62	0.58	—
3,4-di-O-methyl	—	0.50	0.73	0.73	0.69	—
3,5-di-O-methyl	—	0.41	0.62	0.62	0.58	—
2,3,5-tri-O-methyl	—	0.37	0.42	0.42	0.41	—
D-Xylose						
2,3,4-tri-O-methyl	—	—	—	—	—	0.52
L-Rhamnose	—	0.77	1.93	1.92	1.68	—
3-O-methyl	—	0.61	1.26	1.25	1.14	—
4-O-methyl	—	—	—	—	—	0.97
2,4-di-O-methyl	—	—	—	—	—	0.71
3,4-di-O-methyl	—	—	—	—	—	0.79
2,3,4-tri-O-methyl	—	0.49	0.58	0.58	0.57	—
D-Glucose	2.87	—	—	—	—	4.03
2-O-methyl	2.19	—	—	—	—	2.93
3-O-methyl	2.18	—	—	—	—	—
6-O-methyl	1.97	—	—	—	—	—
2,3-di-O-methyl	1.75	—	—	—	—	2.13
2,6-di-O-methyl	1.58	—	—	—	—	1.78
3,4-di-O-methyl	—	—	—	—	—	2.10
3,6-di-O-methyl	1.52	—	—	—	—	—
2,3,4-tri-O-methyl	—	—	—	—	—	1.67
2,3,6-tri-O-methyl	1.29	—	—	—	—	1.32
2,4,6-tri-O-methyl	—	—	—	—	—	1.35
2,3,4-6-tetra-O-methyl	1.00	—	—	—	—	1.01
D-Galactose	—	1.83	4.63	4.61	4.13	—
2-O-methyl	—	1.35	2.89	2.88	2.62	—
3-O-methyl	—	1.35	2.94	2.92	2.64	—
4-O-methyl	—	1.35	2.94	2.92	2.64	—
6-O-methyl	—	1.39	2.99	2.98	2.66	—
2,3-di-O-methyl	—	1.27	2.08	2.09	1.95	—
2,4-di-O-methyl	—	1.14	1.99	1.99	1.84	—
2,6-di-O-methyl	—	1.00	1.79	1.79	1.64	—
3,4-di-O-methyl	—	1.26	2.18	2.19	2.00	—
2,3,4-tri-O-methyl	—	1.20	1.57	1.57	1.50	—
2,3,6-tri-O-methyl	—	1.00	1.34	1.35	1.28	—
2,4,6-tri-O-methyl	—	0.92	1.25	1.25	1.19	—
3,4,6-tri-O-methyl	—	1.22	1.39	1.38	1.32	—
2,3,4,6-tetra-O-methyl	—	1.00	1.00	1.00	1.00	1.00
D-Mannose	—	—	—	—	—	3.90
6-O-methyl	—	—	—	—	—	2.33
2,3-di-O-methyl	—	—	—	—	—	1.88
3,6-di-O-methyl	—	—	—	—	—	1.63

Table GC 22 (continued)
TRIMETHYLSILYL ETHERS OF ALDITOLS DERIVED FROM PARTIALLY METHYLATED SUGARS

Packing	P1	P2	P3	P4	P5	P5
Temperature (°C)	175	140	140	140	140	140
Gas; flow rate (ml/min)	He, 20	He, 60	He, 60	He, 60	He, 60	He, 60
Column						
Length, cm	400	180	180	180	180	180
Diameter (O.D.), cm	0.2 (I.D.)	0.63	0.63	0.63	0.63	0.63
Form	na	coiled	coiled	coiled	coiled	coiled
Material	copper	glass	glass	glass	glass	glass
Detector	FI	FI	FI	FI	FI	FI
Reference	1	2	2	2	2	3
Parent sugar	r^a	r^b	r^b	r^b	r^b	r^b
4,6-di-*O*-methyl	—	—	—	—	—	1.64
2,3,6-tri-*O*-methyl	—	—	—	—	—	1.35
2,4,6-tri-*O*-methyl	—	—	—	—	—	1.19
3,4,6-tri-*O*-methyl	—	—	—	—	—	1.35
2,3,4,6-tetra-*O*-methyl	—	—	—	—	—	1.00

[a] t$_r$ relative to trimethylsilylated 2,3,4,6-tetra-*O*-methyl-D-glucitol (7.8 min).

[b] t$_r$ relative to trimethylsilylated 2,3,4,6-tetra-*O*-methyl-D-galactitol (8, 9, 11, and 5 min for P2, P3, P4, and P5, respectively).

Packing P1 = 2% SE-52 on acid-washed Chromosorb G.
 P2 = 3% OV-17 on Chromosorb W (60 to 80 mesh).
 P3 = 3% OV-101 on Chromosorb W (60 to 80 mesh).
 P4 = 3.8% SE-30 on Supelcoport (60 to 80 mesh).
 P5 = 3% SE-52 on Chromosorb W AW DMCS (80 to 100 mesh).

REFERENCES

1. Patel, S., Rivlin, J., Samuelson, T., Stamm, O. A., and Zollinger, H., A method for elucidation of the structure of products of the reaction of cellulose with formaldehyde, *Helv. Chim. Acta,* 51, 169, 1968.
2. Freeman, B. H., Stephen, A. M., and Van der Bijl, P., Gas-liquid chromatography of some partially methylated alditols as their trimethylsilyl ethers., *J. Chromatogr.,* 73, 29, 1972.
3. Merrifield, E. H., unpublished laboratory data, 1975.

Table GC 23
TRIMETHYLSILYL ETHERS OF PARTIALLY METHYLATED SUGARS AND METHYL GLYCOSIDES

	P1	P2	P3	P4	P5
Packing	T1	150	100	125	136
Temperature (°C)	na	He, na	na	Ar, 20—35	He, 60
Gas; flow rate (ml/min)					
Column					
Length, cm	183	366	244	120	366
Diameter (I.D.), cm	0.63 (O.D.)	0.63 (O.D.)	na	0.4	0.63
Form	straight	straight	na	straight	na
Material	copper	copper	SS	glass	glass
Detector	F1	F1	F1	D1	D2
Reference	1	2	3	4	5,6
Parent compound	r[a]	r[b]	r[c]	r[d]	r[e]
D-Xylose					
2-O-methyl	1.73	—	—	—	—
3-O-methyl	1.57; 1.67	—	—	—	—
4-O-methyl	1.38	—	—	—	—
2,3-di-O-methyl	1.28	—	—	—	—
2,3,4-tri-O-methyl	1.00	—	—	—	—
D-Glucose					
2-O-methyl	—	1.08; 1.50	—	5.98; 9.26	—
3-O-methyl	—	0.76; 1.25	—	4.80; 9.50	—
4-O-methyl	—	0.90; 1.67	—	—	—
6-O-methyl	—	1.52; 1.85	—	7.81; 9.67	—
2,3-di-O-methyl	—	—	0.91; 1.00	2.63; 3.44	—
2,4-di-O-methyl	—	—	0.86; 1.15	—	—
2,6-O-methyl	—	—	—	4.33	—
3,4-O-methyl	—	—	0.78; 1.15	—	—
3,6-di-O-methyl	—	—	—	3.40; 3.99	—
4,6-di-O-methyl	—	—	—	3.91; 4.86	—
2,3,6-tri-O-methyl	—	—	—	1.82; 1.59	—
2,3,4,6-tetra-O-methyl	—	—	—	1.00; 0.91	0.69; 0.76
D-Mannose					
2,4-di-O-methyl	—	—	—	—	0.92

2,6-di-*O*-methyl	—	—	—	1.27; 1.67[f]
3,4-di-*O*-methyl	—	—	—	0.67
3,5-di-*O*-methyl	—	—	—	1.41; 1.70[f]
5,6-di-*O*-methyl	—	—	—	1.45; 1.92[f]
2,3,4-tri-*O*-methyl	—	—	—	0.66
2,5,6-tri-*O*-methyl	—	—	—	0.97
3,4,6-tri-*O*-methyl	—	—	—	0.42
3,5,6-tri-*O*-methyl	—	—	—	1.05;[f] 1.23
2,3,4,6-tetra-*O*-methyl	—	—	—	0.95; 1.04[f]
2,3,5,6-tetra-*O*-methyl	—	—	—	1.12
Methyl glycosides				
Methyl D-mannoside				
2,3-di-*O*-methyl	—	—	—	0.87
2,4-di-*O*-methyl	—	—	—	0.84
2,5-di-*O*-methyl	—	—	—	1.11
2,6-di-*O*-methyl	—	—	—	1.02;[g] 1.30[h]
3,4-di-*O*-methyl	—	—	—	0.62
3,5-di-*O*-methyl	—	—	—	1.20
3,6-di-*O*-methyl	—	—	—	0.78;[h] 1.09[g]
4,6-di-*O*-methyl	—	—	—	0.74
5,6-di-*O*-methyl	—	—	—	1.30
2,3,4-tri-*O*-methyl	—	—	—	0.82
2,3,5-tri-*O*-methyl	—	—	—	1.20
2,3,6-tri-*O*-methyl	—	—	—	1.14
2,4,6-tri-*O*-methyl	—	—	—	0.58
2,5,6-tri-*O*-methyl	—	—	—	0.99
3,4,6-tri-*O*-methyl	—	—	—	0.58
3,5,6-tri-*O*-methyl	—	—	—	0.99
2,3,4-6-tetra-*O*-methyl	—	—	—	1.00

[a] t, relative to trimethylsilylated 2,3,4-tri-*O*-methyl-α-D-xylose (about 16 min).

[b] t, relative to trimethylsilylated α-D-glucose.

[c] t, relative to trimethylsilylated 2,3-di-*O*-methyl-β-D-glucose (about 56 min).

[d] t, relative to trimethylsilylated 2,3,4,6-tetra-*O*-methyl-α-D-glucose.

[e] t, relative to trimethylsilylated 2,3,4,6-tetra-*O*-methyl-α-D-mannoside (29.4 min).

[f] Minor peak.

[g] Furanose.

[h] Pyranose.

Table GC 23 (continued)

TRIMETHYLSILYL ETHERS OF PARTIALLY METHYLATED SUGARS AND METHYL
GLYCOSIDES

Packing   P1   =   3% SE-52 on Gas-Chrom A.
             P2   =   19.5% Carbowax 20M on Chromosorb W (80 to 100 mesh).
             P3   =   5% butanediol succinate polyester on Chromosorb W.
             P4   =   8.7% poly(ethylene glycol succinate) on kieselguhr (60 to 100 mesh).
             P5   =   2% neopentyl glycol succinate polyester on Chromosorb W (80 to 100 mesh).
Temperature T1   =   temperature programmed, 50 → 125°C at 5.6°/min, then isothermal at 125°C for about 20 min.
Detector   D1   =   β-ionization.
             D2   =   thermal conductivity.

REFERENCES

1. Sephton, H. H., Separation and determination of the hydrolysis products of methylated xylan as their trimethylsilyl
    derivatives by vapor phase chromatography, J. Org. Chem., 29, 3415, 1964.
2. Dick, W. E., Baker, B. G., and Hodge, J. E., Preparation and characterization of 1,2,6,2′, 3′, 4′, 6′-hepta-O-acetyl-β-
    maltose, Carbohydr. Res., 6, 52, 1968.
3. Norrman, B., Partial methylation of dextran, Acta Chem. Scand., 22, 1381, 1968.
4. Haworth, S., Roberts, J. G., and Sagar, B. F., Quantitative determination of mixtures of alkyl ethers of D-
    glucose, Carbohydr. Res., 9, 491, 1969.
5. Bhattacharjee, S. S. and Gorin, P. A. J., Identification of di-O-, tri-O-, and tetra-O-methylmannoses by gas-liquid
    chromatography, Can. J. Chem., 47, 1207, 1969.
6. Jones, H. G., Gas-liquid chromatography of methylated sugars, in Methods in Carbohydrate Chemistry, Vol. 6,
    Whistler, R. L. and BeMiller, J. N., Eds., Academic Press, New York, 1972, 25.

Table GC 24

TRIMETHYLSILYL ETHERS AND ESTERS OF URONIC, ALDONIC, AND ALDARIC ACIDS, AND THEIR DERIVATIVES

Packing	P1	P2	P3	P4	P5	P6	P7	P8	P9	P10
Temperature (°C)	190	170	T1	200	175	T2	170	160	160	T3
Gas; flow rate (ml/min)	N₂, 29—32	N₂, 29—32	N₂, 50	N₂, 40	N₂, 40	N₂, 25	Ar, 140	N₂, 30	N₂, 30	N₂, 30
Column										
Length, cm	200	200	250	152	152	122	120	200	200	200
Diameter (I.D.), cm	0.3	0.3	0.32	0.4	0.4	0.32	0.5	0.2	0.2	0.2
Form	na	na	U-tube	coiled	coiled	na	straight	na	na	na
Material	na	na	glass	glass	glass	SS	glass	na	na	na
Detector	FI	FI	FI	FI	FI	FI	D1	FI	FI	FI
Reference	1	1	2	3	3	4	5, 6	7, 8	7, 8	7, 8
Parent compound	r[a]	r[b]	r[b]	r[c]	r[c]	r[d]	r[e]	r[f]	r[f]	r[f]
Hexuronic acids										
D-Glucuronic acid	4.9; 6.3 7.8	5.3; 6.8; 9.2	1.07; 1.27	1.73; 2.10; 2.53	2.52; 3.26; 4.17	—	—	—	—	—
D-Galacturonic acid	4.1; 5.5; 6.0; 7.7	4.5; 5.9; 6.4; 8.3	0.82;[a] 1.18	1.48; 1.89; 2.02; 2.60	2.04; 2.73; 4.06; 4.69	1.00	—	—	—	—
D-Mannuronic acid	4.1; 6.1	4.4; 6.7	—	1.52; 2.11; 2.21; 2.54	2.46; 3.51; 3.96	—	—	—	—	—
L-Guluronic acid	4.4	4.8	—	1.60; 1.91; 2.42; 2.67	2.75; 2.97; 4.06	—	—	—	—	—
L-Iduronic acid	—	—	—	1.48; 1.80; 2.00	2.09; 2.51; 2.94	—	—	—	—	—
Methyl ester methyl glycosides										
Methyl D-glucuronate	3.1;[A] 2.9;[A] 4.9;[i] 4.7[i]	3.0;[A] 4.8[i]	0.45; 0.82	—	—	—	—	—	—	—
Methyl D-galacturonate	3.1;[A] 2.6;[A] 4.1;[i] 4.3[i]	3.0;[A] 2.5;[A] 3.9;[i] 4.0[i]	0.49; 0.56; 0.70	—	—	—	—	—	—	—
Methyl D-mannuronate	2.9;[A] 3.5;[A] 2.7;[i] 3.2;[i]	3.0;[A] 3.5;[A] 2.6;[i] 3.0;[i]	0.52	—	—	—	—	—	—	—
Methyl L-guluronate	—	—	0.30; 0.42; 0.55	—	—	—	—	—	—	—
Methyl L-idurorate	—	—	0.37	—	—	—	—	—	—	—

Table GC 24 (continued)

TRIMETHYLSILYL ETHERS AND ESTERS OF URONIC, ALDONIC, AND ALDARIC ACIDS, AND THEIR DERIVATIVES

Packing	P1	P2	P3	P4	P5	P6	P7	P8	P9	P10
Temperature (°C)	190	170	T1	200	175	T2	170	160	160	T3
Gas; flow rate (ml/min)	N$_2$, 29—32	N$_2$, 29—32	N$_2$, 50	N$_2$, 40	N$_2$, 40	N$_2$, 25	Ar, 140	N$_2$, 30	N$_2$, 30	N$_2$, 30
Column										
Length, cm	200	200	250	152	152	122	120	200	200	200
Diameter (I.D.), cm	0.3	0.3	0.32	0.4	0.4	0.32	0.5	0.2	0.2	0.2
Form	na	na	U-tube	coiled	coiled	na	straight	na	na	na
Material	na	na	glass	glass	glass	SS	glass	na	na	na
Detector	Fl	Fl	Fl	Fl	Fl	Fl	Dl	Fl	Fl	Fl
Reference	1	1	2	3	3	4	5, 6	7, 8	7, 8	7, 8
Parent compound	r^a	r^b	r^c	r^d	r^e	r^f	r^g	r^h	r^i	r^j
Hexuronolactones										
D-Glucurono-6,3-lactone	3.6	3.0; 3.2	—	1.20; 1.27	7.18; 9.13	—	—	—	—	—
D-Mannurono-6,3-lactone	3.6; 4.7	3.1; 4.1	—	1.22; 1.48	10.10; 16.73	—	—	—	—	—
L-Gulurono-6,3-lactone	2.4; 3.2	2.2; 3.8	—	0.86; 1.15	—	—	—	—	—	—
L-Idurono-6,3-lactone	—	—	—	1.00	4.14; 4.32	—	—	—	—	—
Oligogalacturonic acids										
Digalacturonic acid	—	—	—	—	—	4.32	—	—	—	—
methyl ester	—	—	—	—	—	4.04	—	—	—	—
Trigalacturonic acid	—	—	—	—	—	6.51	—	—	—	—
methyl ester	—	—	—	—	—	6.05	—	—	—	—
Aldonic acids										
Glyceronic (glyceric) acid	—	—	—	—	—	—	—	0.052	0.092	0.086
2-C-Methylglyceronic acid	—	—	—	—	—	—	—	0.054	0.082	0.072
L-Threonic acid	—	—	—	—	—	—	—	0.156	0.251	0.237
D-Erythronic acid	—	—	—	—	—	—	—	0.151	0.218	0.212
2-C-Methyl-D-erythronic acid	—	—	—	—	—	—	—	0.184	0.246	0.244
D-Arabinonic acid	—	—	—	—	—	—	—	0.463	0.646	0.632

Compound			V1	V2	V3
D-Xylonic acid	—	—	0.428	0.617	0.570
D-Ribonic acid	—	—	0.438	0.569	0.565
D-Lyxonic acid	—	—	0.457	0.599	0.621
2-Deoxy-D-*erythro*-pentonic acid	—	—	0.252	0.379	0.395
2-Deoxy-D-*threo*-pentonic acid	—	—	0.251	0.373	0.392
3-Deoxy-D-*erythro*-pentonic acid	—	—	0.236	0.380	0.375
3-Deoxy-D-*threo*-pentonic acid	—	—	0.252	0.419	0.418
2-C-Methyl-D-arabinonic acid	—	—	0.554	0.674	0.710
2-C-Methyl-D-ribonic acid	—	—	0.564	0.689	0.742
D-Gluconic acid	6.9	8.1	1.286	1.569	1.639
D-Galactonic acid	6.8	8.2	1.256	1.561	1.570
D-Mannonic acid	6.1	7.2	1.113	1.254	1.385
D-Allonic acid	—	—	1.115	1.276	1.410
D-Altronic acid	—	—	1.247	1.531	1.568
L-Gulonic acid	6.1	6.9	1.087	1.256	1.309
L-Idonic acid	7.4	8.6	1.368	1.743	1.812
D-Talonic acid	—	—	1.235	1.507	1.656
6-Deoxy-L-galactonic acid	—	—	0.615	0.809	0.801
6-Deoxy-L-mannonic acid	—	—	0.528	0.629	0.687
2-Deoxy-D-*arabino*-hexonic acid	—	—	0.659	0.912	1.007
2-Deoxy-D-*lyxo*-hexonic acid	—	—	0.673	0.935	1.035
3-Deoxy-D-*arabino*-hexonic acid	—	—	0.676	0.971	1.004
3-Deoxy-D-*lyxo*-hexonic acid	—	—	0.644	0.894	0.973
3-Deoxy-D-*ribo*-hexonic acid	—	—	0.673	0.946	1.007
3-Deoxy-D-*xylo*-hexonic acid	—	—	0.659	0.995	1.026
2,5-Anhydro-D-gluconic acid	—	—	0.612	1.271	1.928
2,5-Anhydro-D-mannonic acid	—	—	0.477	0.986	1.416
2,5-Anhydro-D-talonic acid	—	—	0.590	1.241	1.776
3-O-Methyl-D-gulonic acid	—	—	0.728	1.038	0.990
Aldonolactones					
L-Threono-1,4-lactone	—	—	0.06	0.17	0.56
D-Erythrono-1,4-lactone	—	—	0.08	0.27	0.92
2-C-Methyl-D-erythrono-1,4-lactone	0.68	—	0.08	0.20	0.68
L-Arabinono-1,4-lactone	0.79	—	0.20	0.49	1.13
D-Xylono-1,4-lactone	—	—	0.22	0.54	1.58
1,5-lactone	—	—	0.26	0.67	1.79
D-Ribono-1,4-lactone	1.00	—	0.25	0.64	2.06
D-Lyxono-1,4-lactone	1.58	—	0.32	0.83	2.47

Table GC 24 (continued)
TRIMETHYLSILYL ETHERS AND ESTERS OF URONIC, ALDONIC, AND ALDARIC ACIDS, AND THEIR DERIVATIVES

	P1	P2	P3	P4	P5	P6	P7	P8	P9	P10
Packing										
Temperature (°C)	190	170	T1	200	175	T2	170	160	160	T3
Gas; flow rate (ml/min)	N_2, 29—32	N_2, 29—32	N_2, 50	N_2, 40	N_2, 40	N_2, 25	Ar, 140	N_2, 30	N_2, 30	N_2, 30
Column										
Length, cm	200	200	250	152	152	122	120	200	200	200
Diameter (I.D.), cm	0.3	0.3	0.32	0.4	0.4	0.32	0.5	0.2	0.2	0.2
Form	na	na	U-tube	coiled	coiled	na	straight	na	na	na
Material	na	na	glass	glass	glass	SS	glass	na	na	na
Detector	Fl	Fl	Fl	Fl	Fl	Fl	DI	Fl	Fl	Fl
Reference	1	1	2	3	3	4	5, 6	7, 8	7, 8	7, 8
Parent compound	r	r	r^b	r^c	r^c	r^d	r^e	r^e	r^e	r^f
2-Deoxy-D-erythro-pentono-1,4-lactone	—	—	—	—	—	—	0.69	0.11	0.33	1.48
2-Deoxy-D-threo-pentono-1,4-lactone	—	—	—	—	—	—	—	0.12	0.44	1.81
3-Deoxy-D-erythro-pentono-1,4-lactone	—	—	—	—	—	—	—	0.10	0.34	0.98
3-Deoxy-D-threo-pentono-1,4-lactone	—	—	—	—	—	—	—	0.11	0.38	1.02
2-C-Methyl-D-arabinono-1,4-lactone	—	—	—	—	—	—	—	0.21	0.42	1.01
2-C-Methyl-D-ribono-1,4-lactone	—	—	—	—	—	—	—	0.21	0.45	1.24
D-Glucono-1,4-lactone	4.5	4.6	—	—	—	—	1.64	0.71	1.48	3.19
1,5-lactone	4.4	4.4	—	—	—	—	—	0.67	1.49	2.99
L-Galactono-1,4-lactone	4.3	4.5	—	—	—	—	1.53	0.69	1.37	2.82
D-Mannono-1,4-lactone	6.3	6.5	—	—	—	—	3.07	1.01	2.21	4.64
1,5-lactone	—	—	—	—	—	—	—	0.65	1.35	3.27
D-Allono-1,4-lactone	—	—	—	—	—	—	—	0.81	1.70	4.80
D-Altrono-1,4-lactone	—	—	—	—	—	—	2.03	0.62	1.19	2.28
L-Gulono-1,4-lactone	5.0	5.0	—	—	—	—	2.36	0.78	1.72	3.58
L-Idono-1,4-lactone	4.9	5.2	—	—	—	—	1.88	0.79	1.63	4.31
D-Talono-1,4-lactone	—	—	—	—	—	—	2.01	0.80	1.70	4.87
6-Deoxy-L-galactono-1,4-lactone	—	—	—	—	—	—	0.76	0.25	0.56	1.31
6-Deoxy-L-mannono-1,4-lactone	—	—	—	—	—	—	1.65	0.38	0.94	2.57

2-Deoxy-D-*arabino*-hexono-1,4-lactone	—	—	—	—	—	—	—	0.42	1.28	3.64
2-Deoxy-D-*lyxo*-hexono-1,4-lactone	—	—	—	—	—	—	—	0.39	1.00	4.17
3-Deoxy-D-*arabino*-hexono-1,4-lactone	—	—	—	—	—	—	—	0.37	1.01	2.24
3-Deoxy-D-*lyxo*-hexono-1,4-lactone	—	—	—	—	—	—	—	0.41	1.09	2.99
3-Deoxy-D-*ribo*-hexono-1,4-lactone	—	—	—	—	—	—	—	0.38	0.92	2.40
3-Deoxy-D-*xylo*-hexono-1,4-lactone	—	—	—	—	—	—	—	0.40	1.13	2.66
D-*glycero*-L-*galacto*-Heptono-1,4-lactone	—	—	—	—	—	2.63	—	—	—	—
D-*glycero*-D-*galacto*-Heptono-1,4-lactone	—	—	—	—	—	2.81	—	—	—	—
D-*glycero*-L-*manno*-Heptono-1,4-lactone	—	—	—	—	—	4.44	—	—	—	—
D-*glycero*-D-*gulo*-Heptono-1,4-lactone	—	—	—	—	—	4.56	—	—	—	—
D-*glycero*-D-*talo*-Heptono-1,4-lactone	—	—	—	—	—	2.92	—	—	—	—
L-*threo*-Hex-2-enono-1,4-lactone[i]	—	—	—	—	—	—	—	0.88	2.01	4.72
Aldaric acids and lactones										
Glyceraric acid	—	—	—	—	—	—	—	0.064	0.152	0.190
Threaric acid	—	—	—	—	—	—	—	0.212	0.432	0.471
Erythraric acid	—	—	—	—	—	—	—	0.174	0.313	0.354
Arabinaric acid	—	—	—	—	—	—	—	0.522	0.864	0.945
Xylaric acid	—	—	—	—	—	—	—	0.545	0.978	1.034
Ribaric acid	—	—	—	—	—	—	—	0.532	0.873	1.037
2-Deoxy-*erythro*-pentaric acid	—	—	—	—	—	—	—	0.278	0.517	0.655
2-Deoxy-*threo*-pentaric acid	—	—	—	—	—	—	—	0.291	0.573	0.736
3-Deoxy-*erythro*-pentaric acid	—	—	—	—	—	—	—	0.280	0.538	0.673
3-Deoxy-*threo*-pentaric acid	—	—	—	—	—	—	—	0.295	0.593	0.774
Glucaric acid	7.4	7.4	—	—	—	—	—	1.295	1.814	1.920
1,4-lactone	5.5	5.6	—	—	—	—	—	—	—	—
6,3-lactone	6.1	6.3	—	—	—	—	—	—	—	—
1,4:6,3-dilactone	2.8	2.2	—	—	—	—	—	—	—	—
Galactaric acid	8.4	9.5	—	—	—	—	—	1.500	2.272	2.347
Mannaric acid	6.2	6.5	—	—	—	—	—	1.074	1.346	1.401
1,4:6,3-dilactone	5.3	4.0	—	—	—	—	—	—	—	—
Allaric acid	—	—	—	—	—	—	—	1.242	1.669	1.904
Altraric acid	—	—	—	—	—	—	—	1.419	2.098	2.195
Idaric acid	—	—	—	—	—	—	—	1.602	2.518	2.589

Table GC 24 (continued)
TRIMETHYLSILYL ETHERS AND ESTERS OF URONIC, ALDONIC, AND ALDARIC ACIDS, AND THEIR DERIVATIVES

	P1	P2	P3	P4	P5	P6	P7	P8	P9	P10
Packing										
Temperature (°C)	190	170	T1	200	175	T2	170	160	160	T3
Gas; flow rate (ml/min)	N$_2$, 29—32	N$_2$, 29—32	N$_2$, 50	N$_2$, 40	N$_2$, 40	N$_2$, 25	Ar, 140	N$_2$, 30	N$_2$, 30	N$_2$, 30
Column										
Length, cm	200	200	250	152	152	122	120	200	200	200
Diameter (I.D.), cm	0.3	0.3	0.32	0.4	0.4	0.32	0.5	0.2	0.2	0.2
Form	na	na	U-tube	coiled	coiled	na	straight	na	na	na
Material	na	na	glass	glass	glass	SS	glass	na	na	na
Detector	Fl	Fl	Fl	Fl	Fl	Fl	Dl	Fl	Fl	Fl
Reference	1	1	2	3	3	4	5, 6	7, 8	7, 8	7, 8
Parent compound	r[a]	r[e]	r[b]	r[c]	r[c]	r[d]	r[e]	r[f]	r[f]	r[f]
2-Deoxy-*arabino*-hexaric acid	—	—	—	—	—	—	—	0.777	1.277	1.565
2-Deoxy-*lyxo*-hexaric acid	—	—	—	—	—	—	—	0.847	1.467	1.751
3-Deoxy-*arabino*-hexaric acid	—	—	—	—	—	—	—	0.767	1.425	1.675
3-Deoxy-*lyxo*-hexaric acid	—	—	—	—	—	—	—	0.789	1.470	1.713
3-Deoxy-*ribo*-hexaric acid	—	—	—	—	—	—	—	0.732	1.290	1.550
3-Deoxy-*xylo*-hexaric acid	—	—	—	—	—	—	—	0.761	1.317	1.639
3-*O*-Methylgularic acid	—	—	—	—	—	—	—	0.909	1.610	1.770

[a] t, relative to trimethylsilylated *meso*-erythritol.
[b] t, relative to trimethylsilylated D-mannitol (about 58 min.)
[c] t, relative to trimethylsilylated ribitol.
[d] t, relative to trimethylsilylated D-galacturonic acid (major peak, 2.2 min).
[e] t, relative to 2,3,5-tri-*O*-(trimethylsilyl)-D-ribono-1,4-lactone (about 10 min).
[f] t, relative to trimethylsilylated D-glucitol (about 12 min for P8, 7 min for P9; for P10 15 min at 120°, 3 min at 160°).
[g] Minor peak.
[h] Furanose.
[i] Pyranose.
[j] L-Ascorbic acid (vitamin C).

Packing   P1 = 2.5% SE-52 on Chromosorb G AW DMCS (80 to 100 mesh).

P2 = 1% SE-30 on Chromosorb G AW DMCS (80 to 100 mesh).

P3 = 3.8% SE-30 on Diatoport S (80 to 100 mesh).

P4 = 10% SE-30 on Celite (100 to 120 mesh).

P5 = 4% XE-60 on Celite (100 to 120 mesh).

P6 = 0.5% SE-30 on acid-washed, silanized Chromosorb W (80 to 100 mesh).

P7 = 10% neopentyl glycol sebacate polyester on acid-washed Chromosorb W (100 to 120 mesh).

P8 = 0.5% OV-1 on Chromosorb G (100 to 120 mesh).

P9 = 0.5% OV-17 on Chromosorb G (100 to 120 mesh).

P10 = 3% DC QF-1 on Gas-Chrom Q (100 to 120 mesh).

Temperature   T1 = temperature programmed, 140 → 200°C at 0.5°/min.

T2 = temperature programmed, 130 → 300°C at 12°/min; isothermal at 300°C for 6 min.

T2 = 120°C for aldonic and aldaric acids, 160°C for lactones

Detector   D1 = ionization detector (^{90}Sr).

REFERENCES

1. **Raunhardt, O., Schmidt, H. W. H. and Neukom, H.,** Gas-chromatographic investigations of uronic acids and uronic acid derivatives, *Helv. Chim. Acta,* 50, 1267, 1967.

2. **Bhatti, T., Chambers, R. E., and Clamp, J. R.,** The gas chromatographic properties of biologically important *N*-acetylglucosamine derivatives, monosaccharides, disaccharides, trisaccharides, tetrasaccharides, and pentasaccharides, *Biochim. Biophys. Acta,* 222, 339, 1970

3. **Kennedy, J. F., Robertson, S. M., and Stacey, M.,** G.l.c. of the *O*-trimethylsilyl derivatives of hexuronic acids, *Carbohydr. Res.,* 49, 243, 1976.

4. **Raymond, W. R. and Nagel, C. W.,** Gas-liquid chromatographic determination of oligogalacturonic acids, *Anal. Chem.,* 41, 1700, 1969.

5. **Perry, M. B. and Hulyalkar, R. K.,** The analysis of hexuronic acids in biological materials by gas-liquid partition chromatography, *Can. J. Biochem.,* 43, 573, 1965.

6. **Morrison, I. M. and Perry, M. B.,** The analysis of neutral glycoses in biological materials by gas-liquid partition chromatography, *Can. J. Biochem.,* 44, 1115, 1966.

7. **Petersson, G.,** Gas-chromatographic analysis of sugars and related hydroxy acids as acyclic oxime and ester trimethylsilyl derivatives, *Carbohydr. Res.,* 33, 47, 1974.

8. **Petersson, G.,** Retention data in GLC analysis; carbohydrate-related hydroxy carboxylic and dicarboxylic acids as trimethylsilyl derivatives, *J. Chromatogr. Sci.,* 15, 245, 1977.

Table GC 25
TRIMETHYLSILYL ETHERS OF AMINODEOXY AND ACETAMIDODEOXY SUGARS, NEURAMINIC ACIDS, AND THEIR DERIVATIVES

	P1	P2	P3	P4	P5	P6	P7	P8
Packing								
Temperature (°C)	T1	187	140	T2	T3	T4	140	220
Gas; flow rate (ml/min)	na, 75	Ar, 15	Ar, 15	N_2, 50	N_2, 30	N_2, 30	He, na	N_2, 40
Column								
Length, cm	183	200	200	250	183	200	91	200
Diameter (I.D.), cm	0.63	0.35	0.35	0.32	0.2	0.2	0.32	0.4
Form	coiled	coiled	coiled	U—tube	coiled	spiral	na	na
Material	SS	glass	glass	glass	glass	SS	glass	glass
Detector	FI	FI	FI	FI	FI	DI	FI	FI
Reference	1	2	2	3	4	5	6, 7, 8	9
Parent compound	r^a	r^a	r^a	r^b	r^c	r^d	r^e	r^f
2-Amino-2-deoxy-D-glucose	0.91; 1.15	0.88; 1.12	1.20; 1.38	0.79; 0.92	—	—	—	—
2-Amino-2-deoxy-D-galactose	0.79	0.77	1.11; 1.22	0.73; 0.83[a]	—	—	—	—
2-Amino-2-deoxy-D-mannose	—	—	—	0.63; 0.81	—	—	—	—
2-Acetamido-2-deoxy-D-arabinose	—	—	—	0.60; 0.66;[a] 0.68	—	—	—	—
2-Acetamido-2-deoxy-D-ribose	—	—	—	0.55;[a] 0.60; 0.68	—	—	—	—
2-Acetamido-2-deoxy-D-glucose	2.27	2.25	8.3; 10.5	1.38	1.00	—	—	—
2-Acetamido-2-deoxy-D-galactose	2.08	2.04	—	1.30	0.95	—	—	—
2-Acetamido-2-deoxy-D-mannose	1.71	—	—	0.97; 1.15	0.68; 0.78	—	—	—
2-Acetamido-2-deoxy-D-allose	—	—	—	1.05;[a] 1.19; 1.24	—	—	—	—
2-Acetamido-2-deoxy-D-altrose	—	—	—	0.96;[a] 1.09;[a] 1.11; 1.21	—	—	—	—
2-Acetamido-2-deoxy-D-gulose	—	—	—	1.05; 1.07	—	—	—	—
2-Acetamido-2-deoxy-D-idose	—	—	—	0.85; 1.00; 1.04; 1.11	—	—	—	—
N-Acetylneuraminic acid	—	—	—	2.41	—	—	—	—
N-Glycolylneuraminic acid	—	—	—	2.80	—	—	—	—

Methyl glycosides						
Methyl 2-amino-2-deoxy-D-glucoside	—	0.70	1.02	—	—	—
Methyl 2-amino-2-deoxy-D-galactoside	—	0.59	1.06	—	—	—
Methyl 2-acetamido-2-deoxy-D-arabinoside	—	—	—	0.53; 0.55	—	—
Methyl 2-acetamido-2-deoxy-D-riboside	—	—	—	0.45; 0.50	—	—
Methyl 2-acetamido-2-deoxy-D-glucoside	—	2.05	—	1.05; 1.17; 1.31; 1.38[a]	0.86;[a] 0.93[a]	2.76;[a] 3.38;[a] 4.00; 5.28[a]
3-O-methyl	—	—	—	—	0.73[a]	—
Methyl 2-acetamido-2-deoxy-D-galactoside	—	1.77	—	1.08; 1.21	0.72;[a] 0.73;[a] 0.87;[a] 0.92[a]	2.92; 3.55
Methyl 2-acetamido-2-deoxy-D-mannoside	—	—	—	0.94; 1.04;[a] 1.18;[a] 1.22[a]	—	—
Methyl 2-acetamido-2-deoxy-D-alloside	—	—	—	0.99; 1.04;[a]	1.12;[a] 1.15	—
Methyl 2-acetamido-2-deoxy-D-altroside	—	—	—	0.84;[a] 0.88; 1.00;[a] 1.08	—	—
Methyl 2-acetamido-2-deoxy-D-guloside	—	—	—	0.88; 1.04; 1.18;[a] 1.20	—	—
Methyl 2-acetamido-2-deoxy-D-idoside	—	—	—	0.86; 0.92;[a] 0.95; 1.00	—	—
Methyl N-acetylneuraminate	7.25	—	—	2.22	—	9.15; 10.10[a]
Methyl N-glycolylneuraminate	—	—	—	—	—	9.15; 10.10
Alditols						
2-Amino-2-deoxy-D-glucitol	—	1.15	1.44	—	—	—
2-Amino-2-deoxy-D-galactitol	—	1.15	1.67	—	—	—
2-Acetamido-2-deoxy-D-glucitol	—	2.12	—	1.33	—	—
2-Acetamido-2-deoxy-D-galactitol	—	2.23	—	1.37	—	—
Disaccharides						
β-D-Galp-(1→4)-D-GlcpNAc	—	—	—	2.04; 2.32	1.00	—
β-D-GlcpNAc-(1→4)-D-GlcpNAc	—	—	—	3.20	1.66	—
β-D-GlcpNAc-(1→6)-D-GlcpNAc	—	—	—	—	—	—
Methylated derivatives						
2-Deoxy-2-methylamino-D-glucose	—	—	—	—	—	1.70; 2.79
3-O-methyl	—	—	—	—	—	1.29; 1.75[a]
4-O-methyl	—	—	—	—	—	1.00; 1.68
6-O-methyl	—	—	—	—	—	1.89; 2.27[a]
3,4-di-O-methyl	—	—	—	—	—	0.89; 1.29
3,6-di-O-methyl	—	—	—	—	—	1.42

Table GC 25 (continued)

TRIMETHYLSILYL ETHERS OF AMINODEOXY AND ACETAMIDODEOXY SUGARS, NEURAMINIC ACIDS, AND THEIR DERIVATIVES

	P1	P2	P3	P4	P5	P6	P7	P8
Packing								
Temperature (°C)	T1	187	140	T2	T3	T4	140	220
Gas; flow rate (ml/min)	na, 75	Ar, 15	Ar, 15	N₂, 50	N₂, 30	N₂, 30	He, na	N₂, 40
Column								
Length, cm	183	200	200	250	183	200	91	200
Diameter (I.D.), cm	0.63	0.35	0.35	0.32	0.2	0.2	0.32	0.4
Form	coiled	coiled	coiled	U—tube	coiled	spiral	na	na
Material	SS	glass	glass	glass	glass	SS	glass	glass
Detector	FI	FI	FI	FI	FI	DI	FI	FI
Reference	1	2	2	3	4	5	6, 7, 8	9
Parent compound	r^a	r^a	r^a	r^a	r^a	r^a	r^a	r^a
4,6-di-O-methyl	—	—	—	—	—	—	1.11; 1.45[a]	—
3,4,6-tri-O-methyl	—	—	—	—	—	—	1.00	—
2-Deoxy-2-methylamino-D-galactose	—	—	—	—	—	—	1.26; 1.77	—
3-O-methyl	—	—	—	—	—	—	1.36; 1.59	—
4-O-methyl	—	—	—	—	—	—	1.30; 1.99	—
6-O-methyl	—	—	—	—	—	—	1.16; 1.57[a]	—
3,4-di-O-methyl	—	—	—	—	—	—	1.70; 2.16	—
3,6-di-O-methyl	—	—	—	—	—	—	1.19; 1.40	—
4,6-di-O-methyl	—	—	—	—	—	—	1.05; 1.67	—
3,4,6-tri-O-methyl	—	—	—	—	—	—	1.40; 1.75	—
2-Deoxy-2-methylamino-D-glucitol	—	—	—	—	—	—	1.29	—
3-O-methyl	—	—	—	—	—	—	1.14	—
4-O-methyl	—	—	—	—	—	—	1.07	—
6-O-methyl	—	—	—	—	—	—	0.93	—
3,4-di-O-methyl	—	—	—	—	—	—	1.12	—
3,6-di-O-methyl	—	—	—	—	—	—	0.90	—
4,6-di-O-methyl	—	—	—	—	—	—	0.79	—
3,4,6-tri-O-methyl	—	—	—	—	—	—	0.93	—
2-Deoxy-2-methylamino-D-galactitol	—	—	—	—	—	—	1.20	—
3-O-methyl	—	—	—	—	—	—	1.26	—

4-*O*-methyl	—	—	—	—	—	1.04	—
6-*O*-methyl	—	—	—	—	—	0.99	—
3,4-di-*O*-methyl	—	—	—	—	—	1.26	—
3,6-di-*O*-methyl	—	—	—	—	—	1.04	—
4,6-di-*O*-methyl	—	—	—	—	—	0.92	—
3,4,6-tri-*O*-methyl	—	—	—	—	—	1.32	—
1,2-*O*-Me-NeuN(Ac,Me)[j]	—	—	—	—	—	—	1.89
4-*O*-methyl	—	—	—	—	—	—	1.70
9-*O*-methyl	—	—	—	—	—	—	1.43
4,9-di-*O*-methyl	—	—	—	—	—	—	1.27
4,7,8-tri-*O*-methyl	—	—	—	—	—	—	1.30
4,7,9-tri-*O*-methyl	—	—	—	—	—	—	1.14
4,8,9-tri-*O*-methyl	—	—	—	—	—	—	1.07
4,7,8,9-tetra-*O*-methyl	—	—	—	—	—	—	1.00

[a] t, relative to trimethylsilyl ether of α-D-glucose (20 min at 140°C, 6.2 min at 160°C, in Reference 1).

[b] t, relative to trimethylsilylated D-mannitol (about 57 min).

[c] t, of acetamidodeoxyhexoses and glycosides of 2-acetamido-2-deoxy-D-glucose relative to TMS ether of 2-acetamido-2-deoxy-D-glucose (37.6 min); t, of 2-acetamido-2-deoxy-D-galactose glycosides relative to TMS ether of 2-acetamido-2-deoxy-D-galactose (7.5 min at 180°C); t, of oligosaccharides relative to TMS ether of β-D-Glc*p*NAc-(1→4)-D-Glc*p*NAc (13.1 min at 243°C).

[d] t, relative to TMS methyl glycoside of racephedrine (6 min).

[e] t, relative to TMS ether of 3,4,6-tri-*O*-methyl-2-deoxy-2-methylamino-D-glucose (7.3 min).

[f] t, relative to 1,2,4,7,8,9-*O*-Me-NeuN(Ac,Me) (about 12 min).

[g] Minor peak.

[h] Pyranoside.

[i] Furanoside.

[j] N,N-Acetyl, methyl-neuraminic acid methyl ester β-D-methyl glycoside.

Packing   P1 = 3% SE-52 on acid-washed, silanized Chromosorb W (80 to 100 mesh).

P2 = 2.2% SE-30 on Gas-Chrom S (100 to 120 mesh).

P3 = 3% QF-1 on Gas-Chrom Q (80 to 100 mesh).

P4 = 3.8% SE-30 on Diatoport S (80 to 100 mesh).

P5 = 3% SE-30 on Chromosorb W HP (80 to 100 mesh).

P6 = 4% SE-30 on Chromosorb W HP (80 to 100 mesh).

P7 = 10% neopentyl glycol sebacate polyester on acid-washed, silanized Chromosorb W (80 to 100 mesh).

P8 = 3.8% SE-30 on Chromosorb W AW DMCS HP (80 to 100 mesh).

Table GC 25 (continued)

TRIMETHYLSILYL ETHERS OF AMINODEOXY AND ACETAMIDODEOXY SUGARS, NEURAMINIC ACIDS, AND THEIR DERIVATIVES

Temperature T1 = 140°C for aminodeoxyhexoses, 160°C for others.

T2 = temperature programmed, 140 → 200°C at 0.5°/min for monosaccharides; isothermal at 220°C for disaccharides.

T3 = temperature programmed, 140 → 160°C at 0.5°/min for acetamidodeoxyhexoses and GlcpNAc glycosides; isothermal at 180°C for GalNAc glycosides; isothermal at 243°C for disaccharides.

T4 = isothermal at 175°C for 30 min, then temperature programmed, 175 → 200°C at 1°/min; held at 200°C for 10 min.

Detector D1 = thermionic nitrogen-phosphorus selective detector.

REFERENCES

1. **Sweeley, C. C., Bentley, R., Makita, M., and Wells, W. W.,** Gas liquid chromatography of trimethylsilyl derivatives of sugars and related substances, *J. Am. Chem. Soc.* 85, 2497, 1963.

2. **Kärkkäinen, J. and Vihko, R.,** Characterisation of 2-amino-2-deoxy-D-glucose, 2-amino-2-deoxy-D-galactose and related compounds, as their trimethylsilyl derivatives by gas-liquid chromatography-mass spectrometry, *Carbohydr. Res.*, 10, 113, 1969.

3. **Bhatti, T., Chambers, R. E., and Clamp, J. R.,** The gas chromatographic properties of biologically important *N*-acetylglucosamine derivatives, monosaccharides, disaccharides, trisaccharides, tetrasaccharides, and pentasaccharides, *Biochim. Biophys. Acta*, 222, 339, 1970.

4. **Coduti, P. L. and Bush, C. A.,** Structure determination of *N*-acetyl amino sugar derivatives and disaccharides by gas chromatography and mass spectroscopy, *Anal. Biochem.*, 78, 21, 1977.

5. **Cahour, A. and Hartmann, L.,** Study of neutral and aminomonosaccharides by gas-liquid differential chromatography: application to three reference glycoproteins, *J. Chromatogr.*, 152, 475, 1978.

6. **Gorin, P. A. J. and Finlayson, A. J.,** Synthesis and chromatographic properties of partially *O*-methylated 2-deoxy-2-methylamino-D-glucoses; standards for methylation studies on polysaccharides, *Carbohydr. Res.*, 18, 269, 1971.

7. **Gorin, P. A. J.,** Syntheses and chromatographic properties of 2-deoxy-2-methylamino-D-galactose and its methyl ethers, *Carbohydr. Res.*, 18, 281, 1971.

8. **Gorin, P. A. J. and Magus, R. J.,** A method for gas-liquid chromatographic identification of *O*-methyl ethers of 2-deoxy-2-methylamino-D-glucose and 2-deoxy-2-methylamino-D-galactose, *Can. J. Chem.*, 49, 2583, 1971.

9. **Haverkamp, J., Kamerling, J. P., Vliegenthart, J. F. G., Veh, R. W., and Schaur, R.,** Methylation analysis determination of acylneuraminic acid residue type 2→8 glycosidic linkage. Application to GT₁ ganglioside and colominic acid, *FEBS Lett.*, 73, 215, 1977.

Section I.II

LIQUID CHROMATOGRAPHY TABLES

The term "liquid chromatography" covers many different types of system (see Section A, Volume II, Section I.III). To facilitate reference, the data on liquid chromatography of carbohydrates presented here are grouped according to the separation mechanism and type of packing involved as follows:

Chromatographic system	Table numbers
High-performance liquid chromatography	
On bonded-phase packings	LC 1—4
On microparticulate silica	LC 5—10
On cation-exchange resins	LC 11, 12
Partition chromatography on ion-exchange resins in aqueous ethanol	LC 13—16
Ion-exchange chromatography	LC 17—27
Gel-permeation chromatography	LC 28—45

The application of affinity chromatography to carbohydrates is reviewed at the end of this section.

The assistance of Dr. Kirsti Granath (Pharmacia AB, Uppsala, Sweden), Professor L. Hough and Dr. R. Sidebotham (Queen Elizabeth College, University of London), and Dr. F. M. Rabel (Whatman Inc., Clifton, New Jersey) in the compilation of this section is gratefully acknowledged.

HIGH-PERFORMANCE LIQUID CHROMATOGRAPHY

The term high-performance liquid chromatography (HPLC) is used here for separations complete in 1 hr or less. Under the conditions stated, monosaccharides can be separated in approximately 20 min and oligosaccharides in approximately 40 min, in the majority of cases. To facilitate comparisons between systems, values of the capacity factor k′ are tabulated, where possible, in preference to actual retention times.

Separations on bonded-phase packings depend upon partition chromatography, those on microparticulate silica on partition and/or adsorption, and the chromatography of sugars on a cation-exchange resin in a salt form (usually Ca^{2+} form) is believed to involve ligand exchange.

Tables LC 1, LC 2, LC 4, LC 7, and LC 11 are largely based on data compiled by Dr. F. M. Rabel (Whatman Inc., Clifton, New Jersey).

Table LC 1
HIGH-PERFORMANCE LIQUID CHROMATOGRAPHY OF SUGARS ON BONDED-PHASE PACKINGS

Packing	P1	P2	P2	P2	P2	P3	P3
Column							
Length	25 cm	30 cm	30 cm	30 cm	30 cm	25 cm	25 cm
Diameter	4.6 mm	3.9 mm	3.9 mm	3.9 mm	3.9 mm	2.1 mm	2.1 mm
Material	SS	SS	SS	SS	SS	SS	SS
Solvent	S1	S2	S2	S3	S4	S5	S3
Flow rate	1 ml/min	2.5 ml/min	1 ml/min	1.5 ml/min	1 ml/min	1 ml/min	1 ml/min
Temperature	25°C	25°C	22°C	25°C	22°C	25°C	25°C
Detection	RI	RI	RI	RI	RI	UV[a]	UV[a]
Reference	1	2	3	2	3	4	4
Compound				Capacity factor (k')			
L-Arabinose	1.42	2.9	3.0	—	—	3.7	—
D-Xylose	1.16	2.4	2.2	—	—	3.0	—
D-Ribose	0.89	—	1.6	—	—	2.0	—
D-Lyxose	1.12	—	2.1	—	—	—	—
D-Fructose	1.53	—	3.5	—	—	—	—
D-Glucose	1.84	5.2	5.1	1.25	0.9	5.0	—
D-Galactose	1.93/2.26[b]	5.9	5.9	—	—	7.9	—
D-Mannose	1.58	4.6	4.4	—	—	—	—
L-Rhamnose	—	1.7	1.7	—	—	—	—
L-Fucose	1.02/1.22[b]	—	2.0	—	—	—	—
2-Deoxy-D-glucose	—	—	1.7	—	—	—	—
6-Deoxy-D-glucose	—	—	2.0	—	—	—	—
Sophorose	—	—	—	—	2.1	—	—
Kojibiose	—	—	—	—	2.1	—	—
Maltose	3.47	—	—	2.75	2.1	—	8.6
Cellobiose	3.47	—	—	—	2.1	—	—
Isomaltose	—	—	—	—	2.5	—	—
Lactose	3.86	—	—	3.25	2.4	—	9.6
Melibiose	4.05	—	—	—	2.7	—	—
Sucrose	2.92	—	—	2.12	1.5	—	6.8

Trehalose	—	7.26	—	—	—	10.6
Raffinose	—	—	—	3.4	—	20.0

a 192 nm.

b Anomer separation.

Packing P1 = Partisil-10 PAC (Whatman®); microparticulate (10 μm) silica with bonded polar amino-cyano phase.

P2 = Waters® μ Bondapak Carbohydrate Analysis Column (10 μm silica with bonded polar phase).

P3 = MicroPak-NH₂ (Varian®); microparticulate silica with bonded amino phase.

Solvent S1 = acetonitrile-H_2O (75:25), pH adjusted to 5.0 by addition of H_3PO_4.

S2 = acetonitrile-H_2O (85:15).

S3 = acetonitrile-H_2O (80:20).

S4 = acetonitrile-H_2O (75:25).

S5 = acetonitrile-H_2O (88:12).

REFERENCES

1. Rabel, F. M., Caputo, A. G., and Butts, E. T., Separation of carbohydrates on a new polar bonded phase material, *J. Chromatogr.*, 126, 731, 1976.
2. Palmer, J. K., A versatile system for sugar analysis via liquid chromatography, *Anal. Lett.*, 8, 215, 1975.
3. Churms, S. C. and Seeman, U. A., unpublished laboratory data, 1977.
4. Hettinger, J. and Majors, R. E., Separation of carbohydrates by high performance liquid chromatography, *Varian Instrum. Appl.*, 10, 6, 1976.

Table LC 2
HIGH-PERFORMANCE LIQUID CHROMATOGRAPHY OF MALTODEXTRINS ON BONDED-PHASE PACKINGS

Packing	P1	P2	P2
Column			
Length	25 cm	30 cm	30 cm
Diameter	4.6 mm	3.9 mm	3.9 mm
Material	SS	SS	SS
Solvent	S1	S2	S3
Flow rate	1 mℓ/min	1 mℓ/min	1.8 mℓ/min
Temperature	25°C	25°C	25°C
Detection	RI	RI	RI
Reference	1	2	3

Degree of polymerization	k'		
1	0.30	0.72	0.9
2	0.55	0.94	1.4
3	0.95	1.25	2.5
4	1.45	1.69	5.2
5	1.95	2.22	6.5
6	2.60	—	—
7	3.30	—	—
8	4.15	—	—
9	5.05	—	—
10	6.20[a]	—	—

[a]   Eluted after about 40 min.

Packing P1 = Partisil-10 PAC (Whatman®).

           P2 = μBondapak Carbohydrate Analysis Column (Waters®).

Solvent   S1 = acetonitrile-0.0025 M sodium acetate (65:35), pH adjusted to 5.0 with acetic acid.

           S2 = acetonitrile-H_2O (65:35).

           S3 = acetonitrile-H_2O (75:25).

REFERENCES

1. Rabel, F. M., Caputo, A. G., and Butts, E. T., Separation of carbohydrates on a new polar bonded phase material, *J. Chromatogr.*, 126, 731, 1976.
2. Palmer, J. K., A versatile system for sugar analysis via liquid chromatography, *Anal. Lett.*, 8, 215, 1975.
3. Linden, J. C. and Lawhead, C. L., Liquid chromatography of saccharides, *J. Chromatogr.*, 105, 125, 1975.

Table LC 3
HIGH-PERFORMANCE LIQUID CHROMATOGRAPHY OF CELLODEXTRINS AND REDUCED DERIVATIVES ON BONDED-PHASE PACKINGS

Packing	P1	P2
Column		
Length	30 cm	25 cm
Diameter	3.9 mm	4.6 mm
Material	SS	SS
Solvent	S1	S2
Flow rate	Programmed[a]	1.5 ml/min
Temperature	25°C	25°C
Detection	RI	RI

Compound	k'	
D-Glucose	1.2	0.5
Cellobiose	2.3	1.0
-triose	4.4	1.6
-tetraose	8.1	2.5
-pentaose	—	3.8
-hexaose	—	5.4
D-Glucitol	1.2	0.6
Cellobiitol	2.6	1.3
-triitol	5.2	2.1
-tetraitol	9.2	3.2
-pentaitol	14.6	4.5
-hexaitol	—	6.5[b]

[a] Waters Model 660 Solvent Programmer: 2 ml/min for 10 min, then flow rate increased to 4.6 ml/min over 45-min period. Under these conditions, cellopentaitol was eluted after 40 min.

[b] Complete separation of oligomers to cellohexaitol in 25 min.

Packing P1 = μ Bondapak Carbohydrate Analysis Column (Waters®).
P2 = Partisil-10 PAC (Whatman®).

Solvent S1 = acetonitrile-H_2O (75:25).
S2 = acetonitrile-H_2O (71:29).

REFERENCE

1. **Gum, E. K. and Brown, R. D.,** Two alternative HPLC separation methods for reduced and normal cellooligosaccharides, *Anal. Biochem.,* 82, 372, 1977.

Table LC 4
HIGH-PERFORMANCE LIQUID CHROMATOGRAPHY OF CHITIN OLIGOSACCHARIDES[a] ON BONDED-PHASE PACKING

Packing	P1
Column	
Length	30 cm
Diameter	3.9 mm
Material	SS
Solvent	S1
Flow rate	1.5 ml/min
Temperature	25°C
Detection	RI

Compound	k'
2-Acetamido-2-deoxy-D-glucose	0.7
Chitobiose	1.3
-triose	2.1
-tetraose	3.4
-pentaose	5.3[b]

[a] β-(1 → 4)-linked oligomers of 2-acetamido-2-deoxy-D-glucose.

[b]   Eluted after about 15 min.

Packing   P1 = μ Bondapak Carbohydrate Analysis Column (Waters®).

Solvent   S1 = acetonitrile-H_2O (70:30).

REFERENCE

1. **Van Eikeren, P. and McLaughlin, H.,** Analysis of the lysozyme-catalysed hydrolysis and transglycosylation of N-acetyl-D-glucosamine oligomers by high-pressure liquid chromatography, *Anal. Biochem.,* 77, 513, 1977.

Table LC 5
CAPACITY FACTORS OF SUGARS AND ALDITOLS ON MICROPARTICULATE SILICA PACKINGS

Packing	P1	P1	P1	P2	P2
Column					
Length	20 cm	20 cm	20 cm	25 cm	25 cm
Diameter	4.6 mm	4.6 mm	4.6 mm	4.6 mm	4.6 mm
Material	SS	SS	SS	SS	SS
Solvent	S1	S2	S3	S3	S4
Flow rate	1.7 ml/min	1.5 ml/min	0.5 ml/min	0.5 ml/min	0.5 ml/min
Temperature	rt	rt	rt	22°C	22°C
Detection	RI	RI	RI	RI	RI
Reference	1	1	1	2	2
Compound			**k$'$**		
L-Arabinose	—	—	1.25	1.06/1.30[a]	0.79/0.97[a]
D-Xylose	—	—	1.20/1.30[a]	1.00/1.13[a]	0.93
D-Ribose	1.0	0.54	1.06	0.80/0.96[a]	0.87
L-Sorbose	1.4	0.70	1.27	—	—
D-Fructose	1.5	0.77	1.50	—	—
D-Glucose	—	—	1.62	1.71	1.23
D-Galactose	—	—	1.70/1.94[a]	2.13[b]	1.58[b]
D-Mannose	—	—	—	1.46	1.00/1.25[a]
D-Glucitol	1.7	1.00	1.88	—	—
Galactitol	—	—	1.92	—	—
D-Mannitol	—	—	1.92	—	—
L-Rhamnose	—	—	—	1.00/1.22[a]	0.75
L-Fucose	—	—	—	1.12/1.33[a]	0.80
Sucrose	2.2	1.2	2.42	—	—
Maltose	2.3	1.4	2.60	2.8	2.0
Cellobiose	2.3	1.4	2.65	—	—
Lactose	2.8	1.8	3.40	3.6	2.8
Trehalose	—	—	3.15	—	—
Raffinose	—	—	4.60	4.8	3.3

[a] Anomer separation.
[b] Broad peak.

Packing P1 = LiChrosorb Si 60 (Merck®; particle size 5 μm).
         P2 = Partisil 10 (Whatman®; particle size 10 μm).

Solvent S1 = ethyl formate-methanol-water (55:20:15).
        S2 = ethyl formate-methanol-water (55:25:10).
        S3 = ethyl formate-methanol-water (60:20:10).
        S4 = ethyl acetate-methanol-water (60:20:10).

REFERENCES

1. **Rocca, J. L. and Rouchouse, A.,** Separation of sugars on microparticulate silica by high performance liquid chromatography, *J. Chromatogr.*, 117, 216, 1976.
2. **Churms, S. C. and Seeman, U. A.,** unpublished laboratory data, 1977.

Table LC 6
HIGH-PERFORMANCE LIQUID CHROMATOGRAPHY OF GLYCOSIDES AND OTHER DERIVATIVES OF SUGARS ON MICROPARTICULATE SILICA

Packing	P1	P1	P1
Column			
Length	25 cm	25 cm	25 cm
Diameter	4.6 mm	4.6 mm	4.6 mm
Material	SS	SS	SS
Solvent	S1	S2	S3
Flow rate	1.2 ml/min	1.2 ml/min	1.2 ml/min
Temperature	rt	rt	rt
Detection	RI	RI	RI

Compound	t_r(min)		
L-Arabinopyranoside, methyl α-	14.2	—	—
D-Xylopyranoside			
methyl α-	12.4	—	—
methyl β-	11.8	—	—
D-Ribopyranoside, methyl β-	11.5	—	—
D-Glucopyranoside			
methyl α-	15.9	—	—
methyl β-	17.7	—	—
phenyl α-	10.0	—	—
phenyl β-	8.8	—	—
D-Galactopyranoside			
methyl α-	20.1	—	—
methyl β-	19.5	—	—
D-Mannopyranoside			
methyl α-	15.3	—	—
methyl β-	18.3	—	—
D-Altropyranoside, methyl α-	15.3	—	—
D-Gulopyranoside,[a] methyl α-	18.3	—	—
D-Glucose			
3-O-methyl-	15.9	—	—
2,3,6-tri-O-methyl-		10.4	—
2,4,6-tri-O-methyl-		10.9	—
D-Glucopyranose			
1,2-O-isopropylidene-		10.4	—
4,6-O-benzylidene-		8.1	—
D-Glucofuranose, 1,2:5,6-di-O-isopropylidene-		—	10.5
D-Mannofuranose, 2,3:5,6-di-O-isopropylidene-		—	8.0

[a]   Hydrate.

Packing   P1   =   Partisil 10 (Whatman®; particle size 10 μm).

Solvent   S1   =   acetonitrile-water (9:1).
           S2   =   acetonitrile-water (18:1).
           S3   =   *n*-hexane-ethyl acetate (1:3).

From McGinnis, G. D. and Fang, P., *J. Chromatogr.*, 153, 107, 1978. With permission.

Table LC 7
HIGH-PERFORMANCE LIQUID
CHROMATOGRAPHY OF 4-NITROBENZOATE
DERIVATIVES OF CARBOHYDRATES ON
MICROPARTICULATE SILICA

Packing	P1	P1	P1
Column			
Length	15 cm	15 cm	25 cm
Diameter	3 mm	3 mm	3 mm
Material	SS	SS	SS
Solvent	S1	S2	S3
Flow rate	2.1 mℓ/min	1.5 mℓ/min	0.8 mℓ/min
Temperature	rt[a]	rt[a]	rt
Detection	UV[b]	UV[b]	RI
Reference	1	1	2

Parent compound	k′		
α-D-Glucopyranose	3.0	1.9	—
β-D-Glucopyranose	5.0	2.3	—
D-Glucopyranoside			
methyl α-	4.2	—	—
methyl β-	8.1	—	—
Sucrose	—	5.9	—
α-Lactose	—	8.2	—
β-Lactose	—	9.8	—
Arabinitol	—	—	2.1
Xylitol	—	—	2.6
D-Mannitol	—	—	3.3
D-Glucitol	—	3.3	3.6
Maltitol	—	—	9.0

[a]   20—22°C

[b]   260 nm.

Packing   P1   =   LiChrosorb Si 60 (Merck; particle size 5 μm).

Solvent   S1 =   *n*-hexane-ethyl acetate (3:1), with 5% dioxane.
            S2 =   *n*-hexane-chloroform-acetonitrile-tetrahydro-furan (10:5:1:0.5).
            S3 =   *n*-hexane-chloroform-acetonitrile (5:2:1).

REFERENCES

1. Nachtmann, F. and Budna, K. W., Sensitive determination of derivatized carbohydrates by high-performance liquid chromatography, *J. Chromatogr.*, 136, 279, 1977.
2. Schwarzenbach, R., Separation of some polyhydric alcohols by high-performance liquid chromatography, *J. Chromatogr.*, 140, 304, 1977.

Table LC 8
HIGH-PERFORMANCE LIQUID CHROMATOGRAPHY OF PERACETYLATED CARBOHYDRATES ON MICROPARTICULATE SILICA

Packing	P1	P2
Column		
Length	25 cm	25 cm
Diameter	3 mm	4.6 mm
Material	SS	SS
Solvent	S1	S2
Flow rate	0.46 ml/min	1.2 ml/min
Temperature	rt	rt
Detection	RI, UV[a]	RI
Reference	1	2

Parent compound	k'	t_r (min)
D-Arabinopyranose		
α-	11.49	—
β-	7.74	—
D-Xylopyranose, α-	8.33	—
D-Lyxopyranose, α-	7.02	—
D-Ribopyranose, β-	10.51	—
L-Rhamnopyranose, α-	5.97	—
D-Glucopyranose		
α-	15.13[b]	17.2
β-	17.7	15.9
D-Galactopyranose		
α-	13.04	15.9
β-	18.3	16.0
D-Mannopyranose		
α-	16.37	—
β-	24.51	—
D-Allopyranose, β-	19.63	—
D-Idopyranose, α-	21.38	—
D-Talopyranose, α-	21.92	—
Erythritol	—	12.6
L-Arabinitol	—	14.0
D-Glucopyranoside		
methyl α-	—	16.3
methyl β-	—	17.3
phenyl α-	—	11.7
phenyl β-	—	13.0

[a]   220 nm.
[b] t_r = about 70 min.

Packing	P1	= LiChrosorb Si60 (Merck®; particle size 5 μm).
	P2	= Partisil 10 (Whatman®; particle size 10 μm).
Solvent	S1	= n-hexane-acetone (10:1).
	S2	= n-hexane-ethyl acetate (1:1).

REFERENCES

1. **Thiem, J., Schwentner, J., Karl, H., Sievers, A., and Reimer, J.,** Separation of peracetylated mono- and disaccharides and quantitative analysis of guaran by high-performance liquid chromatography on silica gel, *J. Chromatogr.*, 155, 107, 1978.
2. **McGinnis, G. D. and Fang, P.,** Separation of substituted carbohydrates by high-performance liquid chromatography II, *J. Chromatogr.*, 153, 107, 1978.

Table LC 9
HIGH-PERFORMANCE LIQUID CHROMATOGRAPHY OF PERACETYLATED OLIGOSACCHARIDES ON MICROPARTICULATE SILICA

Packing	P1	P2
Column		
Length	25 cm	25 cm
Diameter	3 mm	4.6 mm
Material	SS	SS
Solvent	S1	S2
Flow rate	0.39 ml/min	1.2 ml/min
Temperature	rt	rt
Detection	RI	RI
Reference	1	2
Parent oligosaccharide	**k'**	**t, (min)**
α-D-Glucopyranose		
3-*O*-α-D-glucopyranosyl-	7.29	—
3-*O*-β-D-glucopyranosyl-	10.33	—
4-*O*-β-D-glucopyranosyl-	8.96[a]	38.0
4-*O*-β-D-galactopyranosyl-	10.63	—
β-D-Glucopyranose		
4-*O*-α-D-glucopyranosyl-	6.73	—
4-*O*-β-D-glucopyranosyl-	9.47	36.1
6-*O*-β-D-glucopyranosyl-	10.82	—
α-D-Mannopyranose		
4-*O*-β-D-mannopyranosyl-	9.53	
β-D-Mannopyranose		
4-*O*-β-D-mannopyranosyl-	11.53	—
α-D-Glucopyranoside		
α-D-glucopyranosyl-	7.14	—
β-D-Glucopyranoside		
β-D-glucopyranosyl-	10.73	—
β-D-Fructofuranoside		
2-*O*-α-D-glucopyranosyl-	6.14	—

[a] t_r = about 50 min.

Packing	P1	=	LiChrosorb Si60 (Merck®; particle size 5 μm).
	P2	=	Partisil 10 (Whatman®; particle size 10 μm).
Solvent	S1	=	*n*-pentane-acetone (7:2).
	S2	=	*n*-hexane-ethyl acetate (1:1).

REFERENCES

1. **Thiem, J., Schwentner, J., Karl, H., Sievers, A., and Reimer, J.,** Separation of peracetylated mono- and disaccharides and quantitative analysis of guaran by high-performance liquid chromatography on silica gel, *J. Chromatogr.,* 155, 107, 1978.
2. **McGinnis, G. D. and Fang, P.,** Separation of substituted carbohydrates by high-performance liquid chromatography. II, *J. Chromatogr.,* 153, 107, 1978.

Table LC 10
HIGH-PERFORMANCE LIQUID
CHROMATOGRAPHY OF GLYCOLIPIDS ON
MICROPARTICULATE SILICA

Packing	P1	P1
Column		
Length	25 cm	25 cm
Diameter	2.8 mm	2.8 mm
Material	SS	SS
Solvent	S1	S2
Flow rate	na[a]	na[b]
Temperature	rt	rt
Detection	D1	D1

Compound	k'	
Brain gangliosides		
G_{M3}	0.38	—
G_{M2}	0.98	—
G_{M1}	2.16	—
G_{D1a}	2.98[c]	—
G_{D1b}	4.31[c]	—
G_{T1}	6.02[c]	—
Neutral glycosphingolipids		
GL_{1a}	—	0.63
GL_{2a}	—	1.38
GL_3	—	2.21
GL_4	—	4.06

[a]   Analysis time 40 min.
[b]   Analysis time 30 min.
[c]   Doublet peak.

Packing   P1   = Silica Si60 (Merck®), with particle size range 63—200 μm, ground in agate mortar and fractionated by air classifier; fraction having particle size 9 ± 1.5 μm used to pack column.

Solvent   S1   = chloroform-methanol-aqueous HCl (60:35:4); final HCl concentration 0.01 M.

  S2   = chloroform-methanol (3:1).

Detection   D1   = moving-wire system equipped with flame-ionization detector (Pye Unicam LCM2).

REFERENCE

1.   Tjaden, U. R., Krol, J. H., Van Hoeven, R. P., Oomen-Meulemans, E. P. M., and Emmelot, P., High-pressure liquid chromatography of glycosphingolipids (with special reference to gangliosides), *J. Chromatogr.*, 136, 233, 1977.

Table LC 11

CAPACITY FACTORS OF SUGARS AND POLYOLS ON HIGH-PERFORMANCE LIGAND-EXCHANGE CHROMATOGRAPHY: VARIATION WITH CATION ASSOCIATED WITH RESIN[a]

Cation	Tris[b]
Column Length	50 cm
Diameter	2.8 mm
Material	Glass
Solvent	H₂O
Flow rate	0.1 ml/min
Temperature	rt
Detection	RI

Conditions same in all cases

k′

Compound	Tris[b]	Li⁺	Na⁺	K⁺	Rb⁺	Ag⁺	Tl⁺	Mg²⁺	Ca²⁺	Sr²⁺	Ba²⁺	Cd²⁺	La³⁺
D-Xylose		—	—	—	—	1.15	—	—	0.45/0.60[c]	—	—	—	—
D-Fructose		0.45	0.65	1.05	0.90	0.75	1.10	0.35	1.15[d]	1.10[d]	1.00[c]	0.50	0.35
D-Glucose		0.35	0.40	0.70	0.75[c]	0.85	0.70	0.30	0.30/0.45[c]	0.30/0.45[c]	0.40[c]	0.25	0.15
D-Galactose		0.50	0.55	0.80/1.10[c]	0.75/1.00[c]	0.95[e]	0.85	0.40	0.45/0.65[c]	0.40/0.55[c]	0.45[c]	0.35	0.25
D-Mannose		0.45	0.70	—	0.75/1.25[c]	—	—	0.40	0.45/0.65[c]	—	0.70[c]	—	0.30
D-Talose		—	0.50	—	0.95[c]	1.7	—	—	~3.8	—	—	—	~4.0
D-Gulose		—	—	—	1.25	1.15	—	—	1.3	—	—	—	1.0
Sucrose		0.20	0.20	0.40	0.40	0.45	0.45	0.15	0.15	0.20	0.25	0.15	0.10
Glycerol		0.90	0.65	0.80	0.75	1.10	1.00	0.70	1.05	—	0.60	0.80	0.80
D-Mannitol		1.05	0.75	—	0.55	0.80	—	0.40	1.40	—	1.00	—	1.70
Galactitol		—	—	—	0.60	—	—	—	2.1	—	—	—	—

[a] Aminex A-5 (Bio-Rad®; particle size 11 ± 2 μm).

[b] Tris = NH₃C(CH₂OH)₃.

[c] Anomer separation.

[d] Skewed peak.

[e] Asymmetric peak; partial separation of anomers.

From Goulding, R. W., *J. Chromatogr.*, 103, 229, 1975. With permission.

Table LC 12

HIGH-PERFORMANCE LIQUID CHROMATOGRAPHY OF SUGARS, ALDITOLS, AND OLIGOSACCHARIDES ON ION-EXCHANGE RESINS (CALCIUM FORM)

Packing	P1	P1	P2	P2
Column				
Length	60 cm	50 cm	61 cm	61 cm
Diameter	7 mm	7 mm	8 mm	8 mm
Material	na	SS	SS	SS
Solvent	H_2O	H_2O	H_2O	H_2O
Flow rate	1 mℓ/min	0.6 mℓ/min	0.6 mℓ/min	0.55 mℓ/min
Temperature	80°C	85°C	85°C	85°C
Detection	RI	RI[a]	RI[b]	RI[b]
Reference	1	2	3	3
Compound	r[c]	r[d]	r[e]	k'
D-Glucitol	1.63	—	—	—
D-Mannitol	1.35	—	—	—
L-Arabinose	1.19	—	1.26	—
D-Xylose	—	—	1.11	—
D-Galactose	1.10	—	1.13	—
D-Mannose	1.12	—	1.16	—
D-Glucose	1.00	1.00	1.00	1.43[f]
D-Fructose	1.18	—	1.18	—
Sucrose	0.84	—	0.86	—
Maltose	—	0.83	—	—
Cellobiose	—	—	—	1.02
Lactose	—	—	0.89	—
Maltotriose	—	0.74	—	—
Cellotriose	—	—	—	0.70
Cellotetraose	—	—	—	0.49
Cellopentaose	—	—	—	0.35
Cellohexaose	—	—	—	0.25
Celloheptaose	—	—	—	0.18

[a]   Refractometer temperature 45°C.
[b]   Refractometer temperature 30°C.
[c]   t, of D-glucose = about 15 min.
[d]   t, of D-glucose = 19.6 min.
[e]   t, of D-glucose = about 28 min.
[f]   t, of D-glucose = about 35 min.

Packing P1   = Aminex Q15-S (Bio-Rad®; particle size 22 ± 3 μm), converted to Ca^{2+} form before use by treatment with 0.5% aqueous solution of $CaCl_2 \cdot 2H_2O$.

P2   = AG 50W-X4 (Bio-Rad®; particle size 20—30 μm), converted to Ca^{2+} form by treatment, successively, with 0.5%, 2.5%, and 5.0% solutions of $CaCl_2 \cdot 2H_2O$.

REFERENCES

1. **Conrad, E. C. and Palmer, J. K.,** Rapid analysis of carbohydrates by high-pressure liquid chromatography, *Food Technol.,* 30, 84, 1976.
2. American Society of Brewing Chemists, Report of subcommittee on brewing sugars and syrups, *Am. Soc. Brew. Chem. Proc.,* 35, 104, 1977.
3. **Ladisch, M. R., Huebner, A. L., and Tsao, G. T.,** High-speed liquid chromatography of cellodextrins and other saccharide mixtures using water as the eluent, *J. Chromatogr.,* 147, 185, 1978.

PARTITION CHROMATOGRAPHY ON ION-EXCHANGE RESINS IN AQUEOUS ETHANOL

This technique was developed by Professor O. Samuelson (Chalmers Tekniska Högskola, Göteborg, Sweden), whose data published up to 1970 will be found in Section A, Volume I, Section II.II, Table LC 48. The data included here, obtained from more recent work, should be regarded as supplementary to that in the earlier Volume.

Table LC 14 was compiled by Dr. F. M. Rabel (Whatman Inc., Clifton, New Jersey).

Table LC 13
PARTITION CHROMATOGRAPHY ON ION-EXCHANGE RESINS: MONOSACCHARIDES AND COMMON DISACCHARIDES

Packing	P1	P2	P3
Column			
Length	100 cm	26 cm	23 cm
Diameter	4 mm	4 mm	9 mm
Material	glass	glass	glass
Solvent	S1	S2	S3
Flow rate	0.15—0.45 ml/min	0.55 ml/min	0.83 ml/min
Temperature	65°C	88°C	90°C
Detection	D1	D2	D3
Reference	1, 2	3	4
Compound	**k'**	**r^a**	**t, (min)**
L-Arabinose	4.18	0.40	67.8
D-Xylose	3.73	0.48	55.2
D-Lyxose	—	0.35	—
D-Ribose	3.02	0.28	—
D-Galactose	5.33	0.83	—
D-Glucose	4.90	1.00	—
D-Mannose	4.40	0.61	—
D-Gulose	—	0.71	—
L-Rhamnose	2.06	0.20	—
L- Fucose	2.66	0.24	47.8
2-Deoxy-D-ribose	—	0.12	33.6
2-Deoxy-D-galactose	—	0.17	—
6-Deoxy-D-glucose	—	0.32	—
D-Fructose	3.82	0.53	—
L-Sorbose	—	0.58	—
D-Tagatose	—	0.56	—
D-Glucuronic acid	—	—	16.2
D-Galacturonic acid	—	—	18.4
Sucrose	3.80^b	—	—
Maltose	4.73	—	—
Lactose	6.71	—	—

^{a} t, of D-glucose = about 170 min.
^{b} Partially decomposes.

Packing	P1	=	Aminex® A-6 (17.5 ± 2 µm), (CH$_3$)$_3$N$^+$H form.
	P2	=	DAX8 (Durrum®; particle size 11 µm), sulfate form.
	P3	=	M71 cation-exchange resin (Beckman®), Li$^+$ form.
Solvent	S1	=	ethanol-H$_2$O (80:20 w/w).
	S2	=	ethanol-H$_2$O (87:13 w/w).
	S3	=	ethanol-H$_2$O (89:11 v/v).
Detection	D1	=	modified Pye moving wire detector.
	D2	=	copper-bicinchoninate colorimetric method (automated).
	D3	=	tetrazolium blue colorimetric method (automated).

REFERENCES

1. **Hobbs, J. S. and Lawrence, J. G.,** The separation and quantitation of carbohydrates on cation-exchange resin columns having organic counter-ions, *J. Chromatogr.,* 72, 311, 1972.
2. **Lawrence, J. G.,** Analysis of carbohydrates by high-performance liquid chromatography, *Chimia,* 29, 367, 1975.
3. **Mopper, K.,** Improved chromatographic separations on anion-exchange resins. I. Partition chromatography of sugars in ethanol, *Anal. Biochem.,* 85, 528, 1978.
4. **Nackenhorst, R. and Thorn, W.,** Automatic column chromatographic analysis of saccharides and uronic acids, *Res. Exp. Med.,* 172, 63, 1978.

Table LC 14
PARTITION CHROMATOGRAPHY ON ION-EXCHANGE RESINS: ALDITOLS AND REDUCED OLIGOSACCHARIDES

Packing	P1	P1	P2
Column			
Length	68 or 72 cm	68 or 72 cm	117 cm
Diameter	4.4 mm	4.4 mm	4.4 mm
Material	glass	glass	glass
Solvent	S1	S2	S2
Flow rate (linear)	3.2—3.5 cm/min	3.2—3.5 cm/min	3.9 cm/min
Temperature	75°C	75°C	75°C
Detection	D1	D1	D1

Compound	D_v	D_v	D_v
Xylitol	1.41	5.31	4.38
D-Glucitol	2.15	10.5	6.50
D-Mannitol	2.54	14.1	5.83
D-*glycero*-D-*galacto*-Heptitol	4.18	31.6	11.3
D-*glycero*-D-*gulo*-Heptitol	3.25	21.6	9.47
D-*glycero*-D-*manno*-Heptitol	3.60	25.4	8.06
D-Xylitol,4-*O*-β-D-xylopyranosyl-	1.81	13.4	4.70
D-Arabinitol,3-*O*-β-D-galactopyranosyl-	2.81	26.6	9.23
D-Glucitol			
4-*O*-β-D-galactopyranosyl-	3.78	46.1	18.3
6-*O*-α-D-galactopyranosyl-	3.99	49.1	20.9
3-*O*-α-D-glucopyranosyl-	4.80	64.9	12.1
4-*O*-α-D-glucopyranosyl-	3.97	51.5	15.0
4-*O*-β-D-glucopyranosyl-	4.05	52.0	12.0
6-*O*-α-D-glucopyranosyl-	4.26	57.0	15.6
6-*O*-β-D-glucopyranosyl-	4.78	68.0	11.8
D-Mannitol			
4-*O*-β-D-galactopyranosyl-	3.80	46.8	13.4
3-*O*-α-D-glucopyranosyl-	4.75	63.5	12.7
6-*O*-α-D-glucopyranosyl-	4.94	73.9	14.6

Packing P1 = Technicon® T5C (10-17 μm), sulfate form.

            P2 = Aminex® A-6 (15-19 μm), Li⁺ form.

Solvent S1 = ethanol-H_2O (70:30 w/w).

            S2 = ethanol-H_2O (85:15 w/w).

Detection D1 = automated colorimetric assay by orcinol and periodate-acetylacetone (pentane-2,4-dione) methods, simultaneously.

REFERENCE

1. **Havlicek, J. and Samuelson, O.,** Separation of oligomeric sugar alchols by partition chromatography on ion-exchangers, *Chromatographia,* 7, 361, 1974.

Table LC 15
PARTITION CHROMATOGRAPHY ON ION-EXCHANGE RESINS: D-GLUCOSE OLIGOMERS

Packing	P1	P1	P1	P1	P1
Column					
Length	92.5 cm	92.5 cm	92.5 cm	92.5 cm	92.5 cm
Diameter	4 mm	4 mm	4 mm	4mm	4 mm
Material	glass	glass	glass	glass	glass
Solvent	S1	S1	S1	S1	S2
Flow rate (linear)	3.8 cm/min	3.8 cm/min	3.8 cm/min	3.8 cm/min	3.8 cm/min
Temperature	75°C	75°C	75°C	75°C	75°C
Detection	Orcinol	Orcinol	Orcinol	Orcinol	Orcinol

DP	Laminaridextrins[a]	Malto-[b]	Cello-[c]	Isomalto-[d]	
	D_v	D_v	D_v	D_v	D_v
1	3.37	3.37	3.37	3.37	2.24
2	3.89	4.66	4.81	6.71	3.47
3	4.22	6.43	6.34	13.4	5.62
4	4.57	8.87	8.35	26.7	7.76
5	4.95	12.2	11.0	53.3	13.2
6	5.37	16.9	14.5	—	22.4
7	5.82	—	—	—	33.9
8	—	—	—	—	54.9

[a] β-D-(1→3)-linked, from mild acid hydrolysis of laminarin.
[b] α-D-(1→4)-linked series.
[c] β-D-(1→4)-linked series.
[d] α-D-(1→6)-linked series.

Packing   P1   = Dowex® 1-X8 (12-18 μm), sulfate form.
Solvent   S1   = ethanol-H_2O (70:30 w/w).
          S2   = ethanol-H_2O (65:35 w/w).

REFERENCE

1. **Havlicek, J. and Samuelson, O.,** Separation of oligosaccharides by partition chromatography on ion-exchange resins, *Anal. Chem.,* 47, 1854, 1975.

Table LC 16
PARTITION CHROMATOGRAPHY
ON ION-EXCHANGE RESINS:
XYLODEXTRINS

Packing	P1	P2
Column		
Length	60 cm	91 cm
Diameter	4 mm	4 mm
Material	glass	glass
Solvent	S1	S2
Flow rate (linear)	2.8 cm/min	3.0 cm/min
Temperature	75°C	75°C
Detection	Orcinol	Orcinol
Reference	1	1

Xylodextrin[a], DP	D_r	D_r
1	3.16	1.26
2	4.79	1.58
3	7.41	2.24
4	11.2	3.09
5	17.8	4.17
6	26.3	6.03
7	39.8	7.94
8	63.1	11.2
9	—	15.8

[a] β-D-(1→4)-linked, from acid hydrolysis of birch xylan.

Packing	P1	=	Technicon®T5C (14-17 μm), sulfate form.
	P2	=	Dowex®50W-X8 (17-21 μm), Li⁺ form.
Solvent	S1	=	ethanol-H₂O (75:25 w/w).
	S2	=	ethanol-H₂O (80:20 w/w).

REFERENCE

1.   **Havlicek, J. and Samuelson, O.,** Chromatography of oligosaccharides from xylan by various techniques, *Carbohydr. Res.,* 22, 307, 1972.

ION-EXCHANGE CHROMATOGRAPHY

Ion-exchange chromatography of carbohydrates has advanced considerably since the publication of Section A, Volume I (see Section II.II, Tables LC 25, LC 29, and LC 69), and therefore this field is extensively covered here. Anion-exchange resins in the borate form are used in most of the separations reported. The applications of diethyl-aminoethyl-dextran and -cellulose ion-exchangers in chromatography of carbohydrates are reviewed briefly at the end of this section (see Tables LC 26 and LC 27).

Tables LC 17, LC 20, and LC 24 were compiled by Professor L. Hough and Dr. R. Sidebotham (Queen Elizabeth College, University of London). Tables LC 18 and LC 19 include data contributed by Dr. K. Granath (Pharmacia AB, Uppsala, Sweden) and Dr. F. M. Rabel (Whatman Inc., Clifton, New Jersey).

Table LC 17
ANION-EXCHANGE CHROMATOGRAPHY OF SACCHARIDES

Packing	P1	P2	Packing	P1	P2
Column			Column		
Length	75 cm	120 cm	Length	75 cm	120 cm
Diameter	0.6 cm	0.8 cm	Diameter	0.6 cm	0.8 cm
Material	Silica glass	glass	Material	Silica glass	glass
Solvent	S1	S2	Solvent	S1	S2
Flow rate	60 ml/hr	30 ml/hr	Flow rate	60 ml/hr	30 ml/hr
Temperature	62°C	55°C	Temperature	62°C	55°C
Detection	D1	D2	Detection	D1	D2
Reference	1, 2	3	Reference	1, 2	3
Compound	r^a	r^a	**Compound**	r^a	r^a
Trehalose	13.5	—	D-Ribose	60.5	51.0
Sucrose	14.0	11.5	D-Mannose	63.5	60.0
2-Deoxy-D-ribose	15.5	8.5	L-Fucose	70.0	75.0
Raffinose	17.5	—	L-Arabinose	74.0	76.5
Cellobiose	20.0	15.0	D-Fructose	78.5	80.5
Maltose	29.5	21.0	D-Galactose	79.5	83.0
Allosucrose	31.5	—	D-Allose	86.0	—
Lactose	36.5	25.5	D-Xylose	93.0	88.5
L-Rhamnose	42.0	29.5	D-Glucose	100	100

^{a} Relative to D-glucose = 100 (t_r = about 4.5-5.5 hr).

Packing	P1	=	Type-S Chromobeads anion exchange resin (Technicon®); particle size about 20 μm.
	P2	=	LC-R-3 quaternary ammonium ion exchange resin (Jeol).
Solvent	S1	=	Borate-chloride gradient, pH 7.0, borate 0.1–0.4 M, chloride 0–0.2 M.
	S2	=	Potassium tetraborate buffers, 0.13, 0.25, and 0.35 M pH 7.50, 9.08, and 6.0 (step-wise).
Detection	D1	=	cysteine-sulfuric acid colorimetry at 395 nm.
	D2	=	orcinol sulfuric acid colorimetry at 425 and 510 nm.

REFERENCES

1. **Hough, L., Jones, J. V. S., and Wusteman, P.,** On the automated analysis of neutral monosaccharides in glycoproteins and polysaccharides, *Carbohydr. Res.*, 21, 9, 1972.
2. **Hough, L. and Sidebotham, R.,** unpublished data, 1978.
3. **Kennedy, J. F. and Fox, J. E.,** The fully automatic ion exchange and gel-permeation chromatography of neutral monosaccharides and oligosaccharides with a Jeolco JLC-6AH analyser, *Carbohydr. Res.*, 54, 13, 1977.

Table LC 18
HIGH-PERFORMANCE ANION-EXCHANGE CHROMATOGRAPHY OF SUGARS

Packing	P1	P2	P3	P4	P5
Column					
Length	28 cm	19 cm	50.8 cm	47.5 cm	30 cm
Diameter	6 mm	6 mm	6mm	6 mm	4 mm
Material	SS	SS	na	na	glass
Solvent	S1	S2	S3	S3	S4
Flow rate	1 mℓ/min	1.3 mℓ/min	0.82 mℓ/min	0.83 mℓ/min	0.4 mℓ/min
Temperature	60°C	60°C	50°C	50°C	78°C
Detection	D1	D1	D1	D1	D2
Reference	1	1	2	2	3
Compound	r[a]	r[b]	t, (min)	t, (min)	t, (min)
2-Deoxy-D-ribose	8.4	—	—	—	8
Sucrose	9.6	21	—	—	—
Trehalose	11.2	—	40	30	—
Cellobiose	13.2	—	—	—	13
Maltose	26.4	—	—	—	17
L-Rhamnose	30.0	—	76	43	22
Lactose	33.6	—	—	—	20
D-Erythrose	39.9	—	—	—	—
D-Lyxose	42.0	—	—	—	—
D-Ribose	42.8	30	—	—	25
D-Mannose	65.2	50	95	53	36
D-Tagatose	74.0	—	—	—	—
D-Fructose	76.4	—	—	—	43
D-Gulose	76.4	—	—	—	—
Lactulose	77.6	—	—	—	—
D-Allose	77.6	—	—	—	—
L-Arabinose	78.8	62	110	67	47
L-Fucose	78.8	—	—	—	52
D-Talose	85.3	—	—	—	—
D-Galactose	87.6	71	122	75	58
D-Altrose	89.0	—	—	—	—
L-Sorbose	89.2	—	—	—	—
D-Xylose	92.4	80	144	96	72
D-Glucose	100	100	—	125	108
Gentiobiose	114.4	—	—	—	102
Melibiose	125.2	—	—	—	132

[a] Relative to D-glucose = 100 (t, = about 4 hr).   [b] t, of D-glucose = 1 hr.

Packing   P1   = DA-X4 (Durrum®; particle size 20 ± 5 μm); precolumn filled with Dowex® 1-X4 (200 to 400 mesh).

            P2   = DA-X4F (Durrum®; particle size 11 ± 1 μm); precolumn as for P1.

            P3   = Aminex A-15 (Bio-Rad®; particle size 17.5 ± 2 μm).

            P4   = DEAE-Spheron®; diethylaminoethyl derivative prepared in laboratory from Spheron® P-300 (methacrylate gel, particle size 20—40 μm; Lachema, Czechoslovakia).

            P5   = DA-X4, rinsed with 70% (w/w) ethanol after conversion to borate form.

Solvent   S1   = borate buffer gradient (0.1—0.5 MH$_3$BO$_3$, pH 8.0—10.0).

            S2   = 0.4 MH$_3$BO$_3$, pH adjusted to 9.2 with NaOH.

            S3   = borate buffer, pH 8.5 (10.5 g H$_3$BO$_3$ + 42.0 g borax per ℓ).

            S4   = 0.5 MH$_3$BO$_3$, pH adjusted to 8.63 with NaOH.

Detection D1   = orcinol method.

            D2   = copper-bicinchoninate method.

REFERENCES

1. Voelter, W. and Bauer, H., High-performance liquid chromatographic analysis of the isomerization products of carbohydrates, *J. Chromatogr.*, 126, 693, 1976.
2. Chytilova, Z., Mikes, O., Farkas, J., Strop, P., and Vratny, P., Chromatography of sugars on DEAE-Spheron, *J. Chromatogr.*, 153, 37, 1978.
3. Mopper, K., Improved chromatographic separations on anion exchange resins. III. Sugars in borate medium, *Anal. Biochem.*, 87, 162, 1978.

Table LC 19
ANION-EXCHANGE CHROMATOGRAPHY OF METHYLATED SUGARS AND GLYCOSIDES

Packing	P1	P2
Column		
Length	99 cm	28 cm
Diameter	2 mm	6 mm
Material	na	SS
Solvent	S1	S2
Flow rate	na	1 ml/min
Temperature	55°C	60°C
Detection	Orcinol	Orcinol
Reference	1	2
Compound	**k′**	**rᵃ**
D-Xylose	9.2	92.4
2-*O*-methyl-	2.8	—
3-*O*-methyl-	3.5	—
2,3-di-*O*-methyl-	0.5	—
2,3,4-tri-*O*-methyl-	1.1	—
D-Glucose	12.3	100
2-*O*-methyl-	2.0	—
3-*O*-methyl-	7.5	—
3,6-di-*O*-methyl-	2.8	—
2,3,4-tri-*O*-methyl-	1.0	—
2,3,6-tri-*O*-methyl-	0.3	—
2,3,4,6-tetra-*O*-methyl-	1.6	—
D-Mannose	5.3	65.2
2,3-di-*O*-methyl-	1.1	—
2,3,6-tri-*O*-methyl-	0.2	—
2,3,4,6-tetra-*O*-methyl-	1.6	—
D-Galactose	8.0	87.6
2,3,4,6-tetra-*O*-methyl-	1.0	—
D-Arabinopyranoside, methyl β-	—	6.4
D-Xylopyranoside, methyl β-	—	8.0
D-Mannopyranoside, methyl α-	—	12.0
D-Glucopyranoside, methyl β-	—	22.8

ᵃ   Relative to D-glucose = 100 (t, = about 4 hr).

Packing	P1	=	Aminex A-14 (Bio-Rad®; particle size 20 μm).
	P2	=	DA-X4 (Durrum®; particle size 20 ± 5 μm).
Solvent	S1	=	0.11 M potassium tetraborate — 0.17 M boric acid, pH 8.8.
	S2	=	borate buffer gradient (0.1—0.5 M H₃BO₃, pH 8.0—10.0).

REFERENCES

1. **Sinner, M.,** Separation of methyl ethers of xylose, glucose and some other sugars by liquid chromatography, *J. Chromatogr.,* 121, 122, 1976.
2. **Voelter, W. and Bauer, H.,** High-performance liquid chromatographic analysis of the isomerization products of carbohydrates, *J. Chromatogr.,* 126, 693, 1976.

Table LC 20
ANION-EXCHANGE
CHROMATOGRAPHY OF
POLYOLS

Packing	P1
Column	
Length	75 cm
Diameter	0.6 cm
Material	silica glass
Solvent	S1
Flow rate	70 mℓ/hr
Temperature	75°C
Detection	D1

Compound	r[a]
Ethylene glycol[b]	24.0
2-amino-2-deoxy-D-galactitol	54.5
2-amino-2-deoxy-D-glucitol	70.5
Xylitol	81.5
D-Arabinitol	86.0
D-Mannitol	94.0
D-Glucitol	100
Galactitol	109.5

[a] Relative to D-glucitol = 100 (t, = about 4 hr).
[b] Internal standard.

Packing	P1	= Type-S Chromobeads anion exchange resin (Technicon®).
Solvent	S1	= borate-chloride gradient, pH 7.0, borate 0.1—0.4 *M*, chloride 0—0.2 *M*.
Detection	D1	= pentane-2,4-dione, after periodate oxidation.

REFERENCE

1. **Hough, L., Ko, A. M. Y., and Wusteman, P.,** The simultaneous separation and analysis of aminodeoxyhexitols and alditols by automated ion-exchange chromatography, *Carbohydr. Res.*, 44, 97, 1975.

Table LC 21
ANION-EXCHANGE CHROMATOGRAPHY OF D-GLUCOSE OLIGOSACCHARIDES AND REDUCED OLIGOSACCHARIDES

Packing	P1	P1	P1
Column			
Length	15 cm	15 cm	15 cm
Diameter	0.8 cm	0.8 cm	0.8 cm
Material	glass	glass	glass
Solvent	S1	S2	S2
Flow rate	0.51 mℓ/min	0.51 mℓ/min	0.51 mℓ/min
Temperature	55°C	65°C	65°C
Detection	Orcinol	Orcinol	Orcinol
Reference	1	1	2

Compound	r[a]		
D-Glucose	100	100	100
Kojibiose	21	—	—
Kojibiitol	67	—	—
Nigerose	84	—	—
Nigeritol	67	—	—
Maltose	21	—	—
Maltitol	58	—	—
Isomaltose	84	84	84
Isomaltitol	58	—	—
Isomaltotriose	—	69	—
Isomaltotetraose	—	63	—
Isomaltopentaose	—	56	—
Isomaltohexaose	—	49	—
Isomaltoheptaose	—	43	—
Compound 62[b]	—	—	17
B2[b]	—	—	21
B3[b]	—	—	41
63[b]	—	—	57
36[b]	—	—	111
26[b]	—	—	120

Note: The co-operation of Professor M. Torii (Osaka University, Japan) in the compilation of these data is gratefully acknowledged.

[a] Relative to D-glucose (t, = about 4.5-5.5 hr).
[b] Structure appended.

Packing   P1   = LC-R-3 anion-exchange resin, in Jeol Liquid Chromatograph Model JLC-3BC.

Solvent   S1   = (1) 0.13 *M* boric acid, pH adjusted to 7.5 with NaOH.
(2) 0.25 *M* boric acid, pH adjusted to 9.0.
(3) 0.35 *M* boric acid, pH adjusted to 9.6.
Buffers (1), (2), and (3) delivered for 100, 40, and 220 min, respectively.

S2   = as S1, but buffers (1), (2), and (3) delivered for 60, 70, and 230 min, respectively.

REFERENCES

1. Torii, M. and Sakakibara, K., Column chromatographic separation and quantitation of α-linked glucose oligosaccharides, *J. Chromatogr.*, 96, 255, 1974.
2. Torii, M., Sakakibara, K., Misaki, A., and Sawai, T., Degradation of α-linked D-gluco-oligosaccharides and dextrans by an isomalto-dextranase preparation from *Arthrobacter globiformis* T6, *Biochem. Biophys. Res. Commun.*, 70, 459, 1976.

Appendix to Table LC 21: Structures of Numbered Compounds

Compound
62: α-D-Glc*p*-(1→6)-α-D-Glc*p*-(1→2)-D-Glc*p*
63: α-D-Glc*p*-(1→6)-α-D-Glc*p*-(1→3)-D-Glc*p*
36: α-D-Glc*p*-(1→3)-α-D-Glc*p*-(1→6)-D-Glc*p*
26: α-D-Glc*p*-(1→2)-α-D-Glc*p*-(1→6)-D-Glc*p*
α-D-Glc*p*
↓
6
B2: α-D-Glc*p*-(1→2)-α-D-Glc*p*
α-D-Glc*p*
↓
6
B3:   α-D-Glc*p*-(1→3)-D-Glc*p*

Table LC 22
SEPARATION OF URONIC ACIDS
AND ACIDIC OLIGOSACCHARIDES
ON ANION-EXCHANGE RESIN
ACETATE FORM

Packing	P1
Column	
Length	38 cm
Diameter	4 mm
Material	glass
Solvent	S1
Flow rate	30 mℓ/h
Temperature	45°C
Detection	D1

Compound	t_r (min)
D-Glucuronic acid	123
4-*O*-methyl-	87
4-*O*-β-D-glucopyranosyl-	66
D-Galacturonic acid	100
D-Mannuronic acid	130
L-Guluronic acid	114
L-Iduronic acid	138
2-*O*-(GlcA)-D-xylose	70
4-*O*-(GalA)-D-xylose	53
2-*O*-(4-O-methyl-GlcA)-D-xylose	48

Packing P1 = DA-X8F (Durrum®; particle size 11 μm), acetate form.

Solvent S1 = 0.08 *M* sodium acetate, buffered to pH 8.5.

Detection D1 = copper-bicinchoninate method; reagent mixed with equal volume of 95% (w/w) ethanol.

REFERENCE

1. Mopper, K., Improved chromatographic separations on anion-exchange resins. II. Separation of uronic acids in acetate medium and detection with a noncorrosive reagent, *Anal. Biochem.*, 86, 597, 1978.

Table LC 23
AUTOMATED CHROMATOGRAPHY
OF GLYCOSAMINOGLYCANS ON
ANION-EXCHANGE RESIN

Packing	P1
Column	
Length	35 cm
Diameter	6 mm
Material	glass
Solvent	S1
Flow rate	0.7 ml/min
Temperature	rt
Detection	D1

Compound	t_r (min)
Hyaluronic acid	120
Heparan sulfate	244
Chondroitin 6-sulfate[a]	276
Heparin[b]	318

[a]   Eluted together with chondroitin 4-sulfate and dermatan sulfate.
[b]   Broad peak, due to polydispersity.

Packing P1 = AG 1-X2 (Bio-Rad®; 200 to 400 mesh), chloride form.

Solvent S1 = sodium chloride (0.2—1.5 *M*)-magnesium chloride (0—3 *M*) gradient.

Detection D1 = orcinol method (Technicon® automated system), with beaker dialyzer between column outlet and analytical system to reduce concentrations of NaCl and MgCl$_2$ to levels not affecting color response in orcinol-sulfuric acid reaction.

REFERENCE

1. **Radhakrishnamurthy, B., Dalferes, E. R., Ruiz, H., and Berenson, G. S.,** Determination of aorta glycosaminoglycans by automated ion-exchange chromatography, *Anal. Biochem.*, 82, 445, 1977.

Table LC 24
ANION-EXCHANGE CHROMATOGRAPHY OF REDUCED
OLIGOSACCHARIDES DERIVED FROM BLOOD GROUP SUBSTANCE H

Packing	P1	P1	Packing	P1	P1
Column			Column		
Length	30 cm	60 cm	Length	30 cm	60 cm
Diameter	0.6 cm	0.8 cm	Diameter	0.6 cm	0.8 cm
Material	silica glass	silica glass	Material	silica glass	silica glass
Solvent	S1	S1	Solvent	S1	S1
Flow rate	20 mℓ/hr	90 mℓ/hr	Flow rate	20 mℓ/hr	90 mℓ/hr
Temperature	55°C	55°C	Temperature	55°C	55°C
Detection	D1	D1	Detection	D1	D1
Reference	1—4	1, 4	Reference	1—4	1, 4
Reduced oligosaccharide[a]	t_r (min)	t_r (min)	Reduced oligosaccharide[a]	t_r (min)	t_r (min)
17.1	25	21	P	52	—
14.1	28	23	K	57	—
16.1	29	24	13.2	67	53
15.1	31	24	17.3	72	54
13.1 (E)	32	27	11.2	74	48
11.1	33	26	15.4	74	56
15.2 (G)	34	27	16.3	80	62
S	35	—	10.2	86	58
14.2 (M)	36	28	14.4	92	65
14.3	43	33	A	93	—
16.2	45	36	14.5	111	80
10.1	46	35	16.4	123	90
C	47	—	10.3	135	105
17.2	48	36	17.4	136	108
Z	49	—	15.5	150	115
15.3	51	40	10.4	276	240

Note: The co-operation of Professor N. K. Kochetkov (USSR Academy of Sciences, Moscow) in the compilation of these data is gratefully acknowledged.

[a]   Structures appended; designations refer to fractions from gel and paper chromatography.

Packing P1   = DA-X4 anion-exchange resin (Durrum®); particle size 20 ± 5 μm.
Solvent S1   = 0.5 *M* sodium borate, pH 8.5.
Detection D1   = orcinol-sulfuric acid colorimetry at 400 nm.

REFERENCES

1. **Derevitskaya, V. R., Arbatsky, N. P., and Kochetkov, N. K.,** The use of anion-exchange chromatography to separate complex mixtures of oligosaccharides, *Dokl. Akad. Nauk SSSR*, 223, 1137, 1975.
2. **Kochetkov, N. K., Derevitskaya, V. A., and Arbatsky, N. P.,** The structure of pentasaccharide and hexasaccharide from blood group substance H, *Eur. J. Biochem.*, 67, 129, 1976.
3. **Derevitskaya, V. A., Arbatsky, N. P., and Kochetkov, N. K.,** Isolation and elucidation of the structure of higher oligosaccharide from blood group substance H, *Eur. J. Biochem.*, 423, 1978.
4. **Kochetkov, N. K.,** personal communication to L. Hough, 1978.

APPENDIX TO TABLE LC 24: STRUCTURES OF OLIGOSACCHARIDES

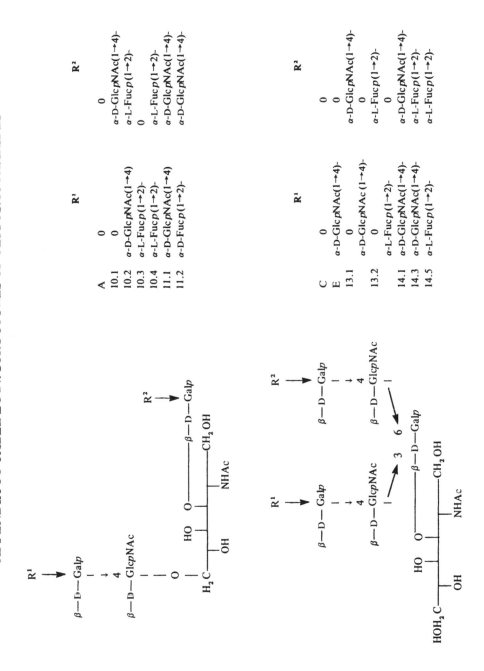

	R¹	R²
A	0	0
10.1	0	α-D-GlcpNAc(1→4)-
10.2	α-D-GlcpNAc(1→4)-	α-L-Fucp(1→2)-
10.3	α-L-Fucp(1→2)-	0
10.4	α-L-Fucp(1→2)-	α-L-Fucp(1→2)-
11.1	α-D-GlcpNAc(1→4)-	α-D-GlcpNAc(1→4)-
11.2	α-D-Fucp(1→2)-	α-D-GlcpNAc(1→4)-

	R¹	R²
C	0	0
E	α-D-GlcpNAc(1→4)-	0
13.1	0	α-D-GlcpNAc(1→4)-
13.2	α-D-GlcpNAc (1→4)-	0
	α-L-Fucp(1→2)-	α-L-Fucp(1→2)-
14.1	α-D-GlcpNAc(1→4)-	0
14.3	α-D-GlcpNAc(1→4)-	α-D-GlcpNAc(1→4)-
14.5	α-L-Fucp(1→2)-	α-L-Fucp(1→2)-

APPENDIX TO TABLE LC 24: STRUCTURES OF OLIGOSACCHARIDES
(continued)

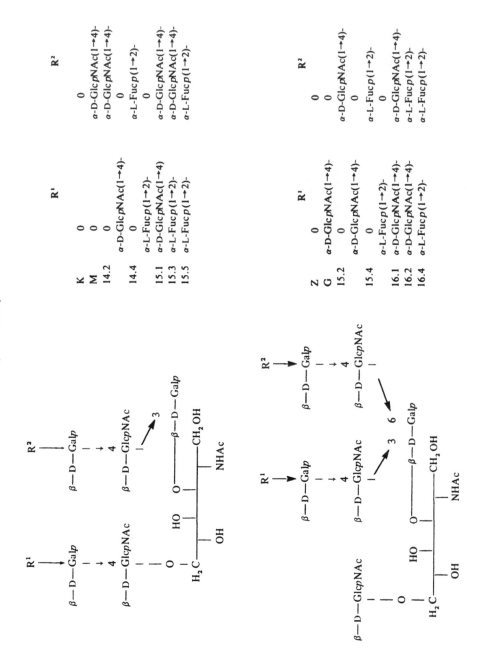

	R¹	R²
K	0	0
M	0	α-D-GlcpNAc(1→4)-
14.2	0	α-D-GlcpNAc(1→4)-
14.4	α-D-GlcpNAc(1→4)-	0
15.1	α-L-Fucp(1→2)-	0
15.3	α-D-GlcpNAc(1→4)-	α-D-GlcpNAc(1→4)-
15.3	α-L-Fucp(1→2)-	α-D-GlcpNAc(1→4)-
15.5	α-L-Fucp(1→2)-	α-L-Fucp(1→2)-

	R¹	R²
Z	0	0
G	α-D-GlcpNAc(1→4)-	0
15.2	α-D-GlcpNAc(1→4)-	α-D-GlcpNAc(1→4)-
15.4	0	α-L-Fucp(1→2)-
16.1	α-D-GlcpNAc(1→4)-	α-D-GlcpNAc(1→4)-
16.2	α-D-GlcpNAc(1→4)-	α-L-Fucp(1→2)-
16.4	α-L-Fucp(1→2)-	α-L-Fucp(1→2)-

	R¹	R²	R³
P	0	0	0
16.3	0	0	α-L-Fucp(1→2)-
	0	α-L-Fucp(1→2)-	0
17.1	α-L-Fucp(1→2)-	0	0
17.2	α-D-GlcpNAc(1→4)-	α-D-GlcpNAc(1→4)-	α-D-GlcpNAc(1→4)-
	0	α-D-GlcpNAc(1→4)-	α-L-Fucp(1→2)
17.3	α-L-Fucp(1→2)-	α-D-GlcpNAc(1→4)-	0
17.4	α-L-Fucp(1→2)-	α-D-GlcpNAc(1→4)-	α-L-Fucp(1→2)-
	0	α-L-Fucp(1→2)-	α-L-Fucp(1→2)-
S	α-L-Fucp(1→2)-	0	α-L-Fucp(1→2)-
	α-L-Fucp(1→2)	α-L-Fucp(1→2)-	0
	0	α-D-GlcpNAc(1→4)-	0

R³
β—D——Galp
↓
4
β—D——GlcpNAc
|

R²
β—D——Galp
↓
4
β—D——GlcpNAc
|
→ 3

R¹
β—D——Galp
↓
4
β—D——GlcpNAc
|
O
|
H₂C——O
|
HO——
|
CH₂OH
OH   NHAc

β—D—Galp
|
→ 6

β—D—GlcpNAc
|

Table LC 25
SEPARATION OF AMINODEOXY SUGARS, AMINODEOXYALDITOLS, AND REDUCED OLIGOSACCHARIDES ON AN AMINO ACID ANALYSER

Packing	P1	P1
Column		
Length	40 cm	40 cm
Diameter	1 cm	1 cm
Material	glass	glass
Solvent	S1	S2
Flow rate	45 ml/hr	45 ml/hr
Temperature	50°C	50°C
Detection	Ninhydrin	Ninhydrin

Compound	t_r (min)	t_r (min)
2-Amino-2-deoxy-D-glucose	170	—
2-Amino-2-deoxy-D-galactose	318	—
2-Amino-2-deoxy-D-mannose	194	—
2-Amino-2-deoxy-D-glucitol	363	366
2-Amino-2-deoxy-D-galactitol	387	387
2-Amino-2-deoxy-D-mannitol	285	—
GlcNβ(1→3)galactitol	—	83[a]
GlcNβ(1→6)galactitol	—	83[a]
GalNα(1→3) galactitol	—	100
Galβ(1→3)2-amino-2-deoxy-D-galactitol	—	146
Galβ(1→3)2-amino-2-deoxy-D-glucitol	—	158
Galβ(1→6)2-amino-2-deoxy-D-glucitol	—	179
Galβ(1→4)2-amino-2-deoxy-D-glucitol	—	192
Galβ(1→4)GlcNβ(1→6)2-amino-2-deoxy-D-galactitol	—	416
GlcNβ(1→4)2-amino-2-deoxy-D-glucitol	—	533
GalNα(1→3)Galβ(1→3)2-amino-2-deoxy-D-glucitol	—	552
GalNα(1→3)Galβ(1→4)2-amino-2-deoxy-D-glucitol	—	608

[a]   Separate if pH of eluting buffer adjusted to 7.0.

Packing	P1	= Locarte No. 12 resin (8% cross-linked sulfonated polystyrene; Locarte, London), Na$^+$ form.
Solvent	S1	= buffer, 0.1 M in Na$^+$, containing trisodium citrate dihydrate (49.0 g), boric acid (1.55 g), 33% (w/v) Brij-35 solution (15 ml), water to 5 l, and HCl to bring pH to 7.5.
	S2	= S1 for 380 min, then buffer changed to solution 0.2 M in Na$^+$, containing trisodium citrate dihydrate (91.88 g), borax (11.9 g), 33% (w/v) Brij-35 solution (15 ml), water to 5 l, and HCl to bring pH to 8.0.

REFERENCE

1. Donald, A. S. R., Separation of hexosamines, hexosaminitols and hexosa-mine-containing di- and trisaccharides on an amino acid analyser, *J. Chromatogr.*, 134, 199, 1977.

CHROMATOGRAPHY OF CARBOHYDRATES ON DIETHYLAMINOETHYL-DEXTRAN ANION-EXCHANGERS

DEAE-Sephadex ion-exchangers (Pharmacia®) are useful in the separation and isolation of acidic carbohydrates, especially oligosaccharides. The isolation of the individual oligomers from a mixture of oligogalacturonic acids produced on enzymic hydrolysis of pectic acid serves as a typical example of the application of this type of chromatography; data are presented in Table LC 26.

Table LC 26
CHROMATOGRAPHY OF OLIGOGALACTURONIC ACIDS ON DEAE-SEPHADEX A-50

Packing	P1
Column	
Length	80 cm
Diameter	10 cm
Material	glass
Solvent	S1
Flow rate	na
Temperature	rt
Detection	D1
Reference	1

Degree of polymerisation	Relative retention volume[a]
1	100
2	115
3	155
4	191
5	236
6	276
7	317
8	379
9	434

[a] Relative to D-galacturonic acid = 100.

Packing P1 = DEAE-Sephadex® A-50, chloride form, equilibrated with water at pH 6.0.
Solvent S1 = 4 *l* water at pH 6.0, then stepwise elution with 4-*l* volumes of sodium chloride solutions, 0.05, 0.10, 0.125, 0.150, 0.175, 0.200, and 0.225 *M*, respectively, followed by 6 *l* each of 0.250 and 0.275 *M* NaCl.
Detection D1 = carbazole method.

REFERENCE

1. Liu, Y. K. and Luh, B. S., Preparation and thin-layer chromatography of oligogalacturonic acids, *J. Chromatogr.*, 151, 39, 1978.

DEAE-Sephadex A-50 may be applied to the fractionation of charged polysaccharides, e.g., agar components differing in sulfate and pyruvate content have been fractionated on the chloride form of this exchanger,[2] by stepwise elution with water and sodium chloride solutions of increasing concentration (0.5—3 M). For polysaccharides containing components having chains terminated by D-mannitol, the use of DEAE-Sephadex A-50 in the molybdate form has been recommended[3] as a means of isolating such components: on chromatography of a laminarin on this packing the components having chains terminated by D-glucose residues are eluted with water, the D-mannitol-terminated chains with 0.25 M sodium chloride.

DEAE-Sephadex A-25 is useful in the examination of hydrolysates of acidic polysaccharides, for the separation of acidic from neutral components, and the fractionation of mixtures of acidic oligosaccharides. For example, Aspinall and co-workers,[4-6] in studying the products of partial acid hydrolysis of plant gums, have used this exchanger in the formate form; neutral sugars present in the hydrolysate are eluted with water, whereas the acidic oligosaccharides, such as the aldobiouronic acid 6-O-(β-D-glucopyranosyluronic acid)-D-galactose, produced from such gums are retained by the exchanger in water but can subsequently be eluted with formic acid solutions (0.50—0.5 M).

REFERENCES

2.   Duckworth, M., Hong, K. C., and Yaphe, W., The agar polysaccharides of *Gracilaria* species, *Carbohydr. Res.*, 18, 1, 1971.
3.   Stark, J. R., A new method for the analysis of laminarins and for preparative-scale fractionation of their components, *Carbohydr. Res.*, 47, 176, 1976.
4.   Aspinall, G. O., Molloy, J. A., and Whitehead, C. C., *Araucaria bidwillii* gum. III. Partial acid hydrolysis of the gum, *Carbohydr. Res.*, 12, 143, 1970.
5.   Aspinall, G. O. and Bhattacharjee, A. K., Plant gums of the genus *Khaya*. IV. Major component of *Khaya ivorensis* gum, *J. Chem. Soc. C*, 361, 1970.
6.   Aspinall, G. O. and Sanderson, G. R., Plant gums of the genus *Sterculia*. IV. Acidic oligosaccharides from *Sterculia urens* gum, *J. Chem. Soc. C*, 2256, 1970.

CHROMATOGRAPHY OF CARBOHYDRATES ON CELLULOSIC ION-EXCHANGERS

Diethylaminoethyl-cellulose affords an efficient means of fractionating charged polysaccharides, and in recent years microgranular DEAE-cellulose has been preferred to ECTEOLA-cellulose in fractionation of glycosaminoglycans, since sharper peaks and virtually quantitative recoveries of these polysaccharides have been obtained with the former exchanger. The data presented in Table LC 27 exemplify this application of DEAE-cellulose.

Table LC 27
CHROMATOGRAPHY OF GLYCOSAMINOGLYCANS OF DEAE-CELLULOSE

Packing	P1 + P2 (in series)	
Column		
Length	40 cm	5 cm
Diameter	1.5 cm	1 cm
Material	glass	glass
Solvent	S1	S2
Flow rate	8 ml/hr	5 ml/hr
Temperature	rt	60°C
Detection		D1
Reference		1

Compound	V_e (ml)
Hyaluronic acid[a]	23
Heparan sulfate[b]	44
Chondroitin 4-sulfate[c]	55
Dermatan sulfate[d]	58
Keratan sulfate[e]	58
Heparin[f]	80

[a] $\overline{M}_w$ about 1×10^6.
[b] $\overline{M}_w$ = 58,000; 0.5 sulfate groups per disaccharide unit.
[c] $\overline{M}_w$ = 19,500; 0.95 sulfate groups per disaccharide unit.
[d] $\overline{M}_w$ = 41,000; about 1 sulfate per disaccharide unit.
[e] $\overline{M}_w$ = 17,500; sulfate as in d.
[f] $\overline{M}_w$ = 13,000; 2.4 sulfate groups per disaccharide unit.

Packing P1 = Sephadex G-50 (Pharmacia®), equilibrated with S1 before use.
P2 = DE 52 (Whatman®), microgranular DEAE-cellulose, chloride form, equilibrated with S1 before use.
Solvent S1 = 0.15 M sodium chloride, containing 0.02% sodium azide.
S2 = lithium chloride gradient (0.2—1.2 M), buffered at pH 4.0 with 0.05 M sodium acetate buffer.
Detection D1 = carbazole method.
Technique The Sephadex column was connected to the DE 52 column, and the glycosaminoglycans (usually in 2 ml NaCl solution) were applied to the former column and transferred to the DE 52 column by elution with S1.

REFERENCE

1. Hallén, A., Chromatography of acidic glycosaminoglycans on DEAE-cellulose, *J. Chromatogr.*, 71, 83, 1972.

A further example of the value of DE 52 in fractionation of acidic carbohydrates is the good resolution obtained on chromatography of the mixture of sialyl oligosaccharides occurring in human milk on a column (45 × 1.5 cm) packed with this exchanger; stepwise elution with acetic acid-pyridine buffers, pH 5.4, of increasing concentration (0.002, 0.012, and 0.060 *M*) yielded fractions from which several of these oligosaccharides, (e.g., 3′- and 6′-sialyllactose) were subsequently isolated.[2]

In studies of acidic plant-gum polysaccharides, DEAE-cellulose in the phosphate form, eluted with phosphate buffers (0.1—0.5 *M*) at pH 6, has been used to fractionate polydisperse gums.[3] The use of this exchanger in the carbonate form, eluted with a 0—0.5 *M* ammonium carbonate gradient, has also been recommended.[4]

Further examples of applications of DEAE-cellulose in the fractionation of acidic polysaccharides will be found in the two review articles[5,6] cited below.

REFERENCES

2. Smith, D. F., Zopf, D. A., and Ginsburg, V., Fractionation of sialyl oligosaccharides of human milk by ion-exchange chromatography, *Anal. Biochem.*, 85, 602, 1978.
3. Aspinall, G. O. and Bhattacharjee, A. K., Plant gums of the genus *Khaya*. IV. Major component of *Khaya ivorensis* gum, *J. Chem. Soc. C*, 361, 1970.
4. Siddiqui, I. R. and Wood, P. J., DEAE-cellulose carbonate form. A useful medium for fractionating polysaccharides, *Carbohydr. Res.*, 16, 452, 1971.
5. Neukom, H. and Kuendig, W., Fractionation of neutral and acidic polysaccharides by ion-exchange column chromatography on diethylaminoethyl (DEAE)-cellulose, in *Methods in Carbohydrate Chemistry*, Vol. 5, Whistler, R. L., BeMiller, J. N., and Wolfrom, M. L., Eds., Academic Press, New York, 1965, 14.
6. Jandera, P. and Churáček, J., Ion-exchange chromatography of aldehydes, ketones, ethers, alcohols, polyols and saccharides, *J. Chromatogr.*, 98, 55, 1974.

GEL-PERMEATION CHROMATOGRAPHY

Gel-permeation chromatography is important in the carbohydrate field, as it is much used in the separation of commercially important oligosaccharides, such as the malto-dextrins, and the fractionation and molecular-weight distribution analysis of polysaccharides. This type of chromatography is, therefore, given extensive coverage here.

For some earlier data on polyacrylamide gel chromatography of sugars, see Section A, Volume I, Section II.II, Table LC 71.

Tables LC 28-37, LC 40, LC 42, and LC 44 were contributed by Dr. Kirsti Granath (Pharmacia AB, Uppsala, Sweden).

Table LC 28
COLUMN CHROMATOGRAPHY ON TIGHTLY CROSS-LINKED GELS: CALIBRATION WITH SELECTED SUGARS AND ALCOHOLS

Packing			P1	P2	P3	P4	P5	P6
Column								
Length			90 cm	90 cm	90 cm	90 cm	90 cm	90 cm
Diameter			1.5 cm	1.5 cm	1.5 cm	1.5 cm	1.5 cm	1.5 cm
Solvent			S1	S1	S1	S1	S1	S1
Flow rate			10 mℓ/hr					
Temperature			rt		Conditions same throughout			
Detection			Rl					
Sample			na					
Compound	M	r (A)[a]	K_d	K_d	K_d	K_d	K_d	K_d
Stachyose	667	5.78	0.17	0.13	0.31	0.31	0.41	0.48
Raffinose	505	5.28	0.24	0.21	0.43	0.39	0.49	0.63
Maltose	342	4.64	0.35	0.32	0.52	0.52	0.60	0.65
Glucose	180	3.56	0.48	0.53	0.63	0.63	0.70	0.82
Glycerol	92	2.89	0.60	0.58	0.76	0.72	0.77	0.89
Propylene glycol	76	2.85	0.67	0.67	0.78	0.76	0.79	0.85
Ethylene glycol	62	2.50	0.67	0.67	0.77	0.75	0.80	0.88
Methanol	32	2.13	0.80	0.80	0.96	0.87	0.87	0.86

[a] Unhydrated radius, derived from Corey-Pauling-Koltun molecular models by method of Goldstein and Solomon (see *J. Gen. Physiol.*, 44, 1, 1960).

Packing   P1   =   Sephadex G-10 (Pharmacia®), allowed to swell in distilled water for at least 24 hr.

         P2   =   Sephadex G-10 (1 *M* HCl). Swollen gel, mixed with equal volume of 1 *M* HCl, placed in boiling-water bath for 2 hr, with stirring. After supernatant decanted, HCl removed by repeated washes with distilled water, followed by vacuum filtration, gel extracted twice with 1 volume chloroform-methanol (1:1), followed by 1 volume diethyl ether, vacuum filtration between each extraction. Gel dried in hood for 24 hr then heated in oven at 100°C for 2 hr. Before use, swollen in distilled water, as P1.

         P3   =   Sephadex G-10 (6 *M* HCl), treated with 6 *M* HCl as described for P2.

         P4   =   Sephadex G-15, prepared as P1.

         P5   =   Sephadex G-15 (1 *M* HCl), treated with 1 *M* HCl as P2.

         P6   =   Bio-Gel® P-2, polyacrylamide gel (Bio-rad®), prepared as P1.

Solvent   S1   =   0.15 *M* NaCl in 0.01 *M* acetic acid (pH 3.3).

REFERENCE

1. **Goodson, J. M., Stefano, V., and Smith, J. C.,** Calibration of tightly cross-linked gel filtration media for determination of the size of low molecular weight, non-interacting solutes, *J. Chromatogr.*, 54, 43, 1971.

Table LC 29

**COLUMN CHROMATOGRAPHY OF SUGARS ON
SEPHADEX G-15; EFFECT OF STRUCTURAL
MODIFICATIONS AND MOLECULAR WEIGHT ON
RELATIVE ELUTION VOLUME**

Packing	Sephadex G-15
Column	
Length (bed)	45 cm
Diameter	2.54 cm
Solvent	H_2O
Flow rate, linear	8.5 cm/hr
Temperature	na
Detection	RI
Sample	4%/0.1 ml

1. Effect of Structural Modifications on Relative Elution (R_g) on Sephadex G-15[1]

Monosaccharide	Corresponding aldopentapyranose[a]	Substituent at indicated C-atom		
		H[b]	OH	OMe
		Carbon-2		
D-Ribose		1.086	1.063	
D-Glucose		1.042	1.000	0.974
D-Galactose		1.029	0.990	
		Carbon-3		
D-Glucose			1.000	0.977
		Carbon-6		
Mannose	1.062	1.040[c]	1.029	
D-Glucose	1.047		1.000	0.957
Galactose	1.024	0.997[c]	0.990	

Note: Data expressed as R_g values (R_g here is V_r for solute relative to V_r for D-glucose).

[a]   Hydroxymethyl substituent on C-5 of the aldohexose replaced by H.
[b]   Deoxy sugars.
[c]   Measurements made with L-monosaccharide.

II. Effect of Site of Linkage (Disaccharides of D-Glucose) on R_g on Sephadex G-15[1]

Disaccharide	Linkage	R_g
Maltose	α-D-(1 → 4)	0.922
Isomaltose	α-D-(1 → 6)	0.889
Cellobiose	β-D-(1 → 4)	0.907
Gentiobiose	β-D-(1 → 6)	0.870

Table LC 29 (continued)
COLUMN CHROMATOGRAPHY OF SUGARS ON SEPHADEX G-15; EFFECT OF STRUCTURAL MODIFICATIONS AND MOLECULAR WEIGHT ON RELATIVE ELUTION VOLUME

Packing	Sephadex G-15
Column	
Length (bed)	45 cm
Diameter	2.54 cm
Solvent	H_2O
Flow rate, linear	8.5 cm/hr
Temperature	na
Detection	RI
Sample	4%/0.1 ml

III. Effect of Configuration at Substituted Anomeric Carbon Atom on R_s of Methyl Glycosides and Disaccharides on Sephadex G-15[1]

Compound	R_s
Methyl α-D-xylopyranoside	1.021
Methyl β-D-xylopyranoside	1.005
Methyl α-D-glucopyranoside	0.971
Methyl β-D-glucopyranoside	0.954
4-0-α-D-Glucopyranosyl-D-glucopyranose	0.922
4-0-β-D-Glucopyranosyl-D-glucopyranose	0.907
6-0-α-D-Glucopyranosyl-D-glucopyranose	0.889
6-0-β-D-Glucopyranosyl-D-glucopyranose	0.870

IV. Relative Elution Volumes of Groups of Isomeric Saccharides on Sephadex G-15[2]

Pentoses M 150	R_s	Hexoses M 180	R_s	Disaccharides M 342	R_s
D-Ribose	1.065	D-Psicose	1.040	Sucrose	0.926
D-Lyxose	1.062	D-Mannose	1.029	Maltose	0.922
D-Xylose	1.047	L-Sorbose	1.026	Turanose	0.920
L-Arabinose	1.024	D-Fructose	1.018	Cellobiose	0.907
		D-Glucose	1.000	Trehalose	0.902
		D-Galactose	0.990	Lactose	0.895
				Isomaltose	0.889
				Melibiose	0.877
				Gentiobiose	0.870

REFERENCES

1. Bertoniere, N. R., Martin, L. F., and Rowland, S. P., Stereoselectivity in the elution of sugars from columns of Sephadex G-15, *Carbohydr. Res.*, 19, 189, 1971.
2. Martin, L. F., Bertoniere, N. R., and Rowland, S. P., The effects of sorption and molecular size of solutes upon elution from polyhydroxylic gels, *J. Chromatogr.*, 64, 263, 1972.

Table LC 30
SEPARATION OF SUBSTITUTED CARBOHYDRATES BY HIGH PERFORMANCE GEL PERMEATION CHROMATOGRAPHY

Packing	P1	P2
Column		
Length	61 cm	122 cm
I.D.	2 mm	2mm
Solvent	S1	S2
Flow rate	0.27 mℓ/min	0.51 mℓ/min
Temperature	rt	rt
Detection	RI, UV	RI, UV
Sample	10 mg/1 mℓ	10 mg/1 mℓ

Compound	r[a]	r[b]	Mol wt
α-Cellobiose octaacetate	1.22	0.92	678.6
Phenyl tetra-O-acetyl β-D-glucopyranoside	2.19	1.03	424.4
α-D-Glucopyranose pentaacetate	2.16	1.00	390.3
β-D-Glucopyranose pentaacetate	2.32	0.99	390.3
α-D-Galactopyranose pentaacetate	2.00	1.01	390.3
β-D-Galactopyranose pentaacetate	2.16	1.01	390.3
Methyl tetra-O-acetyl β-D-glucopyranoside	2.03	1.03	362.3
L-Arabinitol pentaacetate	1.84	1.18	362.3
Erythritol tetraacetate	1.84	1.07	290.2
4,6-O-Benzylidene D-glucopyranose	1.61	1.21	268.3
1,2:5,6-Di-O-isopropylidene α-D-glucofuranose	1.59	1.12	260.3
2,3,6-Tri-O-methyl-D-glucose	1.11	1.15	225.2
2,4,6-Tri-O-methyl-D-Glucose	1.10	1.15	225.2
1,2-O-Isopropylidene α-D-glucofuranose	1.23	1.20	220.2
3-O-Methyl D-glucose	1.00		
D-mannose	0.97		
D-Fructose	1.03		
D-Glucose	1.00		
D-Xylose	1.10		
D-Glucitol	1.00		
D-Arabinitol	1.00		
Methyl α-D-glucopyranoside	1.00		
Methyl β-D-glucopyranoside	1.00		
Phenyl β-D-glucopyranoside	1.22		
Polypropylene glycol standard (M_n = 1220)	0.69		

Note: Commercial liquid chromatograph (Waters Assoc. Model 202) was used.

[a] Relative to D-glucose.
[b] Relative to α-D-glucopyranose pentaacetate.

Packing P1 = EM Gel OR-PVA 500 (Merck®), 50 μm fractionation range, mol wt 0—300. Swollen in methanol for 24 hr. Slurry introduced at top of vertical column under vacuum. High-pressure pump for final packing.

        P2 = Poragel 60 A (Waters®), 37—75 μm fractionation range, mol wt 100—2400. Packed by manufacturer.

Solvent S1 = methanol.
        S2 = chloroform.

From McGinnis, G. D. and Fang, P., *J. Chromatogr.*, 130, 181, 1977. With permission.

Table LC 31
PARTITION COEFFICIENTS (K_d) FOR MALTODEXTRINS, AT DIFFERENT TEMPERATURES, ON POLYACRYLAMIDE GEL

| Packing | P1 | P1 | P1 | P1 | P1 | P1 | | |
| Temperature | 20°C | 30°C | 40°C | 50°C | 60°C | 70°C | | |
Compound	K_d	K_d	K_d	K_d	K_d	K_d	M	d(logK)/dT
Glucose	0.892	0.887	0.893	0.883	0.894	0.877	180	$- 0.86 \times 10^{-4}$
Maltose	0.800	0.789	0.790	0.775	0.784	0.766	342	$- 3.32 \times 10^{-4}$
Maltotriose	0.723	0.703	0.700	0.682	0.683	0.666	504	$- 6.61 \times 10^{-4}$
Maltotetraose	0.647	0.625	0.616	0.599	0.596	0.579	666	$- 9.09 \times 10^{-4}$
Maltopentaose	0.575	0.550	0.539	0.522	0.516	0.500	828	$- 11.51 \times 10^{-4}$
Maltohexaose	0.511	0.486	0.473	0.459	0.448	0.435	990	$- 13.52 \times 10^{-4}$
Maltoheptaose	0.455	0.430	—	0.398	0.391	0.378	1152	$- 15.78 \times 10^{-4}$

Packing   P1 =   Polyacrylamide (Bio-Gel® P-2). From −400 mesh product, 28—60 μm gel fraction was sieved out and further fractionated by repeated settling and decanting. Degassed before packing.

Column   length: 200 cm.
diameter: 1.5 cm.
material: glass, coated with 1% dichlorodimethylsilane in benzene.

Solvent H_2O.
Flow rate   25 ml/hr.
Sample   20 μl, 10% solution.
Detection   automatic analysis of effluent by orcinol method.

REFERENCE

1.   Dellweg, H., John, M., and Trenel, G., Gel permeation chromatography of maltooligosaccharides at different temperatures, *J. Chromatogr.*, 57, 89, 1971; John, M. and Dellweg, H., Gel chromatographic separation of oligosaccharides, *Sep. Purif. Methods*, 2, 231, 1973.

Table LC 32
COLUMN CHROMATOGRAPHY OF MALTODEXTRINS ON POLYACRYLAMIDE GELS; INFLUENCE OF FLOW RATE AND TEMPERATURE ON PARTITION COEFFICIENT K_d

Packing	P1	P2	P1	P1	P1	P1	P1	P1
Column								
Length, cm	100	100	100	100	100	82.0	82.0	82.0
Diameter, cm	1.6	1.6	1.6	1.6	1.6	1.6	1.6	1.6
Solvent	H_2O	H_2O	H_2O	H_2O	H_2O	H_2O	H_2O	H_2O
Flow rate, mℓ/hr	20.0	18.0	19.0	24.6	30.0	28.8	28.8	28.8
Temperature	45°C	45°C	45°C	45°C	45°C	35°C	45°C	55°C
Detection	RI	RI	RI	RI	RI	RI	RI	RI
Sample, $\mu\ell$	200	200	200	200	200	200	200	200
conc	8%	8%	8%	8%	8%	8%	8%	8%
Reference	1	1	2	2	2	2	2	2
Maltodextrin[a]	K_d	K_d	K_d	K_d	K_d	K_d	K_d	K_d
G 1	0.770	0.860	0.768	0.763	0.766	0.896	0.881	0.857
G 2	0.671	0.800	0.672	0.670	0.667	0.788	0.758	0.740
G 3	0.586	0.745	0.583	0.584	0.586	0.699	0.666	0.649
G 4	0.510	0.689	0.509	0.508	0.509	0.611	0.580	0.565
G 5	0.443	0.643	0.436	0.439	0.440	0.527	0.499	0.486
G 6	0.377	0.580	0.378	0.376	0.377	0.462	0.430	0.416
G 7	0.329	0.534	0.326	0.325	0.330	0.399	0.373	0.358
G 8	0.286	0.492	0.282	0.284	0.284	0.347	0.325	0.312
G 9	0.248	0.456	0.245	0.246	0.248	—	—	—
G 10	0.213	0.420	0.213	0.213	0.212	—	—	—
G 11	0.186	0.387	—	—	—	—	—	—
G 12	—	0.358	—	—	—	—	—	—
G 13	—	0.330	—	—	—	—	—	—
G 14	—	0.305	—	—	—	—	—	—
G 15	—	0.280	—	—	—	—	—	—

[a]   Figures denote degree of polymerization of glucose residues (G).

Packing P1 =   Polyacrylamide Bio-Gel® P-2, −400 mesh.
       P2 =   Polyacrylamide Bio-Gel® P-4, −400 mesh. Gel slurry heated at 60° for 10 min degassed.

REFERENCES

1. **Sabbagh, N. K. and Fagerson, I. S.**, Gel permeation chromatography of glucose oligomers using polyacrylamide gels, *J. Chromatogr.*, 86, 184, 1973.
2. **Sabbagh, N. K. and Fagerson, I. S.**, Gel chromatography of glucose oligomers. A study of operational parameters, *J. Chromatogr.*, 120, 55, 1976.

Table LC 33
GEL CHROMATOGRAPHIC ANALYSIS OF α-1,4, α-1,6 AND α-1,4:α-1,6 LINKED OLIGOSACCHARIDES COMPOSED OF D-GLUCOSE

Packing	Bio-Gel P-4, 31—45 μm (Fractionated from −400 mesh)			
Column				
Length	169.0 cm			
Diameter	0.9 cm			
Solvent	H_2O			
Flow rate	10.8 mℓ/hr			
Temperature	65°C			
Detection	Orcinol			
Reference	1, 2	1, 2	1	1

			Pullulan-oligosaccharides[c]	
	Malto-oligo-saccharides[a]	Isomalto-oligo-saccharides[b]	A	B
DP	K_{av}	K_{av}	K_{av}	K_{av}
1	0.778	0.780		
2	0.723	0.700		
3	0.670	0.625	0.670	0.640
4	0.617	0.560		
5	0.566	0.505		
6	0.521	0.453	0.516	0.489
7	0.480	0.408		
8	0.444	0.368		
9	0.408	0.332	0.396	0.376
10	0.377	0.301		
11	0.348	0.273		
12	0.321	0.247	0.308	0.290
13	0.297	0.224		
14	0.274	0.202		
15	0.254	0.182	0.240	0.226
18	—	—	0.187	0.177
21	—	—	0.148	0.138
24	—	—	0.115	0.107
27	—	—	0.090	0.084

[a] Acid hydrolysis of amylose.

[b] Acid hydrolysis of dextran (B 512).

[c] Enzymic hydrolysis of pullulan (linear α-glucan of maltotriose units linked "head to tail" through α-1,6 linkages): A. Hydrolysis by pullulanase (<33.3% α-1,6 linkages). B. Hydrolysis by isopullulanase (>33.3% α-1,6 linkages).

REFERENCES

1.   Schmidt, F. and Enevoldsen, B. S., Gel filtration chromatography of oligosaccharides, *Carlsberg Res. Commun.*, 41, 91, 1976.
2.   Schmidt, F. and Enevoldsen, B. S., Comparative studies of malto- and iso-maltooligosaccharides and their corresponding alditols, *Carbohydr. Res.*, 61, 197, 1978.

Table LC 34
COLUMN GEL CHROMATOGRAPHY OF CELLODEXTRINS

Packing	P1	P2	P3
Column			
Length	60 cm	60 cm	60 cm
Diamter	1.0 cm	1.0 cm	1.0 cm
Solvent	H_2O	H_2O·	H_2O
Flow rate	2 ml/hr	2 ml/hr	2 ml/hr
Temperature	12°C, 25°C, 40°C	25°C	25°C
Detection	RI	RI	RI
Sample (vol, conc)	0.1 ml; 1 mg/ml	0.1 ml; 1 mg/ml	0.1 ml; 1 mg/ml
Reference	1	1	2

Cellodextrin[a]	K_{av} 12°C/25°C/40°C	K_{av}	K_{av}
Glucose	$0.92_6\ 0.92_1\ 0.91_5$	0.82_0	0.61_9
Cellobiose	$0.85_6\ 0.84_7\ 0.83_8$	0.65_5	0.52_7
Cellotriose	$0.78_2\ 0.77_3\ 0.76_1$	0.49_0	0.43_1
Cellotetraose	$0.71_0\ 0.70_0\ 0.68_5$	0.38_5	0.36_2
Cellopentaose	$0.64_9\ 0.63_8\ 0.62_1$	0.33_0	0.30_9
Cellohexaose	$0.60_0\ 0.59_0\ 0.57_2$	0.31_1	0.27_3

[a]   From acid hydrolysis of cellulose.

Packing P1 = Polyacrylamide Bio-Gel P-2, 200 to 400 mesh, swollen for 24 hr in distilled water, degassed under high vacuum.

P2 = Sephadex G-15, 40—120 μm.

P3 = Hydroxyethylcellulose (HEC-100). Laboratory batch. Method of preparation given in Reference 2.

REFERENCES

1.   Brown, W., The separation of cellodextrins by gel permeation chromatography, *J. Chromatogr.*, 52, 273, 1970.
2.   Brown, W. and Chitumbo, K., Properties of hydroxyethylcellulose gels, *Chem. Scr.*, 2, 88, 1972.

Table LC 35
COLUMN GEL CHROMATOGRAPHY OF CELLODEXTRINS IN DIFFERENT SOLVENTS

Packing	P1	P1	P1	P1
Column				
Length (bed)	66.9 cm	57.2 cm	61.1 cm	58.0 cm
Diameter	1.0 cm	1.0 cm	1.0 cm	1.0 cm
Solvent	S1	S2	S3	S4
Flow rate	na	na	na	na
Temperature	25°C	25°C	25°C	25°C
Detection	RI	RI	RI	RI
Sample	0.1 ml; 1 mg/ml	0.1 ml; 1 mg/ml	0.1 ml; 1 mg/ml	0.1 ml; 1 mg/ml
Cellodextrin	K_{av}	K_{av}	K_{av}	K_{av}
Glucose	0.341	0.500	0.513	0.528
Cellobiose	0.190	0.376	0.415	0.427
Cellotriose	0.104	0.293	0.328	0.336
Cellotetraose	0.058	0.213	0.259	0.272
Cellopentaose	0.025	0.144	0.205	0.215
Cellohexaose	Excluded	0.112	0.162	0.189

Packing P1 = Sephadex G-15, 40—120 μm.
Solvent S1 = dimethylsulfoxide.
          S2 = formamide.
          S3 = H_2O.
          S4 = 0.1 M NaCl.

From Brown, W., *J. Chromatogr.*, 59, 335, 1971. With permission.

Table LC 36
COLUMN GEL CHROMATOGRAPHY
OF MANNODEXTRINS

Packing	P1	P2	P3
Column			
Length	60 cm	60 cm	60 cm
Diameter	1.0 cm	1.0 cm	1.0 cm
Solvent	S1	S1	S2
Flow rate	2 ml/hr	2 ml/hr	2 ml/hr
Temperature	25°C	25°C	25°C
Detection	RI	RI	RI
Sample	0.1 ml;	0.1 ml;	0.1 ml;
	1mg/ml	1 mg/ml	1 mg/ml
Reference	1	1	2
Mannodextrin[a]	K_{av}	K_{av}	K_{av}
Mannose	0.80	0.53	0.65
Mannobiose	0.68	0.40	0.52
Mannotriose	0.60	0.30	0.45
Mannotetraose	0.51	0.21	0.36
Mannopentaose	0.45	0.15	0.30
Mannohexaose	0.40	0.12	—
Mannoheptaose	0.33	0.08	0.20

[a] β-D-(1 → 4)-linked; from acid hydrolysis of ivory nut mannan.

Packing P1 = Polyacrylamide Bio-Gel P-2, 200 to 400 mesh, swollen for 24 hr in distilled water, de-gassed under high vacuum.

P2 = Sephadex G-15, 40—120 μm.

P3 = Hydroxyethylcellulose (HEC-100). Laboratory batch.

Solvent S1 = 0.1 M NaCl.

S2 = H_2O.

REFERENCES

1. Brown, W. and Chitumbo, K., Preparation of mannodextrins and their separation by gel chromatography, *J. Chromatogr.*, 66, 370, 1972.
2. Brown, W. and Chitumbo, K., Properties of hydroxyethylcellulose gels, *Chem. Scr.*, 2, 88, 1972.

Table LC 37
COLUMN GEL CHROMATOGRAPHY OF XYLODEXTRINS

Packing	P1	P1	P2	P3
Column				
Length	60 cm	60 cm	60 cm	60 cm
Diameter	1.0 cm	1.0 cm	1.0 cm	1.0 cm
Solvent	S1	S2	S2	S1
Flow rate	2 ml/hr	2 ml/hr	2 ml/hr	2 ml/hr
Temperature	25°C	25°C	25°C	25°C
Detection	RI	RI	RI	RI
Sample	0.1 ml;	0.1 ml;	0.1 ml;	0.1 ml;
	1 mg/ml	1mg/ml	1 mg/ml	1 mg/ml
Reference	1	1	1	2
Xylodextrin[a]	K_{av}	K_{av}	K_{av}	K_{av}
Xylose	0.87	0.81	0.57	0.67
Xylotetraose	0.71	0.58	0.29	0.38
Xylopentaose	0.65	0.51	0.24	0.31
Xylohexaose	0.59	0.47	0.19	0.27
Xyloheptaose	0.55	0.42	0.15	0.22
Xylooctaose	0.53	0.37	0.13	—

[a] β-D-(1 → 4)-linked; from acid hydrolysis of Esparto grass.

Packing P1 = Polyacrylamide Bio-Gel P-2, 200 to 400 mesh.
P2 = Sephadex G-15, 40—120 μm.
P3 = Hydroxyethylcellulose (HEC-100). Laboratory batch.
Solvent S1 = H$_2$O.
S2 = 0.1 M NaCl.

REFERENCES

1. Brown, W. and Andersson, O., Separation of xylodextrins by gel chromatography, *J. Chromatogr.*, 67, 163, 1972.
2. Brown, W. and Chitumbo, K., Properties of hydroxyethylcellulose gels, *Chem. Scr.*, 2, 88, 1972.

Table LC 38
CHROMATOGRAPHIC
SEPARATION OF
XYLODEXTRINS ON BIO-GEL P-2

Packing	Bio-Gel P-2, 12—18 μm
Column	
Length	116 cm
Diameter	1.0 cm
Solvent	H_2O
Flow rate, linear	15 cm/hr
Temperature	65°C
Detection	Orcinol
Sample	5 mg/0.5 ml

Xylodextrin[a], DP	D_v
1	0.631
2	0.575
3	0.525
4	0.479
5	0.436
6	0.407
7	0.380
8	0.347
9	0.316
10	0.288
11	0.269
12	0.245
13	0.224
14	0.204
15	0.186
16	0.174
17	0.159
18	0.148

[a]   From acid hydrolysis of birch xylan.

REFERENCE

1. Havlicek, J. and Samuelson, O., Chromatography of oligosaccharides from xylan by various techniques, *Carbohydr. Res.*, 22, 307, 1972.

Table LC 39
CHROMATOGRAPHY OF CHARACTERIZED DEXTRAN FRACTIONS ON POLYACRYLAMIDE GEL COLUMNS

Packing	P1	P1	P2
Column			
Length	50 cm	85 cm	55 cm
Diameter	6 cm	1.5 cm	1.5 cm
Solvent	M NaCl	M NaCl	M NaCl
Flow rate	na	3 ml/hr	10 ml/hr
Temperature	rt	rt	rt
Detection	Phenol-H_2SO_4	Phenol-H_2SO_4	Phenol-H_2SO_4
Sample	∼ 10 mg/1 ml	5—10 mg/1 ml	∼ 2—5 mg/1 ml
Reference	1	2	2

$\overline{M}_n$	K_d	$\overline{M}_w$	$\overline{M}_n$	K_{av}	$\overline{M}_w$	$\overline{M}_n$	K_{av}
100,000	0.13						
80,000	0.19	71,000	57,200	0.11	11,600	7,800	0.15
50,000	0.38	40,000	33,600	0.26	5,700	4,500	0.31
25,000	0.50	10,000	7,100	0.61	3,900	3,200	0.40
18,000	0.65	3,900	3,200	0.84			
8,000	0.90						

Note: $\overline{M}_n$, $\overline{M}_w$ from manufacturer's data (Pharmacia).

Packing P1 = Bio-Gel P-300, 100 to 200 mesh.
P2 = Bio-Gel P-10, 100 to 200 mesh.

REFERENCES

1. Anderson, D. M. W. and Stoddart, J. F., The use of molecularsieve chromatography in studies on *Acacia senegal* gum, *Carbohydr. Res.,* 2, 104, 1966.
2. Churms, S. C., unpublished data, 1973.

Table LC 40
MOLECULAR WEIGHT DETERMINATION OF DEXTRANS BY CHROMATOGRAPHY ON CALIBRATED SEPHADEX GEL COLUMN

Packing	Sephadex G-200 + G-100[a] (1:2 dry weight ratio)

Column	
Length	75 cm
Diameter	1.4 cm
Solvent	0.3% NaCl
Flow rate	6-7 ml/hr
Temperature	20°C ± 0.1
Detection	Anthrone (Auto Analyzer)
Sample	2 mg/2 ml

Dextran fraction No.	$\overline{M_w}$[b]	$\overline{M_n}$[c]	K_{av}[d]	Stokes radius[e] r (A)
1	147,000	91,000	0.04	82
2	130,000	83,000	0.043	78
3	96,000	66,000	0.064	67.5
4	76,000	53,000	0.085	60.5
5	58,000	46,000	0.110	53.0
6	48,300	38,500	0.148	48.5
7	41,700	33,500	0.173	—
8	36,000	26,500	0.215	42.5
9	32,400	26,500	0.227	—
10	27,800	21,800	0.266	37.7
11	22,400	18,500	0.316	—
12	19,300	16,000	0.380	31.8
13	13,200	9,500	0.476	—
14	10,000	7,100	0.556	23.3
15	7,500	5,100	0.620	—
16	6,100	4,160	0.671	—
17	5,400	4,070	0.719	—

[a]   Sephadex G-200 with water regain (W_r) = 20.1 g/g, Sephadex G-100 with W_r = 9.8 g/g.
[b] $\overline{M_w}$ determined by light scattering method.
[c] $\overline{M_n}$ determined by end group analysis.
[d] K_{av} is not peak-value but corresponds to $\overline{M_w}$ and $\overline{M_n}$-positions on log-normal distribution.
[e]   Stokes radius calculated from r = RT/6 $\pi\eta$ DN, where R = gas constant, T = absolute temperature, η = viscosity of solvent, D = diffusion coefficient, and N = Avogadro's number.

From Granath, K. A. and Kvist, B. E., *J. Chromatogr.*, 28, 69, 1967. With permission.

Table LC 41
MOLECULAR WEIGHT DETERMINATION OF DEXTRANS BY CHROMATOGRAPHY ON CALIBRATED AGAROSE GEL COLUMNS

Packing	P1	P2
Column		
Length	75 cm	30 cm
Diameter	1.4 cm	0.9 cm
Solvent	S1	S2
Flow rate	5 ml/hr	13.8 ml/hr
Temperature	20°C ± 0.1	rt
Detection	Anthrone	Orcinol
Sample	2 mg/2 ml	na
Method of calibration	a	b
Reference	1	2

$\overline{M}_w$	K_{av}	$\overline{M}_w$	V_e/V_o
1,510,000	0.238		
485,000	0.322		
252,000	0.452		
147,000	0.518		
110,000	0.570	113,000	1.173
76,000	0.645	72,600	1.338
41,700	0.720	41,700	1.487
27,800	0.775	22,400	1.666
13,200	0.832	11,200	1.911
6,100	0.870	3,100	2.048

Packing	P1 =	Sepharose® 4B (Pharmacia; 4% agarose).
	P2 =	Bio-Gel® A-0.5m (Bio-Rad®; 10% agarose), 200 to 400 mesh.
Solvent	S1 =	0.9% NaCl.
	S2 =	0.05 M HgCl$_2$.
Method of Calibration		a. From log-normal distribution curves for well-defined dextran fractions ($\overline{M}_w$ by light scattering, $\overline{M}_n$ by end group analysis); final curve by iteration.
		b. Manufacturer's values for $\overline{M}_w$, $\overline{M}_n$ of commercial dextran fractions (Pharmacia, Sigma) used.

REFERENCES

1. Arturson, G. and Granath, K., Dextrans as test molecules in studies of the functional ultrastructure of biological membranes. Molecular weight distribution analysis by gel chromatography, *Clin. Chim. Acta*, 37, 309, 1972.
2. Bathgate, G. N., An automated gel filtration system for determining the molecular weight of polysaccharides, *J. Chromatogr.*, 47, 92, 1970.

Table LC 42
MOLECULAR WEIGHT DETERMINATION OF DEXTRANS BY CHROMATOGRAPHY ON CALIBRATED COLUMNS OF METHACRYLATE GELS AND POROUS GLASS (CPG)

Packing	P1	P2	P3
Column			
Length	2 × 100 cm + 1 × 60 cm	2 × 100 cm	110 cm
Diameter	1.1 cm	0.4 cm	1.5 cm
Material	na	stainless steel	glass
Solvent	H_2O or S1	H_2O or S2	S3
Flow rate	12 mℓ/hr	15 mℓ/hr	ca 8 mℓ/hr
Temperature	na	na	25°C
Sample	5 mg/0.5 mℓ	1.5 mg/50 $\mu\ell$	2—40 mg/0.5 mℓ
Detection	RI	RI + Cassette data logger	RI
Method of calibration	a	b	c
Calculations	manual	computer	computer
Reference	1	2	3

$M_o \times 10^{-3}$	V_e/V_o	$\overline{M}_w$	K_d	$\overline{M}_w$	M_o	K_{av}
170	1.205	589,730	0.037	287,000	287,000	0.042
125	1.241	343,630	0.086	163,000	150,000	0.114
91	1.277	213,880	0.135	74,800	73,500	0.235
70	1.313	140,970	0.183	54,100	52,000	0.306
51	1.349	97,560	0.232	51,400	49,000	0.323
38.5	1.385	70,280	0.281	30,200	30,000	0.425
29	1.420	52,250	0.330	22,400	22,200	0.495
24	1.456	39,790	0.379	16,800	16,500	0.566
16	1.492	30,670	0.428	13,100	13,000	0.615
12	1.528	23,800	0.477	8,540	8,200	0.688
9	1.564	18,410	0.526	5,700	5,500	0.736
6.7	1.600	14,080	0.575	3,330	3,100	0.806
		10,550	0.623	2,650	2,500	0.829
		7,680	0.672	1,860	1,700	0.866
		5,380	0.721	1,000	1,000	0.917
		3,590	0.770			

Packing

P1 = Spheron® P300 + P1000 (Knauer, Berlin). Co-polymer of 2-hydroxyethylmethacrylate with ethylene dimethacrylate.

P2 = Hydrogel® VI (37—75 μm). (Waters Assoc.). Cross-linked polymer of ethylene glycol dimethacrylate.

P3 = Porous glass (CPG®-10, 75—125 μm, Waters). Seven porosities (types 75, 125, 175, 240, 370, 700, and 1250) mixed in ratio of equal pore volumes.

Solvent

S1 = 0.3% NaCl
S2 = 0.02% NaN_3
S3 = 0.5% NaCl

Method of calibration

a. Calibration based on molecular-weight distribution and $\overline{M}_w$, $\overline{M}_n$-data (manufacturers') of commercial dextran fractions (Pharmacia, Sweden). Calibration curve defined by their V_e/V_o vs M_o points used to convert elution curves (conc vs V_e/V_o) of unknown samples to M-distributions. M_o = M of calibration fractions at 50% point.

b. Calibration fractions with $\overline{M}_w$ determined by light scattering. Iteration until calibration curve applied: $M_i = b_5 + \exp[b_4 + b_1(K_d) + b_3(K_d)^3]$ resulted in minimum difference between $\overline{M}_w$, light scattering, and $\overline{M}_w$, calculated from elution profile. Correction for diffusion inserted in computer programme.

c. Calibration with 15 narrow fractions of dextran. $\overline{M}_w$ by sedimentation, $\overline{M}_n$ by end group analysis. Log-normal distribtuion assumed, final curve (4th degree polynome) by iteration. Dispersion correction applied. M_o = molecular weight at peak elution volume.

Table LC 42 (continued)
MOLECULAR WEIGHT DETERMINATION OF DEXTRANS BY CHROMATOGRAPHY ON CALIBRATED COLUMNS OF METHACRYLATE GELS AND POROUS GLASS (CPG)

REFERENCES

1. **Pitz, H.,** Characterization of clinical dextrans by gel chromatography, *Anaesthesist,* 24, 500, 1975.
2. **Alsop, R. M., Byrne, G. A., Done, J. N., Earl, I. E., and Gibbs, R.,** Quality assurance in clinical dextran manufacture by molecular weight characterization, *Process Biochem.,* 12, 15, 1977.
3. **Basedow, A. M., Ebert, K. H., Ederer, H., and Hunger, H.,** Determination of the molecular weight distribution of polymers by permeation chromatography on porous glass, *Makromol. Chem.,* 177, 1501, 1976.

Table LC 43
FRACTIONATION OF DEXTRANS BY COLUMN CHROMATOGRAPHY ON POROUS SILICA

	P1	P1	P2	P3	P4
Packing	P1	P1	P2	P3	P4
Column					
Length	140.5 cm	82.8 cm	30 cm		
Diameter	0.8 cm	1.08 cm	0.46 cm		
Material	glass	glass	stainless steel		
Solvent	H_2O	H_2O	S1	S1	S1
Flow rate	42 ml/hr	9 ml/hr	0.5 ml/min	0.5 ml/min	0.5 ml/min
Temperature	25°C	25°C	rt	rt	rt
Detection	D1	D2	R1	R1	R1
Sample	0.6 mg/0.2 ml	4—10 mg/200 µl	1% solution 20 µl	As P2	As P2
Reference	1	2	3	3	3

$\bar{M}_w$[a]	V_e/V_t	$\bar{M}_w$[b]	V_e/V_t	$\bar{M}_w$[a]	K_{av}	$\bar{M}_w$[a]	K_{av}	$\bar{M}_w$[a]	K_{av}
280,000	0.52	200,000	0.44	75,000	0.04	250,000	0.12	250,000	0.24
210,000	0.55	150,000	0.54	40,000	0.24	170,000	0.25	170,000	0.36
110,000	0.62	110,000	0.58	22,000	0.40	110,000	0.28	110,000	0.45
60,000	0.68	80,000	0.62	10,000	0.62	75,000	0.40	75,000	0.56
20,000	0.75	50,000	0.67			40,000	0.57	40,000	0.70
8,000	0.80	30,000	0.74			10,000	0.80	10,000	0.88

[a] From data supplied by manufacturer (Pharmacia).

[b] From end-group assay of eluted fractions.

Packing   P1 = Porasil-D (Waters®); pore size 40—80 nm), 75 to 100 mesh.
P2 = LiChrospher (Merck®) Si 100, mean particle size 10 µm.
P3 = LiChrospher Si 300.
P4 = LiChrospher Si 500.

Solvent   S1 = 0.5 M sodium acetate (pH 5).

Detection   D1 = automated cysteine-sulfuric acid method.
D2 = AutoAnalyzer with dual systems: D1 and reducing end-group assay (ferricyanide method).

REFERENCES

1. **Barker, S. A., Hatt, B. W., Marsters, J. B., and Somers, P. J.,** Fractionation of dextran and hyaluronic acid on porous silica beads, *Carbohydr. Res.,* 9, 373, 1969.
2. **Barker, S. A., Hatt, B. W., and Somers, P. J.,** Automated continuous analysis for concentration of total carbohydrate and reducing end-group in the fractionation of dextran on porous silica beads, *Carbohydr. Res.,* 11, 355, 1969.
3. **Buytenhuys, F. A. and van der Maeden, F. P. B.,** Gel permeation chromatography on unmodified silica using aqueous solvents, *J. Chromatogr.,* 149, 489, 1978.

Table LC 44
COLUMN CHROMATOGRAPHIC
DATA ON CELLULOSE (RAYON)
AND CM-CELLULOSE IN CADOXEN

Packing	Bio-Gel A-50 m (agarose, 2%)
Column	
Height (bed)	56.0 cm
Diameter	2.5 cm
Solvent	Cadoxen
Flow rate	na
Temperature	na
Detection	Orcinol
Sample	0.35%, vol na

Sample	K_{av}	DPη[a]	Reference
Cuprammonium			
Rayon (Bemberg)	0.35	500	1
Viscose textile			
Rayon yarn	0.51	250	1
CMC 2	0.68	151	1,2
CMC 3	0.46	382	1,2
CMC 5	0.38	434	1,2
Cekol UV	0.42	374	1,2

[a] DPη = degree of polymerization determined by viscosity.

REFERENCES

1. Petterson, B. A., Gel filtration chromatography of regenerated cellulose, *Sven. Papperstidn.*, 72, 14, 1969.
2. Eriksson, K.-E., Pettersson, B. A., and Steenberg, B., Gel filtration chromatography of hemicelluloses and carboxymethyl cellulose in cadoxen solution, *Sven. Papperstidn.*, 71, 695, 1968; Almin, K. E., Eriksson, R.-E., and Pettersson, B. A., Determination of the molecular weight distribution of cellulose on calibrated gel columns. *J. Appl. Polym. Sci.*, 16, 2583, 1972.

Table LC 45
MOLECULAR WEIGHT DETERMINATION OF GLYCOSAMINOGLYCANS BY GEL CHROMATOGRAPHY

Packing	P1	P2	P1	P1	P3	P4
Column						
Length	63.5 cm	79 cm	90 cm	90 cm	58 cm	81.6 cm
Diameter	0.8 cm	1.3 cm	2.5 cm	2.5 cm	0.75 cm	0.8 cm
Solvent	S1	S1	S2	S2	S3	S1
Flow rate	~2 ml/hr	~3 ml/hr	13 ml/hr	13 ml/hr	4—6 ml/hr	na
Temperature	rt	rt	rt	rt	rt	25°C
Detection	D1,D2	D1	D3	D3	D4	D1
Sample	200 μg/200 μl	200—300 μg/200 μl	10—15 mg/1 ml	10—20 mg/1 ml	1.5 mg/1 ml	~3.5 mg/0.9 ml
Calibration	a	a	b	b	c	c
Reference	1,2	2	3	3	4	5

	Chondroitin 4-sulfate						Keratan sulfate		Heparin		Hyaluronic acid	
Polymer	M_w	K_{av}	M_n	K_{av}	M_n	K_{av}	M_n	K_{av}	M_n	K_{av}	M_n	V_e/V_o
	35,800	0.11	36,000	0.39	36,000	0.21	35,000	0.45	29,000	0.15	300,000	0.38
	25,400	0.20	28,000	0.45	28,000	0.30	28,000	0.50	23,000	0.23	150,000	0.43
	19,200	0.28	20,000	0.51	20,000	0.47	25,000	0.54	18,500	0.28	70,000	0.48
	13,300	0.30	18,000	—	18,000	0.50	21,000	0.58	14,000	0.36	45,000	0.53
	16,500	0.34	15,000	0.55	15,000	0.54	18,000	0.61	11,000	0.43	30,000	0.58
	13,900	0.41	10,000	—	10,000	0.66	15,000	0.63	8,200	0.50	22,000	0.62
	12,400	0.45	8,000	0.59	8,000	0.70	12,000	0.68	5,900	0.60	12,000	0.72
	3,600	0.61		0.68			8,000	0.78	5,000	0.69		
	4,800	0.74		0.77			5,000	0.82				
	2,400	0.84		0.85								

Packing P1 = Sephadex G-200, fine grade.
P2 = Sepharose 6B (6% agarose).
P3 = Ultrogel® AcA44 (LKB; agarose-acrylamide gel).
P4 = Porasil-E (Waters®; porous silica, pore size 80—150 nm), 75 to 100 mesh.

Table LC 45 (continued)

MOLECULAR WEIGHT DETERMINATION OF GLYCOSAMINOGLYCANS BY GEL CHROMATOGRAPHY

Solvent   S1 = 0.2 M NaCl.
           S2 = 0.02 M NaCl buffered with 0.02 M imidazole hydrochloride, pH 6.0.
           S3 = 0.3 M NaCl, containing 0.01% of NaN$_3$.

Detection D1 = carbazole method.
           D2 = ^{35}S labeling; activity measured by liquid scintillation counter.
           D3 = reductive labeling (NaB^{3}H$_4$); assay by liquid scintillation counter.
           D4 = RI.

Calibration a. With fractions of chondroitin 4-sulfate isolated by preparative gel chromatography and characterized by ultracentrifugation.
             b. End group analysis (NaB^{3}H$_4$ labeling) used to estimate $\overline{M}_n$ of fractions from gel column.
             c. Viscometry used to estimate mol wt ($\overline{M}_v$) of fractions from gel column.

REFERENCES

1. **Wasteson, Å.,** Properties of fractionated chondroitin sulfate from ox nasal septa, *Biochem. J.,* 122, 477, 1971.
2. **Wasteson, Å.,** A method for the determination of the molecular weight and molecular-weight distribution of chondroitin sulfate, *J. Chromatogr.,* 59, 87, 1971.
3. **Hopwood, J. J. and Robinson, H. C.,** The molecular-weight distribution of glycosaminoglycans, *Biochem. J.,* 135, 631, 1973.
4. **Johnson, E. A. and Mulloy, B.,** The molecular-weight range of mucosal heparin preparations, *Carbohydr. Res.,* 51, 119, 1976.
5. **Barker, S. A., Hatt, B. W., Marsters, J. B., and Somers, P. J.** Fractionation of dextran and hyaluronic acid on porous silica beads, *Carbohydr. Res.,* 9, 373, 1969.

AFFINITY CHROMATOGRAPHY

The application of this comparatively new technique to carbohydrates is reviewed briefly here. No separation data are included because this type of chromatography has been used mainly as a method of isolation and purification for certin polysaccharides and glycoproteins.

APPLICATION TO FRACTIONATION AND PURIFICATION OF POLYSACCHARIDES AND GLYCOPROTEINS

The new, but rapidly expanding technique of affinity chromatography, which involves the use of column packings in which a ligand with a highly specific affinity for molecules containing certain groups, having a particular stereochemistry, is covalently coupled to an insoluble matrix, is being applied to an increasing extent in carbohydrate chemistry, particulary, in studies of glycoproteins. The first such packing suitable for polysaccharide fractionations to become available commercially was Con A-Sepharose (Pharmacia®), in which the lectin (hemagglutinating protein) concanavalin A is covalently bound to the agarose gel Sepharose® 4B by the cyanogen bromide method. This lectin ia s metalloprotein containing Mn^{2+}, and the presence of this cation and CA^{2+} is necessary for the binding activity, which is specific for terminal α-D-glucopyranosyl or α-D-mannopyranosyl residues, internal 2-*O*- linked D-mannopyranosyl residues, or sterically similar configurations. Any polysaccharide containing sufficient residues of the appropriate type will be strongly bound to the packing on chromatography on a column packed with Con A-Sepharose, and may be thus separated from other components of a mixture which are not bound, e.g., dextrans differing in degree of branching may be fractionated.[1] The solvent should be buffered at pH 6 to 7 (phosphate buffers are generally used). Adsorbed polysaccharides may be quantitatively eluted with a solution containing methyl α-D-glucopyranoside or -mannopyranoside,[1,2] but the removal of these carbohydrates from the fractionated material requires a further separation process, and the use of borate buffers for this purpose is now preferred.[3] The borate, which disrupts the lectin-carbohydrate interaction by virtue of its strong complexing action with carbohydrates, is readily removed subsequently by conversion into volatile methyl borate.

Lectins that specifically bind various sugar residues occurring in polysaccharides and glycoproteins are now known and have been used, coupled to an insoluble matrix (usually agarose or polyacrylamide-agarose gel), in affinity chromatography. Examples include the purification of polysaccharides containing terminal nonreducing D-galactose residues, using the galactose-specific lectin from *Ricinus communis* beans, covalently coupled to Sepharose® 4B; this permitted the separation of native fetuin and ceruloplasmin from desialated specimens of these glycoproteins (the presence of which is beleived to be linked to certain diseases), since the latter have exposed D-galactose residues not present in the native glycoproteins.[4] In this case a 0.1 *M* solution of lactose was used for elution of bound material. The isolation of fucosyl-glycoproteins from this mixture of glycoproteins extracted from the synaptosomal plasma membrane of rat brain has been achieved by affinity chromatography on the L-fucose-specific lectin from *Ulex europaeus* (gorse), coupled to Sepharose;[5] the adsorbed material was quantitatively eluted by a solution containing sodium dodecyl sulfate (0.8%) in 0.2 *M* tris · -HCl buffer, pH 9.5. Lectins that specifically bind other sugars forming important constituents of glycoproteins, such as 2-acetamido-2-deoxy-D-glucose (wheat germ-lectin) and 2-acetamido-2-deoxy-D-galactose (lectin from *Helix pomatia* snail), have also been sucessfully applied in affinity chromatography. The use of this technique in the purification of glycoproteins has been comprehensively reviewed by Kristiansen,[6] and by Goldstein and Hayes.[7]

REFERENCES

1. **Lloyd, K. O.,** The preparation of two insoluble forms of the phytohaemagglutinin, concanavalin A, and their interactions with polysaccharides and glycoproteins, *Arch. Biochem. Biophys.,* 137, 460, 1970.
2. **Aspberg, K. and Porath, J.,** Group-specific adsorption of glycoproteins, *Acta Chem. Scand.,* 24, 1839, 1970.
3. **Kennedy, J. F. and Rosevear, A.,** An assessment of the fractionation of carbohydrates on Concana-valin A-Sepharose 4B by affinity chromatography, *J. Chem. Soc. Perkin Trans. 1,* 2041, 1973.
4. **Surolia, A., Ahmad, A., and Bachhawat, B. K.,** Affinity chromatography of galactose-containing biopolymers using covalently coupled *Ricinus communis* lectin to Sepharose 4B, *Biochim. Biophys. Acta,* 404, 83, 1975.
5. **Zanetta, J.-P., Reeber, A., Vincendon, G., and Gombos, G.,** Synaptosomal plasma membrane gly-coproteins. II. Isolation of fucosyl-glycoproteins by affinity chromatography on the *Ulex europaeus* lectin specific for L-fucose, *Brain Res.,* 138, 317, 1977.
6. **Kristiansen, T.,** Group-specific separation of glycoproteins, in *Methods in Enzymology,* Vol. 34, Jakoby, W. B. and Wilchek, M., Eds., Academic Press, New York, 1974, 331.
7. **Goldstein, I. J. and Hayes, C. E.,** The lectin: carbohydrate-binding proteins of plants and animals, *Adv. Carbohydr. Chem. Biochem.,* 35, 127, 1978.

Section I.III

PAPER CHROMATOGRAPHY TABLES

This section supplements the data on paper chromatography of carbohydrates published in Section A, Volume I, Section II.III, Tables PC 12 and PC 70-76.

Table PC 1

CHROMATOGRAPHY[a] OF SUGARS AND DERIVATIVES ON WHATMAN NO. 1 PAPER

Solvent / Compound	S1	S2	S3	S4	S5	S5	S6	S7	S8	S9	S10	S11	S12	S13	S14	S15
	R_G[b] × 100	R_F × 100		R_{xyl}[c] × 100			R_{Glc} × 100									
L-Arabinose	12	54	—	76	92	89	161	136	—	139	116	200	114	121	111	131
D-Xylose	15	44	—	100	100	100	181	—	—	150	121	260	130	150	123	153
D-Lyxose	19	—	—	—	—	—	—	—	—	—	—	—	—	—	—	—
D-Ribose	21	59	—	120	123	117	239	—	—	165	136	370	145	179	131	174
D-Galactose	7	44	—	45	59	59	93	95	—	100	95	85	82	85	92	92
D-Glucose	9	39	—	56	61	63	100	100	—	100	100	100	100	100	100	100
D-Mannose	11	45	—	80	76	75	129	120	—	121	116	145	105	129	111	113
D-Altrose	17	—	—	—	—	—	—	—	—	—	—	—	—	—	—	—
L-Idose	18	—	—	—	—	—	—	—	—	—	—	—	—	—	—	—
D-Talose	19	—	—	—	—	—	—	—	—	—	—	—	—	—	—	—
L-Fucose	21	63	—	110	120	116	—	—	—	—	—	—	—	—	—	—
L-Rhamnose	30	59	—	150	142	146	277	—	—	200	—	420	205	150	214	—
6-Deoxy-D-glucose	28	—	—	—	—	—	—	—	—	—	—	—	—	—	—	—
2-Deoxy-D-ribose	44	—	73	—	—	—	—	—	—	—	—	—	—	—	—	—
D-erythro-Pentulose	25	—	—	130	—	—	—	—	—	—	—	—	—	—	—	—
D-threo-Pentulose	26	—	—	140	—	—	—	—	—	—	—	—	—	—	—	—
D-Fructose	12	51	—	77	89	84	—	130	116	—	—	118	—	—	—	—
L-Sorbose	10	42	—	—	—	—	—	—	—	—	—	—	—	—	—	—
D-Tagatose	12	—	—	—	—	—	—	—	—	—	—	—	—	—	—	—
2,7-Anhydro-D-altro-heptulose	15	67	—	—	—	100	—	—	—	—	—	—	—	—	—	—
2-Amino-2-deoxy-D-glucose	—	62	—	—	—	—	—	—	—	—	—	[d]	51	75	68	—
2-Amino-2-deoxy-D-galactose	—	65	—	—	—	—	—	—	—	—	—	—	—	—	—	—
2-Acetamido-2-deoxy-D-glucose	—	69	—	—	—	—	—	—	—	—	130	127	135	—	150	—
D-Galacturonic acid	2[d]	13	—	6	39	57	—	83	—	79	—	—	11	12	11	—
D-Mannuronic acid	0.8[d]	—	—	—	—	—	—	—	—	—	—	—	—	29	11	—
D-Glucuronic acid	0.8[d]	12	—	2	42	62	110	—	—	89	24	14	12	36	18	—

Compound												
D-Glucuronolactone	—	—	—	—	—	—	—	—	—	—	—	—
L-Ascorbic acid (vitamin C)	24	—	—	—	—	—	—	360	213	213	150	193
D-Mannitol	—	—	—	—	126	110	100	96	97	102	150	—
D-Glucitol	—	—	—	—	122	114	90	91	88	98	—	193
Sucrose	3	39	29	41	34	80	—	72	80	97	102	—
Maltose	2.1	36	—	—	—	—	—	65	68	—	—	—
Isomaltose	—	—	—	—	—	—	—	43	68	—	—	—
Xylobiose	—	—	—	—	65	—	—	51	—	—	—	78
Mannobiose	—	—	—	—	—	—	—	51	—	—	—	—
Raffinose	—	—	—	45	—	—	—	27	—	—	—	—
Maltotriose	—	—	—	—	—	—	—	38	—	—	—	—
Isomaltotriose	—	—	—	—	—	—	—	29	—	—	—	—
Mannotriose	—	—	—	—	—	—	—	18	—	—	—	—

[a]   Descending chromatography, 12 hr or more, at ambient temperature in all cases.

[b] R_G = mobility relative to 2,3,4,6-tetra-O-methyl-D-glucose.

[c] R_{xyl} = mobility relative to D-xylose.

[d]   Streaks.

Solvent   S1   = 1-butanol-ethanol-water (40:11:19).
          S2   = water-saturated phenol + 1% ammonia, in HCN atmosphere.
          S3   = 1-butanol-pyridine-water (10:3:3).
          S4   = ethyl acetate-acetic acid-water (3:3:1).
          S5   = ethyl acetate-acetic acid-formic acid-water (18:3:1:4).
          S6   = ethyl acetate-acetic acid-water (3:1:3).
          S7   = 1-butanol-acetic acid-water (4:1:5).
          S8   = 1-butanol-acetic acid-water (5:1:2).
          S9   = 1-butanol-pyridine-water (6:4:3).
          S10 = ethyl acetate-pyridine-water (8:2:1).
          S11 = ethyl acetate-pyridine-water (10:4:3).
          S12 = benzene-1-butanol-pyridine-water (1:5:3:3).
          S13 = 1-butanol-ethanol-water (2:1:1).
          S14 = 1-butanol-ethanol-water (4:1:1).
          S15 = 1-butanol-1-propanol-water (2:5:3).

From Hough, L. and Jones, J. K. N., in *Methods in Carbohydrate Chemistry*, Vol. 1, Whistler, R. L. and Wolfrom, M. L., Eds., Academic Press, New York, 1962, 21. With permission.

Table PC 2
IMPROVED SEPARATIONS OF
BIOLOGICALLY IMPORTANT SUGARS

	P1	P1	P1	P2
Paper	P1	P1	P1	P2
Solvent	S1	S2	S3	S4
Technique	T1	T2	T2	T3
Reference	1	2	2	3

Compound	$R_{Glc} \times 100$			
D-Glucose	100[a]	100	100	100
D-Galactose	85	80	76	90
D-Mannose	147	124	124	110
D-Fructose	162	133	144	—
D-Xylose	221	200	215	—
D-Arabinose	182	175	180	—
L-Rhamnose	409	330	390	—
L-Fucose	—	225	261	125
D-Glucuronic acid	119	43	37	50
D-Galacturonic acid	107[b]	32	32	—
2-Amino-2-deoxy-D-glucose	—	—	—	74
2-Amino-2-deoxy-D-galactose	—	—	—	64
Sucrose	35	28	28	—
Maltose	20	—	—	—
Lactose	12	—	—	—

[a] R_F of glucose = 0.03.
[b] Gives 2 spots, due to lactone formation.

Paper	P1	=	Whatman® No. 1.
	P2	=	Whatman® No. 3.
Solvent	S1	=	1-butanol-benzene-formic acid-water (100:19:10:25), upper phase; aged over aqueous phase at 23°C for 3 days before use.
	S2	=	ethyl acetate-acetic acid-pyridine-water (10:3:3:2).
	S3	=	ethyl acetate-acetic acid-pyridine-water (12:3:3:2).
	S4	=	ethyl acetate-acetic acid-pyridine-water (5:1:5:3).
Technique	T1	=	paper equilibrated overnight over solvent at bottom of tank before chromatography, then descending development for 60 hr at 23°C (temperature controlled).
	T2	=	descending, 24 hr at 17°C (temperature controlled).
	T3	=	descending; details not specified.

REFERENCES

1. Fuleki, T. and Francis, F. J., A new developing solvent for paper chromatography of various phenolic comounds, sugars and amino acids, *J. Chromatogr.*, 26, 404, 1967.
2. Jarvis, M. C. and Duncan, H. J., Paper chromatography of plant sugars, *J. Chromatogr.*, 92, 454, 1974.
3. Strecker, G. and Lemaire-Poitau, A., Fractionation and characterisation of acidic oligosaccharides and glycopeptides from normal and pathological urines, *J. Chromatogr.*, 143, 553, 1977.

Table PC 3
CHROMATOGRAPHY OF SUGARS AND ALDITOLS ON TUNGSTATE-IMPREGNATED PAPER

Paper	P1	P2	P3	P2
Solvent	S1	S1	S2	S2
Technique	T1	T2	T3	T2
Compound	$R_{Glc}{}^a$(W6) × 100	R_{Glc} × 100	$R_{Glc}{}^b$(W8) × 100	R_{Glc} × 100
Group A[c]				
Erythritol	84	168	103	151
L-Threitol	180	163	44	148
D-Arabinitol	30	136	49	125
1-deoxy-	61	245	99	202
D-Lyxitol, 1-deoxy-	30	245	117	195
Ribitol	48	137	63	127
Xylitol	41	126	13	120
1-deoxy-D-	30	212	51	190
Allitol	26	108	39	112
D-Altritol	25	106	24	107
1-deoxy-	46	172	130	164
1,6-dideoxy-	—	—	130	207
Galactitol	25	102	33	97
1-deoxy-L-	40	188	74	156
1,6-dideoxy-	—	—	112	205
D-Glucitol	24	98	13	100
1-deoxy-	44	168	64	159
2-*O*-α-D-glucopyranosyl-	30	59	14	46
2-*O*-β-D-glucopyranosyl-	13	53	18	50
L-Gulitol, 1-deoxy-	—	—	31	142
1-*O*-α-D-glucopyranosyl-	12	42	5	35
L-Iditol	27	97	13	100
D-Mannitol	24	106	33	106
L-Mannitol, 1-deoxy-	46	190	—	—
1,6-dideoxy-	152	293	175	212
D-Talitol, 1-deoxy-	53	189	72	163
Group B[d]				
D-Lyxose	50—70	145	104	140
D-Ribose	139	157	80	155
D-Gulose	43	110	80	115
D-Mannose	70	119	94	116
L-Gulitol, 3-*O*-α-D-glucopyranosyl-	36	53	7	52
3-*O*-β-D-glucopyranosyl-	37	45	15	50
epi-Inositol	16	48	28	46
Group C[e]				
D-Arabinose	118	120	127	117
D-Xylose	150	141	136	131
D-Galactose	63	87	67	92
D-Glucose	100[f]	100[g]	100[f]	100[h]
D-Glucitol, 3-*O*-β-D-glucopyranosyl-	58	59	59	63
3-*O*-methyl-	—	—	148	152
Kojibiose	51	59	42	48
Sophorose	55	54	51	53
Laminaribiose	63	61	66	62
Maltose	43	46	47	44

Table PC 3 (continued)
CHROMATOGRAPHY OF SUGARS AND ALDITOLS ON TUNGSTATE-IMPREGNATED PAPER

Paper	P1	P2	P3	P2
Solvent	S1	S1	S2	S2
Technique	T1	T2	T3	T2
Compound	$R_{Glc}{}^a$(W6) × 100	R_{Glc} × 100	$R_{Glc}{}^b$(W8) × 100	R_{Glc} × 100
Cellobiose	39	42	45	42
Isomaltose	36	45	27	42
Gentiobiose	34	37	26	39
(+)-Inositol	46	54	48	54
muco-Inositol	72	83	61	75
myo-Inositol	28	38	29	36
scyllo-Inositol	30	42	26	33
Glycerol^f	328	216	312	192
L-Galactose, 6-deoxy-(L-fucose)	187	153	187	140
α-D-Glucopyranoside, methyl^f	233	159	252	150
D-Glucose, 3-*O*-methyl^f	234	168	287	168
Hexose, 2-deoxy-D-*arabino*-	227	172	313	169
2-deoxy-D-*lyxo*-^{i}	200	162	270	154
2-deoxy-D-*ribo*-^{i}	229	182	325	177
α-D-Mannopyranoside, methyl^f	325	209	308	175

^{a} R_{Glc}(W6) = mobility, relative to glucose, on paper impregnated with tungstate solution at pH 6.
^{b} R_{Glc}(W8) = mobility with tungstate solution at pH 8.
^{c} Acyclic compounds with vicinal tetritol system; tungstate strongly complexed.
^{d} Compounds having 3 vicinal OH groups in spatial disposition favoring complex formation; tungstate moderately strongly complexed.
^{e} Compounds not having features described in footnotes ^{c} and ^{d}; tungstate weakly complexed.
^{f} R_f of glucose = 0.10.
^{g} R_f of glucose = 0.18.
^{h} R_f of glucose = 0.23.
^{i} Tungstate probably not complexed.

Paper	P1 =	Whatman® No. 1, impregnated with tungstate at pH 6.
	P2 =	Whatman® No. 1, untreated.
	P3 =	Whatman® No. 1, impregnated with tungstate at pH 8.
Solvent	S1 =	1-butanol-ethanol-water (40:11:19).
	S2 =	acetone-1-butanol-water (5:3:2).
Technique	T1 =	descending chromatography on paper strip (12.5 × 57 cm) pretreated by dipping in 5% aqueous solution of sodium tungstate dihyrate, with pH adjusted to 6 with dilute sulfuric acid; paper blotted in folds of filter paper, then air-dried at room temperature before use.
	T2 =	descending chromatography; no pretreatment.
	T3 =	descending chromatography on paper treated as in T1 but with impregnating solution at pH 8.

From Angus, H. J. F., Briggs, J., Sufi, N. A., and Weigel, H., *Carbohydr. Res.*, 66, 25, 1978. With permission.

Table PC 4
HIGHER OLIGOSACCHARIDES

Type of oligosaccharide	Malto-oligosaccharides					Isomalto-				Cello-	
Paper	P1	P2	P2	P1	P1	P1	P1	P3	P3	P1	P1
Solvent	S1	S2	S2 + S3	S2	S4	S2	S4	S2	S4	S2	S4
Technique	T1	T2	T3	T4	T4	T4	T4	T4	T4	T4	T4
Reference	1	2	2	3	3	3	3	3	3	3	3
Degree of polymerization	$R_{Glc} \times 100$		$R_F \times 100$								
1	100	100	100	45	46	45	46	38	41	45	46
2	65	77	92	27	31	31	34	28	32	26	27
3	38	62	85	17	20	18	25	17	22	13	14
4	21	44	77	8	12	10	15	10	17	5[a]	7
5	15	30	69	4[a]	7	5[a]	10	5.7[a]	12	2.5[a]	3[a]
6	11	22	61	2[a]	4[a]	2.5[a]	6[a]	3.3[a]	9	—	—
7	8	17	54	1[a]	2[a]	1.3[a]	3.7[a]	1.9[a]	6[a]	—	—
8	—	—	47	0.4[a]	0.11[a]	0.6[a]	2.5[a]	1.1[a]	4.2[a]	—	—
9	—	—	40	—	—	—	1.4[a]	0.9[a]	3.0[a]	—	—
10	—	—	34	—	—	—	0.8[a]	—	2.0[a]	—	—
11	—	—	28	—	—	—	—	—	1.4[a]	—	—
12	—	—	23	—	—	—	—	—	0.6[a]	—	—
13	—	—	19	—	—	—	—	—	—	—	—
14	—	—	16	—	—	—	—	—	—	—	—
15	—	—	13	—	—	—	—	—	—	—	—

[a] Measured on paper 160 cm long.

Paper	P1 =	Whatman® No. 1.
	P2 =	Toyo filter paper No. 50.
	P3 =	FN7 (VEB Spezialpapierfabrik, DDR).
Solvent	S1 =	ethyl acetate-pyridine-water (10:4:3).
	S2 =	1-butanol-pyridine-water (6:4:3).
	S3 =	1-butanol-pyridine-water (1:1:1).
	S4 =	1-butanol-pyridine-water (6:5:5).
Technique	T1 =	descending, 24 hr.
	T2 =	descending, 40 hr at 16°C.
	T3 =	after development with S2 as in T2, paper dried then developed 3 times (20 hr each run) with S3.
	T4 =	descending; R_F values ≤0.06 determined using paper strips 160 × 9.5 cm in tank 170 × 10 cm, with solvent trough connected through Teflon tubing to solvent reservoir to maintain level in trough and flow rate; lead band attached to free end of paper to keep it taut.

REFERENCES

1. **Data from Whistler, R. L.**, tabulated by Hough, L. and Jones, J. K. N. in *Methods in Carbohydrates Chemistry*, Vol. 1, Whistler, R. L. and Wolfrom, M. L., Eds., Academic Press, New York, 1962, 21.
2. **Umeki, K. and Kainuma, K.**, Fractionation of maltosaccharides of relatively high degree of polymerisation by multiple descending paper chromatography, *J. Chromatogr.*, 150, 242, 1978.
3. **Teichmann, B.**, Paper chromatographic behaviour of homologous linear polymers. The isomaltodextrin, maltodextrin and cellodextrin series and their 1-(m-substituted phenyl)-flavazoles, *J. Chromatogr.*, 70, 99, 1972.

Table PC 5
SUGAR PHOSPHATES

Paper	P1	P1
Solvent	S1	S2
Technique	T1	T2

Compound	$R_{PO_4}{}^a \times 100$	$R_f \times 100$
Inorganic phosphate	100	72
6-Phosphogluconic acid	49	53
D-Glucose-6-phosphate	41	40
D-Ribose-5-phosphate	55	57
D-*erythro*-Pentulose-5-phosphate	55	61
D-*threo*-Pentulose-5-phosphate	55	52
Erythrose-4-phosphate	57	64
Dihydroxyacetone phosphate	72	65
Glyceraldehyde phosphate	74	60
D-*erythro*-Pentulose-1,5-diphosphate	43	54
Phospho-D-*erythro*-pentulose pyrophosphate	40	36[b]
D-*altro*-Heptulose-7-phosphate	38	43

[a] R_{PO_4} = mobility relative to inorganic phosphate.
[b] Streaks.

Paper	P1 =	Whatman® No. 541.
Solvent	S1 =	1-butanol-1-propanol-acetone-80% (w/v) formic acid-30% (w/v) trichloroacetic acid (8:4:5:5:3).
	S2 =	methanol-90% formic acid-water (16:3:1); tetrasodium salt of EDTA (0.05 g/100 m*l*) added to both solvents.
Technique	T1 =	descending, 2 developments.
	T2 =	descending, single development.

From Wood, T., *J. Chromatogr.*, 35, 352, 1968. With permission.

Section I.IV

THIN-LAYER CHROMATOGRAPHY TABLES

Tables TLC 1, TLC 2, and TLC 4 have been compiled by Professor M. Lato, Dr. M. Ghebregzabher, and Dr. S. Rufini, Instituto di Clinica Pediatrica, Università di Perugia, Italy.

Earlier data will be found in Section A, Volume I, Section II.IV, Tables TLC 23-26 and TLC 126-128.

Table TLC 1
SUGARS, MONO- AND OLIGOSACCHARIDES; ALDITOLS

Compound	$R_F \times 100$													R_{Glc}[a] $\times 100$					R_{Suc}[b] $\times 100$	$R_F \times 100$
Layer	L1	L2	L3	L4	L5	L5	L5	L6	L7	L8	L9	L9	L10	L11	L11	L12	L13	L14	L15	L16
Solvent	S1	S2	S3	S4	S5	S6	S7	S8	S9	S10	S11	S5	S12	S13	S14	S15	S16	S13	S17	S18
Technique	T1	T2	T3	T4	T5	T5	T6	T7	T7	T7	T8	T8	T9	T10	T10	T10	T11	T12	T13	T14
Reference	1	2	3	4	5	5	5	6	6	6	7	7	8	9	9	10	11	12	13	14
	*	*	*	1*	2*	*	**	**	*	*	1*	2*	**	***	***	***	**	***	C	VP
Aldoses																				
Glyoxal	—	—	—	—	—	—	—	64	78	0	—	—	—	—	—	—	—	—	—	—
Glycolaldehyde	—	—	70; 45	—	—	—	—	51	48	47	—	—	—	—	—	—	—	—	—	—
Glyceraldehyde	—	—	59	—	—	—	—	33	60	50	—	—	—	—	—	—	—	—	—	—
Erythrose	—	—	45	—	—	—	—	22	47	45	—	—	—	—	—	—	—	—	—	—
L-Arabinose	67	47	39	39	50	25	42	61	15	37	35	—	44	132	112	131	134	130	—	53
D-Xylose	76	63	47	41	59	38	60	77	18	43	36	33	46	143	113	155	154	134	—	71
D-Lyxose	65	45	—	—	—	—	—	—	—	—	—	—	—	149	114	162	—	—	—	—
D-Ribose	72	60	46	58	39	62	83	12	37	40	20	44	64	161	116	176	—	149	—	—
D-Glucose	65	41	34	28	47	14	26	42	11	28	27	38	34	100	100	100	100	100	—	36
D-Galactose	58	30	30	28	41	10	18	28	10	26	25	29	27	92	98	83	82	93	—	27
D-Mannose	69	50	36	—	—	13	—	—	—	—	32	32	42	120	104	115	115	117	—	44
L-Fucose	75	65	49	53	55	77	91	18	44	42	41	42	—	—	—	162	162	144	—	—
L-Rhamnose	85	83	51	52	63	80	97	98	23	50	48	—	—	180	122	210	192	158	—	84
2-Deoxy-D-glucose	—	—	—	—	—	—	—	—	22	34	—	—	—	—	—	—	—	—	—	—
2-Deoxy-D-Galactose	—	—	—	—	—	—	—	—	28	40	—	—	—	—	—	—	—	—	—	—
6-Deoxy-D-glucose	80	—	—	—	—	—	—	—	—	—	—	—	—	—	—	—	—	—	—	—
2-Deoxy-D-ribose	86	83	—	—	—	—	—	—	20	28	58	56	—	199	—	—	—	—	—	—
Ketoses																				
Dihydroxyacetone	—	—	60; 50	—	—	—	—	33	56	50	—	—	—	—	—	—	—	—	—	—
D-threo-Pentulose	76	—	—	—	—	—	—	—	—	—	4	30	70	—	—	—	—	—	—	—
D-Fructose	64	46	32	—	—	13	33	—	—	—	30	15	30	125	109	115	127	121	—	44
L-Sorbose	69	52	32	—	—	15	36	—	—	—	28	—	—	—	—	129	—	—	—	—
D-Tagatose	60	—	—	—	—	13	33	—	—	—	—	—	32	—	—	—	—	—	—	—

	C1	C2	C3	C4	C5	C6	C7	C8	C9	C10	C11	C12	C13	C14	C15
D-Psicose	—	—	—	—	—	—	7	25	—	—	—	—	—	—	—
D-*altro*-Heptulose	—	—	—	—	—	—	23	28	—	—	—	—	—	—	—
Acids															
Arabinonic acid	—	—	—	0	18	53	—	—	—	—	—	—	—	—	—
D-Gluconic acid	—	—	—	0	9	22	—	—	—	—	—	—	—	—	—
D-Glucuronic acid	61	74	—	22	16	—	—	—	—	—	—	—	101	—	—
D-Galacturonic acid	7	30	—	14	31	—	—	—	—	—	—	—	90	—	—
D-Mannuronic acid	—	—	—	—	—	—	—	—	—	—	—	—	115	—	—
D-Glucuronolactone	—	—	—	30	25; 56	17	—	—	—	—	—	—	—	—	—
D-Galactonolactone	—	—	—	30	15; 45	34	—	—	—	—	—	—	—	—	—
Alditols															
Xylitol	—	—	—	9	35	27	—	—	—	—	—	—	—	—	—
Galactitol	—	—	—	6	25	21	—	—	—	—	—	—	—	—	—
D-Mannitol	—	—	—	6	28	24	—	—	—	—	—	—	—	—	—
D-Glucitol	—	—	—	6	22	22	—	—	—	—	—	—	—	—	—
Inositol	—	—	—	3	18	—	—	—	—	—	—	—	—	—	—
Oligosaccharides															
Cellobiose	56	35	—	6	10	—	—	—	—	69	—	39	—	—	—
Maltose	57	34	—	16	11	—	29	18	—	76	53	48	42	—	11
Isomaltose	44	17	—	4	—	—	—	—	—	—	—	26	—	—	—
α,α-Trehalose	57	31	—	—	9	—	—	—	—	—	—	—	—	—	—
Sucrose	66	47	24	6	14	—	38	20	64	90	66	60	64	—	18
Turanose	—	—	14	3	8	—	—	—	—	87	—	—	—	—	—
Lactose	47	23	—	3	8	—	20	12	15	68	30	29	32	—	5
Lactulose	—	—	—	—	—	—	9	12	—	—	—	—	—	—	—
Melibiose	36	14	—	3	8	—	—	—	—	66	—	—	—	—	—
Gentiobiose	—	20	—	—	—	—	—	—	—	67	—	—	—	—	—
Panose	35	14	—	—	—	—	9	—	—	—	—	—	—	—	—
Palatinose	42	17	—	—	—	—	9	18	—	—	—	—	—	—	—
Raffinose	56	36	—	2	4	8	17	4	—	56	—	13	—	32	—
Meleziose	—	—	—	3	5	10	—	—	—	68	—	—	—	—	—
Gentianose	—	—	—	2	4	8	—	—	—	—	—	—	—	—	—
Maltotriose	46	27	—	—	—	—	—	—	—	51	—	25	—	—	—

Table TLC 1 (continued)
SUGARS, MONO- AND OLIGOSACCHARIDES; ALDITOLS

Compound														$R_f \times 100$			$R_{Glc}^a \times 100$			$R_{Suc}^b \times 100$		$R_f \times 100$	
Layer	L1	L2	L3	L4	L5	L5	L5	L6	L7	L8	L9	L9	L10	L11	L11	L12	L13	L14	L15	L16
Solvent	S1	S2	S3	S4	S5	S6	S7	S8	S9	S10	S11	S5	S12	S13	S14	S15	S16	S13	S17	S18
Technique	T1	T2	T3	T4	T5	T6	T7	T7	T7	T7	T8	T8	T9	T10	T10	T10	T11	T12	T13	T14
Reference	1	2	3	4	5	5	6	6	6	6	7	7	8	9	9	10	11	12	13	14
	*	*	*	1*	2*	*	*	**	*	1*	2*	**		***	***	***	**	***	C	VP
Isomaltotriose	25	6	—	—	—	—	—	—	—	—	—	—	—	—	—	—	—	—	—	—
Isokestose	—	—	—	—	—	—	2	4	11	—	—	—	—	—	—	—	—	—	75	—
neo-Kestose	—	—	—	—	—	—	—	—	—	—	—	—	—	—	—	—	—	—	75	—
1-Kestose	—	—	—	—	—	—	—	—	—	—	—	—	—	—	—	—	—	—	51	—
6-Kestose	—	—	—	—	—	—	—	—	—	—	—	—	—	—	—	—	—	—	41	—
Nystose	—	—	—	—	—	—	—	—	—	—	—	—	—	—	—	—	—	—	29	—
Stachyose	34	—	—	—	—	—	0	2	3	—	—	—	—	—	29	—	—	—	—	—
Maltotetraose	34	—	—	—	—	—	—	—	—	—	—	—	—	—	—	—	—	—	—	—
Isomaltotetraose	17	—	—	—	—	—	—	—	—	—	—	—	—	—	—	—	—	—	—	—
Maltopentaose	27	—	—	—	—	—	—	—	—	—	—	—	—	—	—	—	—	—	—	—
Maltohexaose	23	—	—	—	—	—	—	—	—	—	—	—	—	—	—	—	—	—	—	—

Note: * = single development. ** = double development. *** = threefold development (monodimensional). 1* = two-dimensional development, first run. 2* = two-dimensional development, second run orthogonal to first. C = continuous run. VP = Vapor-programmed TLC.

a R_{Glc} = mobility relative to glucose.
b R_{Suc} = mobility relative to sucrose.

Layer   L1 = precoated Silica Gel 60 (Merck®), conditioned at 105°C for 1 hr, cooled in desiccator.
L2 = precoated Silica Gel 60 (Merck®), impregnated with 0.5 M NaH$_2$PO$_4$.
L3 = 0.25 mm Silica Gel G (Merck®), impregnated with 0.03 M H$_3$BO$_3$; heated at 110°C for 1 hr before use.
L4 = Silica Gel H (Merck®).

L5 = 0.5 mm Kieselguhr G (Merck®), impregnated with 0.15 M NaH$_2$PO$_4$; plates allowed to dry for at least 24 hr at 18—20°C and stored in presence of air at temperature not exceeding 20°C.

L6 = 0.25 mm Silica Gel G 60 (Merck®), dried at 110°C for 2 hr.

L7 = 0.25 mm layer, mixture of Silica Gel G 60 (2 parts), Kieselguhr G (1 part), prepared as for L6.

L8 = 0.25 mm Cellulose MN 300 (Machery-Nagel), allowed to dry overnight.

L9 = 0.30 mm layer, 2:1 mixture of Silica Gel G 60 (Merck®) and Syloid 63 (W. R. Grace), impregnated with 0.03 M sodium tetraborate and 0.05 M sodium tungstate (3:1); dried at 90°C for 30 min.

L10 = precoated 0.25 mm Silica Gel G.

L11 = 0.25 mm Cellulose MN 300 (Machery-Nagel).

L12 = 0.30 mm Cellulose MN 300.

L13 = precoated 0.10 mm Cellulose F (Merck®).

L14 = 0.20 mm cellulose layer (Chromedia CC 41, W. R. Balston), air-dried for 15—20 min; heated at 100°C for 30 min.

L15 = precoated 0.25 mm Silica F$_{254}$ (Merck®).

L16 = 0.30 mm Silica Gel G, impregnated with 0.2 M NaH$_2$PO$_4$; plates air-dried for 15 min, heated at 100°C for 30 min.

Solvent

S1 = 2-propanol-acetone-M lactic acid (2:2:1).

S2 = 2-propanol-acetone-0.1 M lactic acid (2:2:1).

S3 = ethyl acetate-2-propanol-acetic acid-water (4:2:1:1).

S4 = chloroform-methanol-water (16:9:2).

S5 = ethyl acetate-methanol-acetic acid-water (12:3:3:2).

S6 = ethyl acetate-methanol-1-butanol-water (16:3:3:2).

S7 = ethyl acetate-methanol-1-butanol-water (16:3:3:1).

S8 = benzene-ethanol (2:1).

S9 = benzene-acetic acid-ethanol (2:2:1).

S10 = 1-butanol-acetic acid-water (6:1:2).

S11 = ethyl acetate-2-propanol-water (2:2:1).

S12 = pyridine-ethyl acetate-water (13:33:4).

S13 = formic acid-2-butanone-*tert*-butanol-water (3:6:8:3).

S14 = formic acid-2-butanone-*tert*-butanol-water (3:5:7:5).

S15 = ethyl acetate-pyridine-water (20:7:5).

S16 = ethyl acetate-pyridine-water-acetic acid-propionic acid (10:10:2:1:1).

S17 = water-saturated 1-butanol-ethanol (5:2).

S18 = acetone-methanol-water (16:3:1); vapor program in troughs (A = acetone; M = methanol; W = water): trough 3 = A-M-W (7:2:1); trough 6 = A-M-W (3:1:1); trough 8 = A-M-W (4:3:3); trough 10 = A-M-W (1:2:2); troughs 12, 14, 16, and 18 = A-M-W (4:5:11); troughs 1, 2, 4, 5, 7, 9, 11, 13, 15, 17, 19, 20, and 21 = A only.

Table TLC 1 (continued)
SUGARS, MONO- AND OLIGOSACCHARIDES; ALDITOLS

Technique

T1 = ascending development (13 cm; 3 hr); plates dried in stream of warm air (60°C) before spraying.

T2 = ascending development (16 cm; 5 hr); plates dried as above.

T3 = ascending development (10 cm) in pre-equilibrated (for 30 min) chamber.

T4 = two-dimensional development (18 × 18 cm; about 75 min each development).

T5 = samples applied to plates in streaks of about 1 cm, under current of warm air (75—80°C); ascending development (20 cm; 65 min).

T6 = double development (35 cm; 325 min each); after each run plates dried at room temperature for 15 min, then heated at 100°C for 5 min.

T7 = ascending development in filter paper-lined glass chamber.

T8 = two-dimensional development (12 × 12 cm; first run 120 min, second 50 min); plates dried between two runs at 70°C for 4—5 min.

T9 = layer sprayed lightly with solvent before use and dried at 105°C for 2 hr; ascending development (double, 14 cm each) in filter paper-lined chamber, drying between runs at 50°C for 100 min.

T10 = threefold ascending development (about 18.5 cm and 180 min each run), without previous saturation of the chamber.

T11 = ascending development (double about 105 min each run) in presaturated paper-lined chamber; plates air-dried between runs.

T12 = threefold development, 3 hr total time required.

T13 = continuous development for 47—60 hr.

T14 = vapor-programmed (VP) development in 20 cm VP-chamber (Desaga); ambient temperature 24—25°C, saturation time 10 min, development time 3 hr.

REFERENCES

1. **Hansen, S. A.,** Thin-layer chromatographic method for identification of oligosaccharides in starch hydrolysates, *J. Chromatogr.,* 105, 388, 1975.

2. **Hansen, S. A.,** Thin-layer chromatographic method for the identification of mono-, di-, and trisaccharides, *J. Chromatogr.,* 107, 224, 1975.

3. **Nakai, T., Demura, H., and Koyama, M.,** Thin-layer chromatographic detection of glycoladehyde using a fluorescence reaction with o-aminobiphenyl, *J. Chromatogr.,* 66, 87, 1972.

4. **Kartnig, T. and Wegschaider, O.,** A method for identification of sugars from small amounts of glycosides or from sugar mixtures, *J. Chromatogr.,* 61, 375, 1971.

5. **Talukder, M. Q.-K.,** A rapid quantitative thin-layer chromatographic separation of fucose from other neutral monosaccharides, *J. Chromatogr.,* 57, 391, 1971.

6. **Haldorsen, K. M.**, Vanadium pentoxide in sulfuric acid, a general chromogenic spray reagent for carbohydrates, *J. Chromatogr.*, 134, 467, 1977.

7. **Ghebregzabzer, M., Rufini, S., Ciuffini, G., and Lato, M.**, A two-dimensional thin-layer chromatographic method for screening carbohydrate anomalies, *J. Chromatogr.*, 95, 51, 1974.

8. **Szustkiewicz, C. and Demetriou, J.**, Detection of some clinically important carbohydrates in plasma and urine by means of thin-layer chromatography, *Clin. Chim. Acta*, 32, 355, 1971.

9. **Damonte, A., Lombard, A., Tourn, M. L., and Cassone, M. C.**, A modified solvent system and multiple detection technique for the separation and identification of mono- and oligosaccharides on cellulose thin layers, *J. Chromatogr.*, 60, 203, 1971.

10. **Raadsveld, C. W. and Klomp, H.**, Thin-layer chromatographic analysis of sugar mixtures, *J. Chromatogr.*, 57, 99, 1971.

11. **Walkley, J. W. and Tillman, J.**, A simple thin-layer chromatographic technique for the separation of mono- and oligosaccharides, *J. Chromatogr.*, 132, 172, 1977.

12. **Petre, R., Dennis, R., Jackson, B. P., and Jethwa, K. R.**, Thin-layer chromatography of sugars and uronic acids in plant extracts on cellulose, *Planta Med.*, 21, 81, 1972.

13. **Schäffler, K. J. and Morel du Boil, P. G.**, Thin-layer chromatographic separation of oligosaccharides isolated from sucrose-enzyme mixtures, *J. Chromatogr.*, 72, 212, 1972.

14. **De Zeeuw, R. A. and Dull, G. G.**, Rapid analysis of simple carbohydrates by means of vapour-programmed thin-layer chromatography and densitometry and using a new spotting device, *J. Chromatogr.*, 110, 279, 1975.

Table TLC 2
MAIN KETOSES

All entries use Layer L1 and Technique T1. For each solvent, column **a** = $R_f \times 100$ and column **b** = $R_{Fru} \times 100$.

Compound	S1 a	S1 b	S2 a	S2 b	S3 a	S3 b	S4 a	S4 b	S5 a	S5 b	S6 a	S6 b	S7 a	S7 b	S8 a	S8 b	S9 a	S9 b	S10 a	S10 b	S11 a	S11 b	S12 a	S12 b
Dihydroxyacetone	81	231	82	186	80	210	71	254	71	237	71	215	74	195	68	189	70	233	76	190	64	305	69	157
L-*glycero*-Tetrulose	75	214	78	177	74	195	67	239	63	210	65	197	70	184	64	178	66	220	71	178	54	257	66	150
D-*threo*-Pentulose	70	200	76	173	72	189	63	225	61	203	63	191	69	182	64	178	66	220	69	173	46	219	66	150
D-*erythro*-Pentulose	60	171	64	145	60	136	50	179	49	163	51	155	55	145	52	144	50	167	59	148	38	181	56	127
D-Tagatose	50	143	60	136	54	123	43	154	42	137	45	136	51	134	48	133	47	157	54	135	28	133	56	127
D-Psicose	49	140	55	125	50	114	39	139	41	137	43	130	48	126	45	125	41	137	49	123	29	138	50	114
L-Sorbose	42	120	51	116	46	121	34	121	35	117	38	115	44	116	42	117	37	123	48	120	24	114	50	114
D-Fructose	35	100	44	100	38	100	28	100	30	100	30	100	38	100	36	100	30	100	40	100	21	100	44	100
D-*altro*-Heptulose	42	120	54	123	47	124	35	125	34	113	37	112	43	113	42	117	40	133	48	120	21	100	50	114
D-*manno*-Heptulose	32	91	43	98	39	103	27	96	29	97	32	97	38	100	38	106	33	110	41	103	18	86	47	107
D-*ido*-Heptulose	33	94	41	93	36	95	28	100	30	100	31	94	35	92	35	97	27	90	35	88	19	90	41	93
D-*allo*-Heptulose	24	69	37	84	32	84	21	75	25	83	28	85	33	87	33	92	27	90	33	83	14	67	39	89
D-*gluco*-Heptulose	22	63	33	75	29	76	20	71	23	77	26	79	31	82	32	89	24	80	32	80	14	67	39	89
L-*galacto*-Heptulose	17	49	25	57	22	58	15	54	19	63	19	58	23	61	23	64	16	53	23	58	12	57	29	66
-*epi*-Inosose-2	10	29	9	20	9	24	6	21	6	20	6	18	7	18	6	17	4	13	10	25	6	29	9	20
scyllo-Inosose	4	11	4	9	4	11	2	7	3	10	3	9	4	11	3	8	2	7	5	13	3	14	5	11

$a = R_f \times 100.$

$b = R_{Fru} \times 100.$

Layer L1 = 0.25 mm Silica Gel 60 (Merck®).

Solvent
S1 = acetone–water (9:1).
S2 = 2-propanol–acetone–water (4:5:1).
S3 = 1-propanol–acetone–water (4:5:1).
S4 = 1-butanol–acetone water (4:5:1).
S5 = 2-propanol–ethyl acetate–water (4:5:1).
S6 = 2-propanol–ethyl acetate–water (5:4:1).
S7 = 2-propanol–ethyl acetate–water (6:3:1).
S8 = 2-propanol–ethyl acetate–water (7:2:1).
S9 = 2-propanol–ethyl acetate–water (83:11;6).
S10 = 1-butanol–acetone–methanol–water (33:36:18:9).
S11 = ethyl acetate-1-butanol–methanol–water (16:3:3:2).
S12 = ethanol-2-butanol–water (6:3:1).

Technique
T1 = ascending development (15 cm; 60 min); developing chamber equilibrated with solvent system for 1 hr before use.

From Papin, J.-P. and Udiman, M., *J. Chromatogr.*, 132, 339, 1977. With permission.

Table TLC 3
MAIN SUGAR CONSTITUENTS OF GLYCOPROTEINS AND GLYCOLIPIDS

Layer	L1	L1	L1	L1	L1	L2	L2	L3	L4
Solvent	S1	S2	S3	S4	S5	S6	S7	S8	S9
Technique	T1	T1	T1	T2	T3	T4	T4	T5	T6
Reference	1	1	1	1	1	2	2	3	4
					$R_F \times 100$				$R_{ADG}{}^a \times 100$
Compound	*	*	*	*	***	1**	2**	*	*
D-Mannose	—	—	—	—	—	—	—	69	—
D-Galactose	54	55	100	72	30	—	—	58	—
D-Glucose	62	61	100	75	42	—	—	65	—
L-Fucose	—	—	—	—	—	66	50	75	—
2-Amino-2-deoxy-D-glucose	21	26	54	31	10	—	—	34	100
2-Acetamido-2-deoxy-D-glucose	54	59	100	74	33	64	53	79	116
2-Amino-2-deoxy-D-galactose	13	19	43	22	7	—	—	—	85
2-Acetamido-2-deoxy-D-galactose	45	51	100	66	26	52	57	—	107
N-Acetylneuraminic acid	3	5	44	2	1	—	—	—	—
N-Glycolylneuraminic acid	—	—	41	—	—	—	—	—	—

Note: * = single development. ** = double developments. *** = threefold development. 1 refers to first run in 2-dimensional development. 2 refers to second run, orthogonal to first, in 2-dimensional development.

a R_{ADG} = mobility relative to 2-amino-2-deoxy-D-glucose.

Layer L1 = 0.25 mm Silica Gel G (Merck®); plates air-dried for 1 hr, heated at 110°C for 30 min before use.

L2 = 0.25 mm Silica Gel F$_{254}$ (Merck®), impregnated with 0.2 M NaH$_2$PO$_4$; heated at 110—120°C for 30 min before use.

L3 = precoated Silica Gel 60 (Merck®), conditioned at 105°C for 1 hr.

L4 = ChromAR 500 (Mallinckrodt) sheets (about 70% neutral silicic acid and 30% fiber glass) cut into 83° wedge, base 2.5 cm wide, sides about 12 cm long.

Solvent S1 = 1-propanol-water (7:1).

S2 = 1-propanol-water (14:3).

S3 = methanol-water (5:2).

S4 = 1-propanol-ethyl acetate-water (5:1:4).

S5 = methyl acetate-2-propanol-water (18:1:1).

S6 = 1-butanol-acetone-water (4:5:1).

S7 = phenol-water (3:1 w/v).

S8 = 2-propanol-acetone-M lactic acid (2:2:1).

S9 = 1-propanol-concentrated ammonia (4:1).

Technique T1 = ascending development, 4 hr.

T2 = ascending development, 5 hr.

T3 = threefold development, 50 min each run; plates dried after each development.

T4 = two-dimensional development, twice to 15 cm in each direction; plate air-dried between runs.

T5 = ascending development (13 cm; 3 hr).

T6 = sheets uniformly sprayed with methanolic solution of CuSO$_4$ (0.05%), dried in stream of warm air for 30 min before sample application; ascending development (15 cm) in Gelman chamber, without presaturation.

REFERENCES

1. **Gal, A. E.**, Separation and identification of monosaccharides from biological materials by thin-layer chromatography, *Anal. Biochem.*, 24, 452, 1968.
2. **Hotta, K. and Kurokawa, M.**, Separation of fucose and acetylhexosamine by two-dimensional thin-layer chromatography, *Anal. Biochem.*, 26, 472, 1968.
3. **Hansen, S. A.**, Thin-layer chromatographic method for identification of oligosaccharides in starch hydrolysates, *J. Chromatogr.*, 105, 388, 1975.
4. **Martz, M. D. and Krivis, A. F.**, Thin-layer chromatography of hexosamines on copper-impregnated sheets, *Anal. Chem.*, 43, 790, 1971.

Table TLC 4
METHYL, ARYL, AND ACETYL DERIVATIVES OF SUGARS

Layer	L1	L2	L3	L4	L4	L5	L5	L6	L7	L8
Solvent	S1	S2	S3	S4	S5	S6	S7	S8	S9	S9
Technique	T1	T1	T2	T3	T3	T4	T5	T6	T7	T8
Reference	1	1	2	3	3	4	4	4	4	5
	$R_f \times 100$		$R_c' \times 100$	$R_f \times 100$						$R_{Ara}^b \times 100$
Compound	*	*	*	*	*	**	**	**	**	*
D-Glucopyranoside										
methyl β-	20	51	—	—	—	—	—	—	—	—
methyl 4,6-O-benzylidene-α-	62	81	—	—	—	—	—	—	—	—
methyl 2,3,4-tri-O-methyl-α-	—	—	—	63	31	—	—	—	—	—
methyl 2,3,4-tri-O-methyl-β-	—	—	—	—	35	—	—	—	—	—
methyl 2,3,4,6-tetra-O-methyl-α-	—	—	—	100	49	—	—	—	—	—
methyl 2,3,4,6-tetra-O-methyl-β-	—	—	—	—	62	—	—	—	—	—
phenyl 2,3,4,6-tetra-O-acetyl-α-	—	—	176	—	—	—	—	—	—	—
phenyl 2,3,4,6-tetra-O-acetyl-β-	—	—	140	—	—	—	—	—	—	—
2-chlorophenyl 2,3,4,6-tetra-O-acetyl-α-	—	—	148	—	—	—	—	—	—	—
2-chlorophenyl 2,3,4,6-tetra-O-acetyl-β-	—	—	120	—	—	—	—	—	—	—
4-chlorophenyl 2,3,4,6-tetra-O-acetyl-α-	—	—	200	—	—	—	—	—	—	—
4-chlorophenyl 2,3,4,6-tetra-O-acetyl-β-	—	—	152	—	—	—	—	—	—	—
2-methylphenyl 2,3,4,6-tetra-O-acetyl-α-	—	—	158	—	—	—	—	—	—	—
2-methylphenyl 2,3,4,6-tetra-O-acetyl-β-	—	—	117	—	—	—	—	—	—	—
4-methylphenyl 2,3,4,6-tetra-O-acetyl-α-	—	—	173	—	—	—	—	—	—	—
4-methylphenyl 2,3,4,6-tetra-O-acetyl-β-	—	—	117	—	—	—	—	—	—	—
4-nitrophenyl 2,3,4,6-tetra-O-acetyl-α-	—	—	150	—	—	—	—	—	—	—
4-nitrophenyl 2,3,4,6-tetra-O-acetyl-β-	—	—	104	—	—	—	—	—	—	—
4-phenylphenyl 2,3,4,6-tetra-O-acetyl-α-	—	—	135	—	—	—	—	—	—	—
4-phenylphenyl 2,3,4,6-tetra-O-acetyl-β-	—	—	104	—	—	—	—	—	—	—
D-Xylopyranoside, methyl β-	30	58	—	—	—	—	—	—	—	—
D-Glucopyranose										
4,6-O-ethylidene-α-	40	68	—	—	—	—	—	—	—	—
1,2,3,4,6-penta-O-acetyl-α-	75	90	—	—	—	—	—	—	—	—
1,2,3,4,6-penta-O-acetyl-β-	—	—	100	—	—	—	—	—	—	—

D-Arabinose								
2-O-methyl-	—	—	—	—	—	—	—	24
3-O-methyl-	—	—	—	—	—	—	—	15
4-O-methyl-	—	—	—	—	—	—	—	11
5-O-methyl-	—	—	—	—	—	—	—	40
2,3-di-O-methyl-	—	—	—	—	—	—	—	56
2,4-di-O-methyl-	—	—	—	—	—	—	—	43
2,5-di-O-methyl-	—	—	—	—	—	—	—	81
3,4-di-O-methyl-	—	—	—	—	—	—	—	26
3,5-di-O-methyl-	—	—	—	—	—	—	—	81
2,3,5-tri-O-methyl-	—	—	—	—	—	—	—	100
D-Xylose								
2-O-methyl-	—	—	—	44	49	68	35	—
3-O-methyl-	—	—	—	44	49	46	36	—
2,3-di-O-methyl-	—	—	—	68	73	78	51	—
2,3,4-tri-O-methyl-	—	—	—	87	93	84	59	—
D-Glucose								
2-O-methyl-	—	—	—	32	33	53	10	—
3-O-methyl-	—	—	—	32	33	44	10	—
3,6-di-O-methyl-	—	—	—	54	52	65	28	—
2,3,4-tri-O-methyl-	—	—	—	79	80	79	53	—
2,3,6-tri-O-methyl-	—	—	—	77	—	77	52	—
2,3,4,6-tetra-O-methyl-	—	—	—	—	—	85	79	—
D-Galactose,2,3,4,6-tetra-O-methyl-	—	—	—	80	—	70	59	—
D-Mannose								
2,3-di-O-methyl-	—	—	—	72	—	65	41	—
2,3,6-tri-O-methyl-	—	—	—	72	—	73	43	—
2,3,4,6-tetra-O-methyl-	—	—	—	—	—	79	61	—

[a] R_G = mobility relative to that of penta-O-acetyl-β-D-glucopyranose.

[b] R_{Ara} = mobility relative to that of 2,3,5-tri-O-methyl-D-arabinose.

* Single development.

** Double development.

Layer L1 = 0.25 mm Silica Gel 60 (Merck®), dried at 110°C for 2 hr.

L2 = 0.25 mm layer, mixture of Silica Gel G 60 (2 parts) and Kieselguhr G (1 part), prepared as for L1.

L3 = Silica gel G (Merck®), air-dried then heated at 110°C for 1 hr.

Table TLC 4 (continued)
METHYL, ARYL, AND ACETYL DERIVATIVES OF SUGARS

L4 = preparative (0.5 mm) Silica Gel H (Merck®).
L5 = cellulose.
L6 = silica gel buffered with 0.1 MH_3BO_3 activated at 110°C.
L7 = silica gel.
L8 = 0.25 mm Silica Gel F_{254} (Merck®).

Solvent
S1 = benzene-ethanol (2:1).
S2 = benzene-acetic acid-ethanol (2:2:1).
S3 = 2-butanone-petroleum ether, 40—60° (1:3).
S4 = chloroform-methanol (47:3).
S5 = benzene-ethanol-water (170:47:15, upper layer).
S6 = 1-butanol-ethanol-water-ammonia (40:10:49:1).
S7 = ethyl acetate-acetic acid-water (18:7:8).
S8 = 1-butanol-acetone-water (4:5:1).
S9 = 2-butanone saturated with 3% aqueous ammonia.

Technique
T1 = ascending development in filter paper-lined glass chamber.
T2 = ascending development in saturated chamber, distance = 15 cm.
T3 = not specified.
T4 = double development (12.5 and 12.7 cm), 6.5 hr total time.
T5 = double development (14.6 and 15.1 cm), 4.5 hr total time.
T6 = double development (13.6 and 14.4 cm), 2.5 hr total time.
T7 = double development (15.1 and 13.8 cm), 7 hr total time.
T8 = single development, standard procedure.

REFERENCES

1. **Haldorsen, K. M.,** Vanadium pentoxide in sulfuric acid, a general chromogenic spray reagent for carbohydrates, *J. Chromatogr.,* 134, 467, 1977.
2. **Audichya, T. D.,** Thin-layer chromatography of anomeric aryltetra-O-acetyl-D-glucopyranosides, *J. Chromatogr.,* 57, 161, 1971.
3. **Brennan, P. J.,** Application of thin-layer chromatography to the purification and characterisation of some methyl glucosides, *J. Chromatogr.,* 59, 231, 1971.
4. **Sinner, M.** Separation of methyl ethers of xylose, glucose and some other sugars by liquid chromatography, *J. Chromatogr.,* 121, 122, 1976.
5. **Mied, P. A. and Lee, Y. C.,** Preparation and separation of methyl ethers of D-arabinose, *Anal. Biochem.,* 49, 534, 1972.

Table TLC 5
HIGHER OLIGOSACCHARIDES

	Malto-oligosaccharides											Isomalto-		Fructans		Oligogalacturonic acids			
	$R_F \times 100$						**$R_{M3} \times 100$**					**$R_{IM3} \times 100$**		**$R_F \times 100$**		**$R_{EA} \times 100$**			
Layer	L1	L1	L1	L1	L1	L1	L2	L2	L2	L2	L2	L3	L3	L1	L1	L4	L5	L5	L5
Solvent	S1	S2	S3	S4	S1	S2	S5	S6	S7	S7	S7	S8	S8	S9	S10	S11	S12	S11	S12
Technique	T1	T1	T1	T1	T2	T2	T3	T4	T5	T6	T7	T8	T8	T9	T9	T10	T11	T11	T11
Reference	1	1	1	1	1	1	2	2	2	2	2	3	3	4	4	5	5	5	5
Degree of polymerization	*	*	*	*	*	*	*	*	*	*	*	C	C	*	*	*	*	**	**
1	89	92	93	93	91	94	—	141	110	110	107	C	C	85	96	65	59	100	100
2	78	90	89	90	87	91	—	119	100	100	100	100	100	78	94	58	48	88	76
3	60	84	84	89	83	87	100	100	100	100	100	86	81*	52	87	52	38	80	58
4	39	77	78	84	81	81	78	89	93	90	81*	69	66*	38	83	48	25	60	37
5	20	71	70	81	77	75	66	77	84	85	66*	52	48*	24	77	31	13	47	22
6	—	64	63	77	73	70	53	68	72	75	48*	—	37*	15	70	19	7	27	12
7	—	57	55	73	68	65	44	60	65	72	37*	28	27*	10	62	13	5	17	6
8	—	49	44	68	62	55	37	52	55	65	27*	17	20*	7	55	8	2	13	3
9	—	39	36	62	55	44	31	45	45	57	20*	—	14*	—	48	4	—	8	2
10	—	31	30	55	49	33	25	40	37	48	14*	—	—	—	43	1	—	4	1
11	—	23	22	49	42	25	—	35	30	38	—	—	—	—	38	—	—	—	—
12	—	21	17	42	36	19	—	30	—	31	—	—	—	—	34	—	—	—	—
13	—	19	14	36	29	14	—	26	—	—	—	—	—	—	29	—	—	—	—
14	—	15	9	29	24	10	—	23	—	—	—	—	—	—	23	—	—	—	—
15	—	12	6	24	20	8	—	20	—	—	—	—	—	—	19	—	—	—	—
16	—	9	5	20	16	4	—	17	—	—	—	—	—	—	15	—	—	—	—
17	—	7	3	16	—	—	—	—	—	—	—	—	—	—	12	—	—	—	—
18	—	—	2	—	—	—	—	—	—	—	—	—	—	—	10	—	—	—	—
19	—	—	—	—	—	—	—	—	—	—	—	—	—	—	8	—	—	—	—
20	—	—	—	—	—	—	—	—	—	—	—	—	—	—	7	—	—	—	—

Note: * = single development. ** = double development. C = continuous run.

Table TLC 5 (continued)
HIGHER OLIGOSACCHARIDES

^a R_{MI} = mobility relative to maltotriose, calculated from published data.

^b R_{IMI} = mobility relative to isomaltose, calculated from published chromatogram.

^c R_{GA} = mobility relative to α-D-galacturonic acid.

^d Sucrose.

^e Single α-(1→3) branch.

Layer   L1 = 0.25 mm Kieselguhr G (Merck®).

L2 = precoated silica gel high-performance TLC plates (Merck®, Cat. No. 5633).

L3 = precoated 0.25 mm silica gel plates (Merck®, Cat. No. 5721).

L4 = Cellulose MN-300, Polygram Cel-300 plates.

L5 = Eastman E-13255 cellulose plates.

Solvent   S1 = 1-butanol-pyridine-water (13:4:3).

S2 = 1-butanol-pyridine-water (6:4:3).

S3 = 1-butanol-pyridine-water (5:4:2).

S4 = 1-butanol-pyridine-water (20:19:11).

S5 = 2-propanol-acetone-water (8:7:5).

S6 = 2-propanol-water (7:3).

S7 = ethanol-acetone-water (9:6:5).

S8 = 1-propanol-nitromethane-water (5:2:3).

S9 = 1-propanol-ethyl acetate-water (4:5:1).

S10 = 1-propanol-ethyl acetate-water (3:1:1).

S11 = ethyl acetate-acetic acid-water (2:1:2).

S12 = ethyl acetate-acetic acid-water (4:2:3).

Technique   T1 = plates 5 × 20 cm; oligosaccharides in 90% dimethylsulfoxide solution, applied as narrow band 1 cm long; ascending development.

T2 = plates 20 × 45 cm; oligosaccharide mixtures applied as narrow band 16 cm long.

T3 = oligosaccharides, dissolved in 70% aqueous ethanol, spotted on plate; developed for 30 min at 60°C.

T4 = developed for 2 hr at 60°C.

T5 = developed for 30 min at 26°C.

T6 = developed for 30 min at 40°C.

T7 = developed for 30 min at 58°C.

T8 = continuous development for 19 hr at 30°C.

T9 = single, ascending development (12 cm) at ambient temperature.

T10 = single, ascending development (95 min) at 24°C.
T11 = double development (2.5 hr each) at 24°C.

REFERENCES

1. Shannon, J. C. and Creech, R. G., Thin-layer chromatography of malto-oligosaccharides, *J. Chromatogr.*, 44, 307, 1969.
2. Nurok, D. and Zlatkis, A., Separation of malto-oligosaccharides by high-performance thin-layer chromatography at moderate temperature, *J. Chromatogr.*, 142, 449, 1977.
3. Covacevich, M. T. and Richards, G. N., Continuous quantitative thin-layer chromatography of oligosaccharides, *J. Chromatogr.*, 129, 420, 1976.
4. Collins, F. W. and Chandorkar, K. R., Thin-layer chromatography of fructo-oligosaccharides, *J. Chromatogr.*, 56, 163, 1971.
5. Liu, Y. K. and Luh, B. S., Preparation and thin-layer chromatography of oligogalacturonic acids, *J. Chromatogr.*, 151, 39, 1978.

Table TLC 6
GLYCOSAMINOGLYCANS, GLYCOPEPTIDES, AND GLYCOLIPIDS

Layer	L1	L1	L2	L3	L4
Solvent	S1	S1	S2	S3	S4
Detection	D1	D1	D2	D3	D4
Technique	T1	T2	T3	T4	T5
Reference	1	1	2	3	4

Compound	$R_F \times 100$			$R_{Chs}{}^a \times 100$	$R_{Lb}{}^b \times 100$
Glycosaminoglycans					
Chondroitin 6-sulfate	32	35	—	—	—
Chondroitin 4-sulfate	31	36	—	—	—
$\overline{M}_n$ 13,000	—	—	—	60	—
$\overline{M}_n$ 31,000	—	—	—	80	—
$\overline{M}_n$ 61,000	—	—	—	100	—
Dermatan sulfate	15	23	—	72	—
Heparin	14	16	—	55	—
N-Acetylheparan sulfate	21	22	—	—	—
Keratan sulfate	43	75	—	—	—
Glycopeptides					
Glucosylgalactosylhydroxylysine	—	—	—	—	50
Ovalbumin GP IV, $\overline{M}_n$ 1,500	—	—	—	—	56
Ovalbumin GP III, $\overline{M}_n$ 1,900	—	—	—	—	58
Transferrin GP, $\overline{M}_n$ 2,600	—	—	—	—	65
Fetuin GP, $\overline{M}_n$ 3,300	—	—	—	—	68
Glycolipids					
Brain gangliosides					
G_{M3}	—	—	73	—	—
G_{M2}	—	—	63	—	—
G_{M1}	—	—	54	—	—
G_{D3}	—	—	38	—	—
G_{D1a}	—	—	29	—	—
G_{D1b}	—	—	16	—	—
G_{T1}	—	—	6	—	—
G_{Q1}	—	—	1	—	—

Note: GP = glycopeptide

[a] R_{Chs} = mobility relative to chondroitin 4-sulfate, $\overline{M}_n$ 61,000, calculated from published data
[b] R_{Lb} = mobility relative to a-lactalbumin CNBr peptide I.

Layer	L1 = 0.10 mm MN-Polygram Cellulose 400 (Machery-Nagel).
	L2 = precoated silica gel high-performance plates (Merck).
	L3 = 0.5 mm Sephadex G-200, Superfine (Pharmacia).
	L4 = 1.25 mm Sephadex G-50, Superfine (Pharmacia).
Solvent	S1 = 0.04 M ammonium formate-methanol (9:11), containing 1.1 m M complexone-III.
	S2 = methyl acetate-2-propanol-0.033 M KCl (9:6:4).
	S3 = 2 M NaCl.
	S4 = 6 M guanidine hydrochloride-0.1 M phosphate buffer (pH 7.0).
Detection	D1 = Azure A dye (50 mg) in acetone-methanol-2% acetic acid (20:60:20); after treatment for 2—3 min, plate washed with 1% acetic acid.
	D2 = plate sprayed with orcinol-hydrochloric acid.
	D3 = plate covered with filter paper; paper dried and stained with 0.1% toluidine blue in 2% acetic acid for 30 min, then washed with 2% acetic acid.

Table TLC 6 (continued)
GLYCOSAMINOGLYCANS, GLYCOPEPTIDES, AND GLYCOLIPIDS

D4 = paper replica of plate stained with fluorescamine in acetone (2 mg/ml), heated at 60°C for 5 min and examined under U.V.

Technique   T1 = plates impregnated with solution containing lead acetate (3 g) in 80 ml 96% ethanol, with 20 ml 40% formalin + 1 ml glacial acetic acid added; development in presaturated chamber for 3 hr.

T2 = as T1; development for 24 hr.

T3 = gangliosides dissolved in chloroform-methanol (2:1) at concentrations 0.1 mg/ml sialic acid each; 1 μl applied in 5 mm streak; development for 30 min.

T4 = plate inclined at 15°, with filter paper bridge to solvent; solvent allowed to flow for at least 24 hr before application of samples (1—3 μl, containing 50—100 μg), absorbed from filter paper strips laid on gel surface; development continued until Blue Dextran 2000 marker migrated 10 cm (2—3 hr).

T5 = plate inclined at about 8° in Pharmacia TLG apparatus; samples dissolved in solvent at least 2 hr before application (in volume 5 μl, containing 75—100 μg); cytochrome *c* as marker; development for 16—18 hr.

REFERENCES

1. Havass, Z. and Szabo, L., Thin-layer chromatographic separation of glycosaminoglycans, *J. Chromatogr.*, 71, 580—584, 1972.
2. Zanetta, J.-P., Vitiello, F., and Robert, J., Thin-layer chromatography of gangliosides, *J. Chromatogr.*, 137, 481, 1977.
3. Taniguchi, N., Thin-layer gel filtration of urinary mucopolysaccharides in Hunter's syndrome and in normal individuals, *Clin. Chim. Acta*, 30, 801, 1970.
4. Hung, C.-H., Strickland, D. K., and Hudson, B. G., Estimation of molecular weights of peptides and glycopeptides by thin-layer gel filtration, *Anal. Biochem.*, 80, 91, 1977.

Section I.V

ELECTROPHORETIC DATA

This section is included because paper electrophoresis serves as a useful complement to paper and thin-layer chromatography in examination of sugars and derivatives. Comprehensive data obtained by this method are presented in Tables EL 1 to EL 3.

The application to carbohydrates of other electrophoretic techniques (glass-fiber, cellulose acetate and gel electrophoresis) is briefly reviewed at the end of this section (Tables EL 4 to EL 6 and accompanying text).

PAPER ELECTROPHORESIS

Tables EL 1 to EL 3 have been compiled mainly from data originally published by Dr. H. Weigel (Royal Holloway College, University of London) in *Advances in Carbohydrate Chemistry,* 18, 61—97, 1963. These data are reprinted by permission of Dr. Weigel and Academic Press.

The list of papers, electrolytes, techniques, and references applicable to all three tables will be found after Table EL 3.

Table EL 1
PAPER ELECTROPHORESIS OF SUGARS[a]

Paper	P1	P2	P3	P4	P4	P2	P2	P5	P5
Electrolyte	E1	E2	E3	E4	E5	E6	E7	E8	E9
Technique	T1	T2	T3	T4	T5	T6	T2	T7	T7
Reference	1	2	3	4	5	2	2	6,7	7,8
Compound	$M_G{}^b \times 100$					$M_R{}^c \times 100$		$M_s{}^d \times 100$	
Aldoses									
D-Erythrose	—	—	—	—	—	—	—	90	110'
L-Threose	—	—	—	—	—	—	—	60	5
L-Arabinose	96	91	—	150	240	30	7	0	0
D-Xylose	100	101	—	140	180	17	8	0	0
D-Lyxose	—	71	—	190	230	42	30	110	104'
D-Ribose	77	75	—	210	470	100	100	40	20
D-Glucose	100	100	100	100	100	16	6	0	0
D-Galactose	93	93	—	130	180	28	10	0	0
D-Mannose	72	69	—	140	110	35	41	90'	110'
D-Allose	—	83	—	180	—	75	33	—	—
D-Altrose	—	97	—	—	580	77	10	0	—
D-Gulose	—	82	—	—	—	53	31	110	110
L-Idose	—	102	—	—	—	115	42	—	—
D-Talose	—	87	—	—	—	119	110	70	—
L-Rhamnose	52	49	—	130	50	32	28	60'	110'
L-Fucose	89	83'	—	—	—	22	6	0	0
2-Acetamido-2-deoxy-D-glucose	23	—	—	10	—	—	—	—	—
2-Acetamido-2-deoxy-D-galactose	35	—	—	—	—	—	—	—	—
D-*glycero*-L-*gluco*-Heptose	—	104	—	—	—	23	7	20	0
D-*glycero*-D-*galacto*-Heptose	—	—	—	—	—	—	—	40	94'
D-*glycero*-L-*galacto*-Heptose	—	—	—	—	—	—	—	40	21
D-*glycero*-L-*manno*-Heptose	—	—	—	—	—	—	—	80	100'
D-*glycero*-D-*allo*-Heptose	—	—	—	—	—	—	—	90	48
D-*glycero*-D-*gulo*-Heptose	—	—	—	—	—	—	—	110	98'
D-*glycero*-D-*ido*-Heptose	—	—	—	—	—	—	—	100	73
Ketoses									
D-*erythro*-Pentulose	—	90	—	—	—	209	73	—	—
D-*threo*-Pentulose	—	75	—	—	—	194	41	—	—
D-Fructose	90	89	—	210	930	75	22	50	25'
L-Sorbose	95	97	—	200	850	73	16	30	20'
D-Tagatose	—	95	—	240	860	103	65	105	110'
D-Psicose	—	76	—	—	—	188	91'	—	—
L-*galacto*-Heptulose	89	—	—	—	—	—	—	—	—
D-*gluco*-Heptulose	—	—	—	—	—	—	—	100	—
D-*manno*-Heptulose	87	—	—	—	—	—	—	40	—
Uronic acids									
D-Glucuronic acid	120	—	126	—	—	—	—	—	—
D-Galacturonic acid	—	—	104	—	—	—	—	—	—
D-Mannuronic acid	—	—	88	—	—	—	—	—	—
L-Guluronic acid	—	—	75	—	—	—	—	—	—

Table EL 1 (continued)
PAPER ELECTROPHORESIS OF SUGARS[a]

Paper	P1	P2	P3	P4	P4	P2	P2	P5	P5
Electrolyte	E1	E2	E3	E4	E5	E6	E7	E8	E9
Technique	T1	T2	T3	T4	T5	T6	T2	T7	T7
Reference	1	2	3	4	5	2	2	6,7	7,8
Compound	$M_G{}^b \times 100$					$M_R{}^c \times 100$		$M_S{}^d \times 100$	
Oligosaccharides									
Sophorose	24	—	—	—	—	—	—	0	0
Nigerose	69	—	—	130	—	—	—	0	0
Laminaribiose	69	—	—	110	—	—	—	0	0
Maltose	32	30	—	40	—	15	7	0	0
Cellobiose	23	22	—	30	—	15	10	0	0
Isomaltose	69	—	—	90	—	—	—	0	0
Gentiobiose	75	—	—	100	—	—	—	0	0
Lactose	38	37	—	70	—	24	13	0	0
Melibiose	80	77	—	140	—	32	10	0	0
Turanose	—	64	—	—	—	30	7	10	0
Sucrose	18	16	—	10	—	14	4	0	0
Raffinose	28	26	—	40	—	25	7	—	—

[a]   Explanation of terms and references at end of Table EL 3.
[b] M_G = mobility relative to D-glucose.
[c] M_R = mobility relative to D-ribose.
[d] M_S = mobility relative to D-glucitol.
[e]   Streaks.

Table EL 2
PAPER ELECTROPHORESIS OF ALDITOLS, CYCLITOLS, AND REDUCED DISACCHARIDES[a]

Paper	P1	P2	P4	P4	P2	P2	P5	P5
Electrolyte	E1	E2	E4	E5	E6	E7	E8	E9
Technique	T1	T2	T4	T5	T6	T2	T7	T7
Reference	1	2	4	5	2, 7	2	6, 7, 9	7, 8
Compound	$M_G \times 100$			$M_M{}^b \times 100$	$M_R \times 100$		$M_S \times 100$	
Alditols								
Erythritol	75	75	100	10	53	3	100	90[c]
L-Threitol	—	75	—	30	96	11	50	24
D-Arabinitol	90	87	—	—	124	14	110	104
L-Arabinitol	—	—	180	60	—	—	—	—
Xylitol	—	79	170	90	155	25	110	104
Ribitol	—	85	120	30	76	4	110	103
D-Glucitol	—	83	190	130	161	47	100	100
Galactitol	98	97	210	100	145	32	100	100
D-Mannitol	90	91	190	100	130	23	100	100
Allitol	—	90	—	—	92	9	94	97[c]
D-Altritol	—	—	—	—	—	—	99	97[c]
L-Iditol	—	81	—	140	173	57	—	100
D-Talitol	—	89	—	—	138	17	—	—
L-Rhamnitol	—	—	—	80	—	—	—	—
D-*glycero*-D-*gluco*-Heptitol	—	88	—	—	171	53[c]	—	—
D-*glycero*-L-*gluco*-Heptitol	92	95	—	—	176	59[c]	—	—
D-*glycero*-D-*galacto*-Heptitol	100	98	—	—	140	51[c]	—	—
D-*glycero*-D-*altro*-Heptitol	—	92	—	—	144	27	—	—

Table EL 2 (continued)
PAPER ELECTROPHORESIS OF ALDITOLS, CYCLITOLS, AND REDUCED DISACCHARIDES[a]

Paper	P1	P2	P4	P4	P2	P2	P5	P5
Electrolyte	E1	E2	E4	E5	E6	E7	E8	E9
Technique	T1	T2	T4	T5	T6	T2	T7	T7
Reference	1	2	4	5	2, 7	2	6, 7, 9	7, 8
Compound	$M_G \times 100$		$M_M{}^b \times 100$		$M_R \times 100$		$M_S \times 100$	
D-*glycero*-D-*talo*-Heptitol	—	93	—	—	140	34	—	—
D-*glycero*-D-*ido*-Heptitol	—	85	—	—	168	71[c]	—	—
(*meso*)-*glycero*-*ido*-Heptitol	—	78	—	—	182	79[c]	—	—
(*meso*)-*glycero*-*gulo*-Heptitol	—	85	—	—	160	72[c]	—	—
(*meso*)-*glycero*-*allo*-Heptitol	—	95	—	—	100	11	—	—
Cyclitols								
allo-Inositol	88	85	—	—	50	62	40	8
cis-Inositol	—	79	—	—	40	116	—	—
epi-Inositol	73	76	180	180	29	74[c]	110	100
(+)-Inositol	63	—	100	0	—	—	0	0
(−)-Inositol	63	59	—	—	23	20	—	—
muco-Inositol	96	97	—	—	36	41	0	0
myo-Inositol	53	49[c]	70	0	16	75[c]	20	0
neo-Inositol	—	59	—	—	43	64	—	—
scyllo-Inositol	<5	2	20	—	7	[d]	0	0
Reduced disaccharides								
D-glucitol								
2-*O*-α-D-gluco-pyranosyl-	—	—	—	—	—	—	69	76
2-*O*-β-D-gluco-pyranosyl-	—	—	—	120	—	—	70	73
3-*O*-α-D-gluco-pyranosyl-	—	—	—	—	39	—	0	0
3-*O*-β-D-gluco-pyranosyl-	—	—	—	10	—	—	0	0
4-*O*-α-D-gluco-pyranosyl-	—	—	—	—	—	—	40	—
4-*O*-β-D-gluco-pyranosyl-	—	—	120	110	—	—	40	—
D-glucitol								
6-*O*-α-D-gluco-pyranosyl-	—	—	—	—	—	—	80	—
6-*O*-β-D-gluco-pyranosyl-	—	—	—	60	—	—	80	—
6-*O*-α-D-galacto-pyranosyl-	—	—	150	—	—	—	80	—
4-*O*-β-D-galacto-pyranosyl-	—	—	150	—	—	—	40	—
D-mannitol								
2-*O*-α-D-manno-pyranosyl-	—	—	—	—	—	—	80	—
3-*O*-α-D-manno-pyranosyl-	—	—	—	—	—	—	0	0
2-*O*-α-D-gluco-pyranosyl-	—	—	—	—	—	—	72	70
L-gulitol								
1-*O*-α-D-galacto-pyranosyl-	—	—	—	—	100	—	—	68
3-*O*-β-D-galacto-pyranosyl-	—	—	—	—	63	—	—	10
1-*O*-α-D-gluco-pyranosyl-	—	—	—	—	—	—	78	76
1-*O*-β-D-gluco-pyranosyl-	—	—	—	—	—	—	69	66
2-*O*-α-D-gluco-pyranosyl-	—	—	—	—	—	—	72	70
3-*O*-α-D-gluco-pyranosyl-	—	—	—	—	66	—	46	17
3-*O*-β-D-gluco-pyranosyl-	—	—	—	—	63	—	37	10

[a] Explanation of terms and references at end of Table EL 3.
[b] M_M = mobility relative to D-mannitol.
[c] Streaks.
[d] Adsorbed.

Table EL 3
PAPER ELECTROPHORESIS OF DERIVATIVES OF SUGARS AND POLYOLS

Paper	P1	P2	P4	P4	P2	P2	P5	P5
Electrolyte	E1	E2	E4	E5	E6	E7	E8	E9
Technique	T1	T2	T4	T5	T6	T2	T7	T7
Reference	1	2	4	5	2	2	6, 7, 9	7, 8
Compound	$M_G \times 100$				$M_R \times 100$		$M_S \times 100$	
Glycosides								
D-Arabinofuranoside								
methyl α-, β-	4	—	—	—	—	—	—	—
D-Arabinopyranoside								
methyl α-, β-	38	—	—	—	—	—	0	—
L-Arabinofuranoside								
methyl α-	—	—	0	—	—	—	0	—
L-Arabinopyranoside								
methyl α-	—	—	70	—	—	—	—	—
methyl β-	—	—	60	—	—	—	—	—
D-Xylofuranoside								
methyl α-	56	—	—	230	—	—	0	—
methyl β-	33	—	5	—	—	—	—	—
D-Xylopyranoside								
methyl α-	0	—	0	—	—	—	—	—
methyl β-	0	—	0	0	—	—	—	—
D-Lyxopyranoside								
methyl α-	45	—	—	—	—	—	0	0
methyl β-	27	—	—	—	—	—	0	—
D-Ribopyranoside								
methyl α-	—	—	—	—	—	—	10	0
methyl β-	53	—	—	—	—	—	—	—
D-Glucofuranoside								
methyl α-	73	—	—	—	—	—	—	—
methyl β-	—	—	—	200	—	—	—	—
D-Glucopyranoside								
methyl α-	11	10	0	0	9	1	0	0
methyl β-	19	15	0	—	6	3	—	—
D-Galactofuranoside								
methyl α-	41	—	—	—	—	—	—	—
methyl β-	31	—	20	40	—	—	—	—
D-Galactopyranoside								
methyl α-	38	35	60	0	19	5	—	—
methyl β-	38	34	50	—	19	5	—	—
D-Mannofuranoside								
methyl α-	—	—	140	1600	—	—	—	—
D-Mannopyranoside								
methyl α-	42	39	50	—	17	0	0	0
methyl β-	31	—	40	0	—	—	0	—
D-Altropyranoside								
methyl α-	—	58	—	—	—	1	—	—
D-Gulopyranoside								
methyl α-	59	—	—	—	—	—	—	—
methyl β-	72	—	—	—	—	—	—	—
L-Rhamnopyranoside								
methyl α-	31	—	—	—	—	—	—	—
methyl β-	14	—	—	—	—	—	—	—
D-Fructopyranoside								
methyl α-	71	—	—	—	—	—	—	—
methyl β-	59	—	—	—	—	—	—	—

Table EL 3 (continued)
PAPER ELECTROPHORESIS OF DERIVATIVES OF SUGARS AND POLYOLS

Paper	P1	P2	P4	P4	P2	P2	P5	P5
Electrolyte	E1	E2	E4	E5	E6	E7	E8	E9
Technique	T1	T2	T4	T5	T6	T2	T7	T7
Reference	1	2	4	5	2	2	6, 7, 9	7, 8
Compound	$M_G \times 100$				$M_R \times 100$		$M_S \times 100$	
Ethers								
D-Xylose								
2-O-methyl-	—	—	0	0	—	—	—	—
3-O-methyl-	—	—	170	290	—	—	—	—
4-O-methyl-	—	—	30	0	—	—	—	—
5-O-methyl-	—	—	—	1300	—	—	—	—
3,5-di-O-methyl-	—	—	—	930	—	—	—	—
D-Glucose								
2-O-methyl-	23	—	0	0	—	—	—	—
3-O-methyl-	80	76	140	130	13	0	0	0
4-O-methyl-	24	—	30	0	—	—	0	0
6-O-methyl-	80	—	96	50	—	—	0	0
2,3-di-O-methyl-	12	—	0	—	—	—	—	—
2,4-di-O-methyl-	<5	—	—	—	—	—	—	—
3,4-di-O-methyl-	31	—	—	—	—	—	—	—
2,3,4-tri-O-methyl-	0	—	—	—	—	—	0	0
2,3,6-tri-O-methyl-	—	0	—	—	0	0	0	—
2,4,6-tri-O-methyl-	—	0	—	—	0	0	0	0
3,5,6-tri-O-methyl-	71	65	—	—	134	8	—	—
2,3,4,6-tetra-O-methyl-	0	0	—	—	0	0	0	—
D-Galactose								
2-O-methyl-	32	—	60	—	—	—	—	—
3-O-methyl-	—	—	140	—	—	—	—	—
4-O-methyl-	27	—	40	—	—	—	—	—
6-O-methyl-	—	—	120	—	—	—	—	—
2,3-di-O-methyl-	—	—	10	—	—	—	—	—
2,4-di-O-methyl-	—	—	0	—	—	—	—	—
2,6-di-O-methyl-	—	—	50	—	—	—	—	—
D-Mannose								
3-O-methyl	—	—	80	—	—	—	—	—
3,4-di-O-methyl-	—	—	—	—	—	—	0	0
D-Rhamnose								
2,3-di-O-methyl-	<5	—	—	—	—	—	—	—
2,4-di-O-methyl-	<5	—	—	—	—	—	—	—
3,4-di-O-methyl-	36	—	—	—	—	—	—	—
L-Arabinitol								
2-O-methyl-	—	—	—	10°	—	—	—	—
4-O-methyl-	—	—	—	50°	—	—	—	—
Xylitol								
2-O-methyl-	—	—	120	40°	—	—	—	—
3-O-methyl-	—	—	50	10°	—	—	—	—
5-O-methyl-	—	—	—	60°	—	—	—	—
D-Glucitol								
2-O-methyl-	—	—	150	120°	—	—	—	—
3-O-methyl-	—	—	80	10°	—	—	0	0
4-O-methyl-	—	—	140	130°	—	—	—	—
6-O-methyl-	—	—	—	60°	—	—	—	—
2,3-di-O-methyl-	—	—	—	—	—	—	0	0

Table EL 3 (continued)
PAPER ELECTROPHORESIS OF DERIVATIVES OF SUGARS AND POLYOLS

Paper	P1	P2	P4	P4	P2	P2	P5	P5
Electrolyte	E1	E2	E4	E5	E6	E7	E8	E9
Technique	T1	T2	T4	T5	T6	T2	T7	T7
Reference	1	2	4	5	2	2	6, 7, 9	7, 8
Compound	$M_G \times 100$				$M_R \times 100$		$M_S \times 100$	
Galactitol								
2-*O*-methyl-	—	—	170	50[a]	—	—	—	—
3-*O*-methyl-	—	—	140	100[a]	—	—	—	—
6-*O*-methyl-	—	—	160	80[a]	—	—	—	—
D-Mannitol								
2-*O*-methyl-	—	—	—	—	—	—	100	98
1,2-di-*O*-methyl-	—	—	—	—	—	—	100	95
(+)-Inositol								
3-*O*-methyl-	66	59	100	—	35	2	0	—
(−)-Inositol								
2-*O*-methyl-	—	24	—	—	27	14	—	—
myo-Inositol								
1-*O*-methyl-	—	11	—	—	20	26	—	—
5-*O*-methyl-	15	15	50	—	25	22	—	—

Paper	P1 =	Whatman® No. 3.
	P2 =	Whatman® No. 4.
	P3 =	Schleicher and Schuell 2043b.
	P4 =	Whatman® No. 1.
	P5 =	Whatman® No. 3MM.
Electrolyte	E1 =	0.2 *M* sodium borate, pH 10.
	E2 =	0.05 *M* borax, pH 9.2.
	E3 =	0.01 *M* borax, pH 9.2, containing 0.005 *M* CaCl₂.
	E4 =	0.05 *M* germanate buffer, pH 10.7 (50% aqueous NaOH added to suspension of germanium dioxide in water until clear solution obtained).
	E5 =	0.05 *M* sulfonated phenylboronic acid buffer, pH 6.5.
	E6 =	0.1 *M* sodium arsenite, pH 9.6 (19.8 g As₂O₃ in 1 *l* 0.13 *M* NaOH).
	E7 =	5.8% aqueous solution solution of basic lead acetate, pH 6.8.
	E8 =	2% aqueous solution of sodium molybdate, pH 5 (adjusted by addition of conc H₂SO₄).
	E9 =	2% aqueous solution of sodium tungstate, pH 5 (adjusted as E8).
Technique	T1 =	1.5 hr at applied potential 500 V (15 V/cm).
	T2 =	1.5 hr at 1300—1400 V(20 to 25 V/cm), with room temperature 20—25°C, cooling water 18—20°C.
	T3 =	applied current 0.5 mA/cm, further details na.
	T4 =	1.5 hr at 1500 V (25—30 V/cm), with circulating water at 40°C.
	T5 =	3—6 hr at 500 V (about 10 V/cm), at 40°C as in T4.
	T6 =	as T2; connection to electrodes through salt bridges advised, owing to possible evolution of arsine if platinum electrodes inserted directly into arsenite solution.
	T7 =	1—2 hr at 30—60 V/cm.

REFERENCES

1. **Foster, A. B.**, Zone electrophoresis of carbohydrates, *Ad. Carbohydr. Chem.*, 12, 81, 1957.
2. **Frahn, J. L. and Mills, J. A.**, Paper ionophoresis of carbohydrates, *Aust. J. Chem.*, 12, 65, 1959.
3. **Haug, A. and Larsen, B.** Separation of uronic acids by paper electrophoresis, *Acta Chem. Scand.*, 15, 1395, 1961.
4. **Lindberg, B. and Swan, B.**, Paper electrophoresis of carbohydrates in germanate buffer, *Acta Chem. Scand.*, 14, 1043, 1960.
5. **Garegg, P. J. and Lindberg, B.**, Paper electrophoresis of carbohydrates in sulfonated phenylboronic acid buffer, *Acta Chem. Scand.*, 15, 1913, 1961.

Table EL 3 (continued)
PAPER ELECTROPHORESIS OF DERIVATIVES OF SUGARS AND POLYOLS

6. Bourne, E. J., Hutson, D. H., and Weigel, H., Paper ionophoresis of sugars and other cyclic poly-hydroxy-compounds in molybdate solution, *J. Chem. Soc.*, 4252, 1960.
7. Weigel, H., Paper electrophoesis of carbohydrates, *Ad. Carbohydr. Chem.*, 18, 61, 1963.
8. Angus, H. J. F., Bourne, E. J., Searle, F., and Weigel, H., Complexes between polyhydroxy-compounds and inorganic oxy-acids. Tungstate complexes of sugars and other cyclic polyhydroxy-compounds, *Tetrahedron Lett.*, 55, 1964.
9. Bourne, E. J., Hutson, D. H., and Weigel, H., Complexes between molybdate and acyclic polyhydroxy-compounds, *J. Chem. Soc.*, 35, 1961.

ELECTROPHORESIS ON GLASS-FIBER STRIPS

Table EL 4
Electrophoresis of Sugars and Polyols on Glass-Fiber Strips

Paper   Whatman® GF[a], 38 × 10 cm
Electrolyte 0.2 *M* sodium borate, pH 10
Technique   1.5 hr at 15 V/cm
Reference   1

Compound	$M_G{}^b$ × 100
L-Arabinose	94
D-Xylose	100
D-Ribose	74
D-Glucose	100
D-Galactose	91
D-Mannose	67
L-Rhamnose	49
L-Fucose	88
2-Deoxy-D-ribose	28
D-Fructose	88
L-Sorbose	95
L-*galacto*-Heptulose	86
D-*manno*-Heptulose	83
Sophorose	28
Nigerose	65
Maltose	30
Cellobiose	26
Gentiobiose	69
Sucrose	15
Raffinose	23
Melezitose	21
D-Glucuronic acid	123
D-Glucose-1-phosphate	110
Erythritol	70
D-Arabinitol	86
D-Glucitol	82
Galactitol	95
D-Mannitol	85
D-*glycero*-D-*galacto*-Heptitol	95
D-*glycero*-D-*manno*-Heptitol	87
myo-Inositol	48

[a]   Type not specified.
[b]   Mobility relative to D-glucose.

Table EL 4 (continued)
Electrophoresis of Sugars and Polyols on Glass-Fiber Strips

Strip	P1 =	Millipore cellulose acetate membrane.
	P2 =	cellulose acetate strips from Membranfiltergesellschaft (Gottingen).
	P3 =	Separax or Oxoid strips.
	P4 =	Sepraphore III (Gelman).
Electrolyte	E1 =	0.1 M pyridine-0.566 M formic acid buffer, pH 3.0.
	E2 =	0.1 M barium acetate, pH 8.
	E3 =	0.3 M calcium acetate, pH 7.25.
	E4 =	0.1 M cupric acetate, pH 5.3.
	E5 =	0.1 M zinc acetate, pH 6.3.
	E6 =	0.1 M potassium phosphate buffer, pH 6.7.
	E7 =	0.1 M pyridine-0.47 M formic acid, pH 3.
Detection	D1 =	Alcian blue, 1% in 5% acetic acid; membrane stained for 2 min, then washed with 5% acetic acid.
	D2 =	Alcian blue, 0.05% in 95% ethanol saturated with NaCl; strips stained for 15 min, then washed for 30 min in ethanol solution.
	D3 =	Alcian blue or toluidine blue, 0.5% in 3% acetic acid; strips stained for 20 min, then washed with 1% acetic acid for 1 min, tap water for 10 min.
	D4 =	Alcian blue, 0.1% in 0.1% acetic acid; strips stained for 10 min, then washed with 0.1% acetic acid for 20 min.
Technique	T1 =	13 min at 10 V/cm.
	T2 =	3 hr at 5 V/cm.
	T3 =	3 hr at 1.0 mA/cm.
	T4 =	4.5 hr at 1 mA/cm; fresh solution for each run.
	T5 =	1.5—2 hr at 0.5 mA/cm; fresh solution for each run.
	T6 =	1.5 hr at 1 mA/cm; fresh solution for each run.
	T7 =	1 hr at 1 mA/cm; solution can be used repeatedly.
	T8 =	80—100 min at 1 mA/cm; solution can be used repeatedly.
	T9 =	two-dimensional electrophoresis on 10 × 10 cm strip; after run with E7 for 100 min at 1 mA/cm run in orthogonal direction with E2, conditions as T4.

POLYSACCHARIDES

Glass-fiber electrophoresis has been applied to polysaccharides,[2] especially pectic substances,[3] but with limited success owing to high rates of endosmotic flow. This can be obviated by trimethylsilylation of the glass-fiber support.[4] Whatman® GF/C strips, after removal of organic material by heating at 400°C for 2 hr, are immersed for 18 hr in a CCl₄ solution of dimethyldichlorosilane (2%), then rinsed in toluene and dried. It is necessary to add a surfactant to the electrolyte solution in order to wet trimethyl-silylated glass-fiber strips prior to use in electrophoresis. Glass-fiber strips thus modi-fied have been successfully applied to analytical electrophoresis of a variety of polysac-charides derived from plant cell walls, e.g., de-esterified pectins from potato tubers are separated in 10 min by electrophoresis, at potential gradient 13 V/cm, in 0.1 M acetate buffer (pH 5.0) containing 0.02 M EDTA.[4]

REFERENCES

1. Bourne, E. J., Foster, A. B., and Grant, P. M., Ionophoresis of carbohydrates. IV. Separations of carbohydrates on fibre-glass sheets, *J. Chem. Soc.*, 4311, 1956.
2. Northcote, D. H., Zone electrophoresis: qualitative, quantitative, and preparative electrophoretic separations of neutral polysaccharides, in *Methods in Carbohydrate Chemistry*, Vol. 5, Whistler, R. L., BeMiller, J. N., and Wolfrom M. L., Eds., Academic Press, New York, 1965, 49.
3. Haaland, E., Water-soluble polysaccharides from the leaves of *Tussilago farfara* L., *Acta Chem. Scand.*, 23, 2546, 1969.
4. Jarvis, M. C., Threlfall, D. R., and Friend, J., Separation of macromolecular components of plant cell walls: electrophoretic methods, *Phytochemistry*, 16, 849, 1977.

ELECTROPHORESIS ON CELLULOSE ACETATE STRIPS

Table EL 5
CELLULOSE ACETATE ELECTROPHORESIS OF GLYCOSAMINOGLYCANS

Strip	P1	P2	P3	P4	P4	P4	P4	P4	P4
Electrolyte	E1	E2	E3	E2	E4	E5	E6	E7	E2
Detection	D1	D2	D3	D4	D4	D4	D4	D4	D4
Technique	T1	T2	T3	T4	T5	T6	T7	T8	T9
Reference	1	2	3	4	4	4	4	4	4
Compound	$R_{chs}{}^a \times 100$							1*	2*
Chondroitin 6-sulfate	100	109	109	108	91	103	100	100	107
Chondroitin 4-sulfate	100	100	100	100	100	100	100	100	100
Dermatan sulfate	72	68	77	84	85	86	100	91	84
Keratan sulfate	66	—	—	89	109[b]	96	77[b]	85	92[b]
Hyaluronic acid	48	64	—	80	39	70	68	63	80
Heparin	129	44	—	66	28[b]	91[b]	110	110	66
Heparitin sulfate (heparin monosulfate)	90	44	—	66	64[b]	78	86	87	66

Note: 1* = two-dimensional electrophoresis, first run.

2* = two-dimensional electrophoresis, second run, orthogonal to first.

[a] R_{chs} = mobility relative to chondroitin 4-sulfate.

[b] Broad spot.

CELLULOSE ACETATE ELECTROPHORESIS: OTHER POLYSACCHARIDES

Choy and Dutton[5] have successfully applied this technique to the acidic capsular polysaccharides from bacteria of the genus *Klebsiella;* the pyridine-formic acid buffer (pH 3) used in electrophoresis of glycosaminoglycans (see Table EL 5) and also a veronal-Tris buffer (pH 8.8) have been used in this work. The distances of migration of these polysaccharides on electrophoresis for 30 min at 300 V were found to be approximately in inverse proportion to their equivalent weights. Alcian blue is a suitable staining reagent for these polysaccharides on cellulose acetate strips.

Neutral polysaccharides may be examined by electrophoresis on cellulose acetate strips in a borate buffer; Dudman and Bishop[6] advocate the use of 0.1 *M* sodium tetraborate-0.1 *M* sodium chloride, pH 9.3. These authors have reported electrophoretic studies of a number of such polysaccharides, which were dyed with Procion blue 3GS or Procion red 2BS before application to the strips, to facilitate observation of their behavior. Optimal conditions were 3 to 5 min electrophoresis at 25 to 35 V/cm, with the dyed polysaccharide in a 1% aqueous solution applied as a thin streak.

REFERENCES

1. Herd, J. K., Identification of acid mucopolysaccharides by microelectrophoresis, *Anal. Biochem.*, 23, 117, 1968.
2. Wessler, E., Analytical and preparative separation of acidic glycosaminoglycans by electrophoresis in barium acetate, *Anal. Biochem.*, 26, 439, 1968.
3. Seno, N., Anno, K., Kondo, K., Nagase, S., and Saito, S., Improved method for electrophoretic separation and rapid quantitation of isomeric chondroitin sulfates on cellulose acetate strips, *Anal. Biochem.*, 37, 197, 1970.
4. Hata, R. and Nagai, Y., A rapid and micro method for separation of acidic glycosaminoglycans by two-dimensional electrophoresis, *Anal. Biochem.*, 45, 462, 1972.
5. Choy, Y. M. and Dutton, G. G. S., The structure of the capsular polysaccharide from *Klebsiella* K-type 21, *Can. J. Chem.*, 51, 198, 1973.
6. Dudman, W. F. and Bishop, C. T., Electrophoresis of dyed polysaccharides on cellulose acetate, *Can. J. Chem.*, 46, 3079, 1968.

GEL ELECTROPHORESIS

Table EL 6
GEL ELECTROPHORESIS OF GLYCOSAMINOGLYCANS

	G1	G2	G2	G2
Gel	G1	G2	G2	G2
Electrolyte	E1	E2	E3	E4
Detection	D1	D2	D2	D2
Technique	T1	T2	T2	T2
Reference	1 2	3	3	3
Compound	$R_{chs}{}^a \times 100$			
Chondroitin 6-sulfate	56[b]	100	97	96
Chondroitin 4-sulfate	100[b]	100	100	100
Dermatan sulfate	73[b]	100	87	91
Keratan sulfate	—	69	89	0
Hyaluronic acid	0[b]	52	83	—
Heparin	110[b]	115	76	0
Heparitin sulfate	—	77	82	81

[a] R_{chs} = mobility relative to chondroitin 4-sulfate.

[b] Molecular-weight averages, from viscosity measurements, of samples used were: chondroitin 6-sulfate 40,000, 4-sulfate 12,000; dermatan sulfate 27,000; hyaluronic acid 230,000; heparin 11,000.

Table EL 6 (continued)
GEL ELECTROPHORESIS OF GLYCOSAMINOGLYCANS

Gel	G1 =	Slab (18 × 10 × 0.5 cm) of 6% polyacrylamide gel, prepared by dissolving 6 g Cyanogum-41 (American Cyanamide Company) and 0.2 g ammonium persulfate in 100 ml buffer, filtering, then adding 1.0 ml dimethylaminopropionitrile.
	G2 =	Slide (5.0 × 7.5 × 0.1 cm) of 0.9% agarose gel.
Electrolyte	E1 =	0.1 M Na$_3$PO$_4$ — 0.1 M Na$_2$HPO$_4$, pH 11.5, containing HCOONa (0.125 M in troughs, 0.250 M in gel)
	E2 =	0.06 M barbital buffer, pH 8.6.
	E3 =	0.05 M 1,3-diaminopropane; acetic acid added to bring pH to 8.5.
	E4 =	0.025—0.05 M 1,10-diaminodecane (maintained at 60°C); acetic acid added to bring pH to 10.3.
Detection	D1 =	Alcian blue, 0.5% in 3% acetic acid, to stain gel; destaining with 3% acetic acid.
	D2 =	staining with toluidine blue, 0.1% in acetic acid-ethanol-water (0.1:5:5); solvent used in destaining.
Technique	T1 =	samples, dissolved in buffer (about 10 μl), applied on strips (1 × 0.3 cm) of Whatman® No. 3 paper inserted in gel; electrophoresis for 7 hr at 100 V, horizontally; gel water-cooled.
	T2 =	samples (1—10 μg) dissolved in buffer (3—5 μl) applied to gel as in T1; electrophoresis at 5°C at 150 V until dye indicator (cresol red) had migrated 3 cm (with E2) or 4.5 cm (with E3 or E4); after electrophoresis gel immersed in 0.1% Cetavlon solution for 3 hr, then covered with strip of Whatman® 3MM paper wetted with Cetavlon solution and placed under 250 W infrared lamp for 2 hr, in current of air, prior to staining

GEL ELECTROPHORESIS:
OTHER POLYSACCHARIDES

Pavlenko and Ovodov[4,5] have examined a number of polysaccharides, both neutral and acidic, by disc electrophoresis in polyacrylamide gel with the following buffers.

1. Boric acid (0.6 g) + EDTA sodium salt (1 g) + Tris (10 g) in 1 *l* water (pH 9.2)
2. Sodium tetraborate (2 g) + EDTA sodium salt (1 g) + Tris (10 g) in 1 *l* water (pH 9.3)

The gel (4.6% acrylamide) was in silica tubes (5.5 × 0.7 cm). The sample solution (20 to 30 μl, containing 15 to 200 μg polysaccharide) was mixed with 15% Cyanogum-41 (0.7 m*l*) and buffer (0.3 m*l*) and applied as a layer on top of the gel. Electrophoresis for 20 to 25 min at 200 V (1 mA per tube), then for 2.5 to 3 hr at 400 V (7 mA per tube), gave optimal results. Staining[4] with ethanolic HI for 1 hr followed by Schiff's reagent (0.5 to 1 hr) or iodine solution for amylopectin was insufficiently sensitive, and therefore, dyeing polysaccharides (cyanuric chloride Amido Black 10B was recommended) before application to the gel is considered a superior method of detection.[5]

REFERENCES

1. Hilborn, J. C. and Anastassiadis, P. A., Acrylamide gel electrophoresis of acidic mucopolysaccharides, *Anal. Biochem.*, 31, 51, 1969.
2. Hilborn, J. C. and Anastassiadis, P. A., Estimation of the molecular weights of acidic mucopolysaccharides by polyacrylamide gel electrophoresis, *Anal. Biochem.*, 39, 88, 1971.
3. Dietrich, C. P., McDuffie, N. M., and Sampaio, L. O., Identification of acidic mucopolysaccharides by agarose gel electrophoresis, *J. Chromatogr.*, 130, 299, 1977.
4. Pavlenko, A. F. and Ovodov, Yu. S., Disc electrophoresis in polyacrylamide gel of neutral and uronic acid-containing polysaccharides, *Izv. Akad. Nauk SSSR Ser. Khim.*, 10, 2315, 1969; *Anal. Abstr.*, 19, 4170, 1970.
5. Pavlenko, A. F. and Ovodov, Yu. S., Polyacrylamide gel electrophoresis and gel filtration of dyed polysaccharides, *J. Chromatogr.*, 52, 165, 1970.

Section II

Detection Techniques

Section II

Section II.I

DETECTION AND IDENTIFICATION OF CARBOHYDRATES IN GAS CHROMATOGRAPHY

GC DETECTORS

The flame ionization detector (FID) is now universally employed in gas chromatography of carbohydrates, as for other organic compounds. If strongly electronegative atoms are introduced on derivatization, as is the case with trifluoroacetylation, the more sensitive and selective electron capture detector can be used. Detection of carbohydrates at picogram level by this method has been reported (see Table GC 12, Reference 3; Table GC 14, References 2 and 3).

Compounds containing nitrogen, such as aminodeoxy sugars and neuraminic acid derivatives, can be more selectively detected in the presence of other sugars by use of a thermionic nitrogen-phosphorus selective detector (NPSD). For optimal selectivity, the correct choice of operating conditions is critical. Under appropriate conditions (see Table GC 25, Reference 5) a selectivity factor of 100 has been claimed, with the sensitivity of detection of aminodeoxy sugars by the NPSD enhanced about fivefold in comparison with that possible with the FID. The NPSD will also be applicable to sugars containing phosphorus.

IDENTIFICATION BY MASS SPECTROMETRY

The use of mass spectrometry in the identification of substances separated by gas chromatography is now a well-established technique, which is being applied to an increasing extent in the carbohydrate field. Of the various carbohydrate derivatives prepared for gas chromatography, most have been studied by mass spectrometry. Those giving the simplest chromatograms, without multiple peaks due to the formation of anomeric mixtures, are the best choice for analysis of complex mixtures (e.g., of partially methylated sugars), since the mass spectra of such derivatives are also simple. The alditol acetates have been extensively used in GC-MS analyses, though the formation of the same alditol by two different sugars in some cases is a disadvantage, not shared by the aldononitrile acetates, which also give simple mass spectra. The ambiguity introduced on reduction of methylated sugars to alditols can, however, be obviated by deuterium labeling if sodium borodeuteride is used as the reducing agent.

For oligosaccharides, the permethylated or trimethylsilylated derivatives of the reduced oligosaccharides are the best choice for analysis by GC-MS.

Electron-impact (EI) mass spectrometry has to date been the technique used almost exclusively in GC-MS of carbohydrates. However, it has recently been reported that some partially methylated alditol acetates that cannot be differentiated by EI mass spectrometry may be distinguished on the basis of their chemical-ionization (CI) mass spectra (see Reference 9); a similar conclusion has been reached after a study of the per(trimethylsilyl) ethers of the common hexoses by the latter method (Reference 37). It is probable, therefore, that CI mass spectrometry will assume increasing importance in this field in the future.

A detailed treatment of the mass spectrometry of carbohydrates is obviously beyond the scope of this book. Several of the papers cited in connection with the GC data presented in Section I.I contain MS data also. These are listed below, together with some excellent review articles and papers dealing with MS studies of specific types of derivatives, to which the reader is referred for comprehensive information on this subject.

REFERENCES

Review Articles

1. **Hanessian, S.,** Mass spectrometry in the determination of structure of certain natural products containing sugars, in *Methods of Biochemical Analysis*, Vol. 19, Glick, D., Ed., John Wiley & Sons, New York, 1971, 105.
2. **Kochetkov, N. K. and Chizhov, O. S.,** Mass spectrometry of carbohydrates, in *Methods in Carbohydrate Chemistry*, Vol. 6, Whistler, R. L. and BeMiller, J. N., Eds., Academic Press, New York, 1972, 540.
3. **Lönngren, J. and Svensson, S.,** Mass spectrometry in structural analysis of natural carbohydrates, *Adv. Carbohydr. Chem. Biochem.*, 29, 41, 1974.
4. **Dutton, G. G. S.,** Applications of gas-liquid chromatography to carbohydrates. II, *Adv. Carbohydr. Chem. Biochem.*, 30, 9, 1974.

Selected Papers
Alditol Acetates Derived from Partially Methylated Sugars

5. **Björndal, H., Hellerqvist, C. G., Lindberg, B., and Svensson, S.,** Gas-liquid chromatography and mass spectrometry in methylation analysis of polysaccharides, *Angew. Chem. Int. Ed. Engl.*, 9, 610, 1970.
6. **Björndal, H., Lindberg, B., Pilotti, A., and Svensson, S.,** Mass spectra of partially methylated alditol acetates. II. Deuterium labelling experiments, *Carbohydr. Res.*, 15, 339, 1970.
7. **Borén, H. B., Garegg, P. J., Lindberg, B., and Svensson, S.,** Mass spectra of partially methylated alditol acetates. III. Labelling experiments in the mass spectrometry of partially methylated deoxy-alditol acetates, *Acta Chem. Scand.*, 25, 3299, 1971.
8. **Choy, Y.-M., Dutton, G. G. S., Gibney, K. B., Kabir, S., and Whyte, J. N. C.,** Comments on gas-liquid chromatography and mass spectrometry of methylated alditol acetates, *J. Chromatogr.*, 72, 13, 1972.
9. **McNeil, M. and Albersheim, P.,** Chemical-ionization mass spectrometry of methylated hexitol acetates, *Carbohydr. Res.*, 56, 239, 1977.

Aldononitrile Acetates

10. **Dmitriev, B. A., Backinowsky, L. V., Chizhov, O. S., Zolotarev, B. M., and Kochetkov, N. K.,** Gas-liquid chromatography and mass spectrometry of aldononitrile acetates and partially methylated aldononitrile acetates, *Carbohydr. Res.*, 19, 432, 1971.
11. **Szafranek, J., Pfaffenberger, C. D., and Horning E. C.,** The mass spectra of some per-*O*-acetylaldononitriles, *Carbohydr. Res.*, 38, 97, 1974.
12. **Seymour, F. R., Plattner, R. D., and Slodki, M. E.,** Gas-liquid chromatography-mass spectrometry of methylated and deuteriomethylated per-*O*-acetyl-aldononitriles from D-mannose, *Carbohydr. Res.*, 44, 181, 1975.
13. **Seymour, F. R., Slodki, M. E., Plattner, R. D., and Jeanes, A.,** Six unusual dextrans: methylation structural analysis by combined g.l.c.-m.s. of per-*O*-acetylaldononitriles, *Carbohydr. Res.*, 53, 153, 1977.

Boronates

14. **Reinhold, V. N., Wirtz-Peitz, F., and Biemann, K.,** Synthesis, gas-liquid chromatography and mass spectrometry of per-*O*-trimethylsilyl carbohydrate boronates, *Carbohydr. Res.*, 37, 203, 1974.

Isopropylidene Derivatives

15. **Morgenlie, S.,** Analysis of mixtures of the common aldoses by gas chromatography-mass spectrometry of their *O*-isopropylidene derivatives, *Carbohydr. Res.*, 41, 285, 1975.

Methyl Glycosides of Partially Methylated Sugars

16. **Heyns, K., Kiessling, G., and Müller, D.,** The mass spectrometric determination of linkages in polysaccharides containing amino sugars. Mass spectra of derivatives of 2-amino-2-deoxy-D-galactose, *Carbohydr. Res.*, 4, 452, 1967.
17. **Heyns, K., Sperling, K. R., and Grützmacher, H. F.,** Combination of gas chromatography and mass spectrometry for analysis of partially methylated sugar derivatives. The mass spectra of partially methylated methyl glucosides, *Carbohydr. Res.*, 9, 79, 1969.
18. **Kováčik, V., Bauer, S., Rosik, J., and Kováč, P.,** Mass spectrometry of uronic acid derivatives. III. The fragmentation of methyl ester methyl glycosides of methylated uronic and aldobiouronic acids, *Carbohydr. Res.*, 8, 282, 1968.

19. Kováčik, V., Bauer, S., and Rosik, J., Mass spectrometry of uronic acid derivatives. IV. The fragmentation of methyl ester methyl glycosides of methylated aldotriouronic acids, *Carbohydr. Res.*, 8, 291, 1968.

20. Kováčik, V. and Kováč, P., Identification of methyl *O*-methyl-D-xylofuranosides by mass spectrometry, *Carbohydr. Res.*, 24, 23, 1972.

21. Kováčik, V., Mihalov, V., and Kováč, P., Identification of methyl (methyl *O*-acetyl-*O*-methylhexopyranosid)uronates by mass spectrometry, *Carbohydr. Res.*, 54, 23, 1977.

22. Rauvala, H. and Kärkkäinen, J., Methylation analysis of neuraminic acids by gas chromatography-mass spectrometry, *Carbohydr. Res.*, 56, 1, 1977.

Permethylated Derivatives of Reduced Oligosaccharides

23. Kärkkäinen, J., Analysis of disaccharides as permethylated disaccharide alditols by gas-liquid chromatography-mass spectrometry, *Carbohydr. Res.*, 14, 27, 1970.

24. Kärkkäinen, J., Structural analysis of trisaccharides as permethylated trisaccharide alditols by gas-liquid chromatography-mass spectrometry, *Carbohydr. Res.*, 17, 11, 1971.

25. Mononen, I., Finne, J., and Kärkkäinen, J., Analysis of permethylated hexopyranosyl-2-acetamido-2-deoxyhexitols by g.l.c.-m.s., *Carbohydr. Res.*, 60, 371, 1978.

26. Rauvala, H. and Kärkkäinen, J., Methylation analysis of neuraminic acids by gas chromatography-mass spectrometry, *Carbohydr. Res.*, 56, 1, 1977.

Trifluoroacetylated Alditols

27. Chizhov, O. S., Dmitriev, B. A., Zolotarev, B. M., Chernyak, A. Y., and Kochetkov, N. K., Mass spectra of alditol trifluoroacetates, *Org. Mass Spectrom.*, 2, 947, 1969.

Trimethylsilyl Ethers

28. Chizhov, O. S., Molodtsov, N. V., and Kochetkov, N. K., Mass spectrometry of trimethylsilyl ethers of carbohydrates, *Carbohydr. Res.*, 4, 273, 1967.

29. De Jongh, D. C., Radford, T., Hribar, J. D., Hanessian, S., Bieber, M., Dawson, G., and Sweeley, C. C., Analysis of trimethylsilyl derivatives of carbohydrates by gas chromatography and mass spectrometry, *J. Am. Chem. Soc.*, 91, 1728, 1969.

30. Haverkamp, J., Kamerling, J. P., Vliegenthart, J. F. G., Veh, R. W., and Schaur, R., Methylation analysis determination of acylneuraminic acid residue type 2 → 8 glycosidic linkage, *FEBS Lett.*, 73, 215, 1977.

31. Kamerling, J. P., Vliegenthart, J. F. G., Vink, J. and de Ridder, J. J., Mass spectrometry of pertrimethylsilylated aldosyl oligosaccharides, *Tetrahedron*, 27, 4275, 1971.

32. Kamerling, J. P., Vliegenthart, J. F. G., Vink, J., and de Ridder, J. J., Mass spectrometry of 2-acetamido-2-deoxyglycose containing oligosaccharides, *Tetrahedron*, 27, 4749, 1971.

33. Kamerling, J. P., Vliegenthart, J. F. G., Vink, J., and de Ridder, J. J., Mass spectrometry of pertrimethylsilylated oligosaccharides containing fructose units, *Tetrahedron*, 28, 4375, 1972.

34. Kärkkäinen, J., Determination of the structures of disaccharides as *O*-trimethylsilyl derivatives of disaccharide alditols by gas-liquid chromatography-mass spectrometry, *Carbohydr. Res.*, 11, 247, 1969.

35. Kärkkäinen, J. and Vihko, R., Characterization of 2-amino-2-deoxy-D-glucose, 2-amino-2-deoxy-D-galactose, and related compounds, as their trimethylsilyl derivatives by gas-liquid chromatography-mass spectrometry, *Carbohydr. Res.*, 10, 113, 1969.

36. Kochetkov, N. K., Chizhov, O. S., and Molodtsov, N. V., Mass spectrometry of oligosaccharides, *Tetrahedron*, 24, 5587, 1968.

37. Laine, R. A. and Sweeley, C. C., *O*-Methyl oximes of sugars. Analysis as *O*-trimethylsilyl derivatives by gas-liquid chromatography and mass spectrometry, *Carbohydr. Res.*, 27, 199, 1973.

38. Murata, T. and Takahashi, S., Characterization of *O*-trimethylsilyl derivatives of D-glucose, D-galactose and D-mannose by gas-liquid chromatography chemical-ionization mass spectrometry with ammonia as reagent gas, *Carbohydr. Res.*, 62, 1, 1978.

39. Petersson, G., Mass spectrometry of alditols as trimethylsilyl derivatives, *Tetrahedron*, 25, 4437, 1969.

40. Petersson, G., Gas-chromatographic analysis of sugars and related hydroxy acids as acyclic oxime and ester trimethylsilyl derivatives, *Carbohydr. Res.*, 33, 47, 1974.

41. Petersson, G., Samuelson, O., Anjou, K., and von Sydow, E., Mass spectrometric identification of aldonolactones as TMS ethers, *Acta Chem. Scand.*, 21, 1251, 1967.

19. Kovacik, V., Bauer, S., and Rosík, J., Mass spectrometry of uronic acid derivatives. IV. The fragmentation of methyl ester methyl glycosides of permethylated aldobiouronic acids, *Carbohyd. Res.*, 8,
291, 1968.

20. Kovacik, V., et al., ... , ... , *Carbohyd. Res.*, ... , ... , 197 ...

21. Kocourek, V., Mikhael, ... , ... , ... , *Carbohyd. Res.*, ... , ... , ...

22. Petuely, F. and Lindner, ... , ... , ... , *Monatsh.*, ... , ... , 1971.

Section II.II

DETECTION METHODS FOR LIQUID CHROMATOGRAPHY

HPLC DETECTORS

The detector most used in HPLC of carbohydrates to date has been the differential refractometer, though the introduction of UV detectors capable of operating at 190 nm makes possible the application of this method to HPLC of sugars.[1] For monosaccharides, the sensitivity of this UV detector is approximately 12 times greater than that of most RI detectors (detection limit approximately 20 μg sugar[2]), but for oligosaccharides sensitivity decreases with increasing DP of the sugar. Derivatization of sugars (benzoylation[3] or 4-nitrobenzoylation[4]) permits the use of UV detectors operating at 260 nm; the resulting increase in sensitivity makes possible detection of carbohydrates in the nanogram range.[4]

DETECTION SYSTEMS FOR OTHER COLUMN METHODS

RI and UV Methods

The differential refractometer is a suitable detector for liquid chromatography systems not involving eluents containing salts in high concentration. Its use in anion-exchange chromatography of sugars in borate medium has been reported,[5] but sensitivity is low, especially for the monosaccharides, detector response being highly dependent upon the concentration of inorganic ions in the eluate. A similar conclusion was reached following an attempt to apply UV detection (approximately 200 nm) to the anion-exchange separation of D-glucose and D-fructose;[6] the sensitivity of UV detection was found to be approximately 100 times lower than that of the orcinol-sulfuric acid colorimetric method (see below). For D-glucitol, D-mannitol, and gluconic and glucaric acids, the response of the UV detector is higher, and, therefore, its use in anion-exchange chromatography of compounds of these types seems feasible.[6] A UV photometer operating at 254 and 306 nm has been used as a detector in chromatography of a mixture of 16 standard sugars (0.05 μmol of each) in a system in which the solution emerging from the anion-exchange column (flow rate 11 mℓ/hr) was dynamically mixed with concentrated sulfuric acid (flow rate 19 mℓ/hr; reactor temperature 100°C), so that the sugars were converted to compounds (furfural derivatives) having UV-absorbing chromophores.[7]

Another variant is the incorporation in an automated anion-exchange chromatography system of a cerate oxidative detector, which is based upon the reduction of Ce[4+] to the fluorescent Ce[3+] by sugars (or other reducing agents) emerging from the column.[8,9] The reagent, containing Ce[4+] and sulfuric acid in concentrations of 5×10^{-4} and 1.5 mol/ℓ, respectively, with sodium bismuthate (20 mg/ℓ) added for stabilization, is dynamically mixed (flow rate 5 mℓ/hr) with the solution eluted from the column (flow rate 10 mℓ/hr)[9] and, after passing through a heating coil (100°C), the mixture flows through a fluorometer (260 nm excitation, 350 nm emission). This detector, which is compatible with the borate buffers used in anion-exchange chromatography of sugars provided that chloride ion is absent,[9] is approximately 10 times more sensitive than that using concentrated sulfuric acid,[7] described above, or the phenol-sulfuric acid colorimetric method (see below).

Colorimetric Methods

Several of the standard colorimetric methods for the quantitative determination of carbohydrates[10-12] are adaptable to use in automated analytical systems, such as Technicon AutoAnalyzer®, in conjunction with the various column systems employed in liquid chromatography of carbohydrates (Section I.II) These are reviewed below. Those applicable to carbohydrates in general are discussed first, followed by analytical methods for specific classes of carbohydrates.

General Methods

Most of these involve the use of sulfuric acid to form furfural derivatives, which react with compounds such as phenols, aromatic amines, or heterocyclic hydrocarbons to form colored products.

Anthrone Method

Anthrone (9,10-dihydro-9-oxo-anthracene) reacts with most carbohydrates in concentrated sulfuric acid to produce a characteristic blue-green color. Aminodeoxy sugars do not give any color and *N*-acetylneuraminic acids give only a faint yellow-green. In manual analysis of carbohydrate solutions by this method, the recommended procedure[11,13] is to pipette anthrone reagent (10 mℓ, 0.2 g anthrone per 100 mℓ concentrated sulfuric acid) into test tubes immersed in a cold-water bath (10 to 15°C) and carefully overlay each anthrone solution with 5 mℓ of sample (20 to 40 μg/mℓ). The tubes are then shaken vigorously, while remaining immersed in the cold-water bath, until the contents are thoroughly mixed, after which they are heated in a boiling-water bath for 16 min and then cooled before the optical densities of the solutions are read at 625 nm.

The anthrone method as automated by Jenner[14] has been much used in analysis of dextran fractions from gel columns.[15] The sample plate from the fraction-collector connected to the column is transferred to the AutoAnalyzer® sampler module and each sample, pumped at 0.42 mℓ/min, is dynamically mixed with anthrone reagent (1.5 g anthrone dissolved in 750 mℓ concentrated sulfuric acid; water added to total volume 1000 mℓ), pumped at 1.19 mℓ/min, with air (1.20 mℓ/min) segmenting the stream. A pulse suppressor is necessary where the mixing occurs. After mixing, each sample passes through a heating coil at 95°C for 10 min, followed by a cooling coil and a debubbler, and the color is then measured at 625 nm in a colorimeter fitted with a 15-mm flowcell of 0.15-mℓ volume. This method is applicable to dextran solutions of concentration up to 100 μg/mℓ.

Chloride ions interfere in the anthrone method and therefore, in the analysis of fractions from gel columns eluted with sodium chloride solutions, standards used to calibrate the analytical system should contain chloride in the same concentration as the fractions from the column. Interference by substances used as preservatives for sugar solutions (e.g., phenyl mercuric acetate), carbonyl compounds (e.g., formaldehyde), and some solvents used in paper chromatography of sugars (e.g., 1-butanol) have also been reported.[13]

A further disadvantage of this method is the instability of the anthrone reagent, which requires standardization each day and deteriorates approximately 1 week after preparation.[13]

Cysteine-Sulfuric Acid Method

Most carbohydrates react with L-cysteine in the presence of sulfuric acid, with the production of a color that varies with the type of sugar involved. In the procedure recommended[10,12] for manual analysis, 4 mℓ concentrated sulfuric acid is added, with cooling in tap water, to 1 mℓ sample solution (containing 5 to 50 μg sugar); this addition is followed by vigorous shaking to ensure mixing. The reaction mixture is then left for 1 hr at room temperature, with occasional shaking, after which 0.1 mℓ of a 3% aqueous solution of L-cysteine hydrochloride monohydrate (reagent grade) is added with shaking. Spectrophotometric readings of the optical density may be taken 15 to 20 min after the addition of L-cysteine where the sugars to be determined are pentoses or hexuronic acids (yellow color, absorption maximum 390 nm), or hexoses (yellow, maximum 412 to 414 nm). Heptoses give initially an orange color (maximum approximately 430 nm), but this changes to purple (510 nm) after a few hours. 6-Deoxy-hexoses show after 15 min only a small absorption (maximum at 400 nm); this increases continuously during 24 hr at room temperature. 2-Deoxypentoses have no absorption for approximately 30 min after addition of L-cysteine, but after 24 hr a very sharp absorption maximum appears at 375 nm. It is thus possible, by careful selection of wavelength and time of analysis, to detect the presence of several different classes of carbohydrates in the same solution by the cysteine-sulfuric acid method.

Automation of this method[16,17] for use in conjunction with anion-exchange systems for chromatography of sugars has proved particularly successful. The reagent, a solution (0.07% w/v) of L-cysteine hydrochloride in sulfuric acid (86% v/v), is dynamically mixed with the solution emerging from the column in a ratio of approximately 5:1, and, after passing through a heating coil (97°C, residence time 3 min), cooling, and debubbling, the color of the solution is measured by a colorimeter fitted with 15-mm flow cells. The response at 395 nm is used for quantitation of the chromatogram given by a mixture of sugars. Hough and co-workers[17] measure the absorbance also at 425 nm, at which wavelength only hexoses respond significantly at 405 nm. From the ratio of the heights of the peaks at 405 nm to those at 395 nm (0.50 to 0.57 for pentoses, 0.69 to 0.76 for 6-deoxyhexoses) the different classes of monosaccharides can be distinguished.

Further advantages of the cysteine-sulfuric acid method are the stability of the reagent and its sensitivity; sugars in amounts below 100 nmol (approximately 18 μg) can be determined accurately by the automated procedure.[17] The use of sulfuric acid diluted to 86%, in a reagent mixed before use, eliminates the pulsing that presents problems in systems involving dynamic mixing of concentrated sulfuric acid with aqueous solutions. This method is highly recommended as a detection system for automated anion-exchange chromatography of sugars.

Orcinol-Sulfuric Acid Method

An automated procedure based on the color developed (absorbance maximum at 420 to 425 nm) upon reaction of orcinol with the products of sulfuric acid degradation of carbohydrates has been widely used as a detection method in anion-exchange chromatography of sugars,[18,19] partition chromatography on ion-exchange resins,[20] and gel-permeation chromatography with water as eluent.[19,21] The composition of the orcinol reagent as originally described by Kesler[18] was 1.0 g orcinol per liter 70% (v/v) sulfuric acid, but Kennedy and Fox[19] have successfully used a reagent consisting of orcinol (1.5 g/l) dissolved in concentrated sulfuric acid. For partition chromatography of sugars on ion-exchange resins with aqueous ethanol eluents, Samuelson[20] advocates a system in which an aqueous solution (1.6%) of orcinol is dynamically mixed with the column effluent and then with 60% (v/v) sulfuric acid. In Kennedy's carbohydrate analyzer,[19] the reagent is pumped at 24 ml/hr and the sample stream at 12 ml/hr, but proportions vary in the different systems according to the composition of the reagent. After mixing, the solution passes through a heating coil at 95°C (residence time 15 min). The absorbance is usually measured at 420 to 425 nm; Kennedy[19] measured that at 510 nm also, as hexuronic acids give a product having maximum absorption at 518 nm, and the relative absorbances at 425 and 510 nm of the chromophores produced by the 6-deoxy-hexoses are different from those of the other monosaccharides.

With a colorimeter fitted with 2-mm flow cells, as little as 0.1 μg of carbohydrate can be detected by the orcinol-sulfuric acid method.[19] Disadvantages of the method include instability of the orcinol reagent (which lasts only 2 to 3 weeks) and precipitation of purple material in the heating coil, resulting in base line "noise".

Phenol-Sulfuric Acid Method

The phenol-sulfuric acid method for the determination of carbohydrates, originally developed by Smith and co-workers,[22] is applicable to a wide variety of carbohydrates, including methyl ethers with free, or potentially free, reducing groups. A characteristic orange-yellow color is produced, the absorption maximum being at 490 nm for hexoses and their methylated derivatives, and 480 nm for pentoses, 6-deoxyhexoses, and hexuronic acids. Some methylated pentoses and their methyl glycosides (e.g., 2,3,5-tri-O-methyl-L-arabinose and its methyl glycoside) give products showing an absorption maximum at 415 nm, and such compounds should therefore be determined at that wavelength. However, in general, this colorimetric assay is carried out at 480 to 490 nm. Only aminodeoxy sugars do not react.

In the procedure recommended[11,22] for manual analysis by the phenol-sulfuric acid method, 1 mℓ of a 5% solution of phenol in water is added to 2 mℓ of sample solution (containing 10 to 80 μg carbohydrate) and, after mixing, 5 mℓ concentrated sulfuric acid is introduced from a fast-flowing pipette so that the stream hits the liquid surface directly, producing good mixing and heat distribution. The temperature required for color development is reached as a result of the exothermic reaction of sulfuric acid and water, and, therefore, no further heating is necessary. After the reaction mixture has stood for 30 min, the color, which remains stable for several hours, may be read at the wavelength appropriate to the class of compound.

Automation of the phenol-sulfuric acid colorimetric method has proved difficult, as the system must be very carefully designed to control the excessive liberation of heat and pulsing produced when the concentrated sulfuric acid is mixed with the aqueous solutions. Nevertheless, automated anion-exchange chromatography systems for sugars in which this detection method is employed have been reported.[23-25]

Although not readily adaptable to automation, the phenol-sulfuric acid method is highly recommended for manual analysis of carbohydrate solutions, because of its high accuracy and extreme simplicity. Amino acids and proteins do not interfere, and, therefore, this method is particularly suitable for the analysis of hydrolysates from glycoproteins and glycopeptides.[25] There is also no interference from solvents, such as 1-butanol, ethanol, or 2-butanone, used in paper chromatography of sugars and their methylated derivatives.[22] Some metallic cations, particularly Cu^{2+} and Fe^{3+}, cause deepening of the color intensity,[26] while the presence of borate decreases the color. The latter is a disadvantage insofar as the application of this method to analysis of sugars eluted with borate buffers from anion-exchange columns is concerned. There is little interference from chloride ion, however, when the analysis is performed manually in an open test tube, so that gas generated (a further problem in an automated system) is freely evolved. Therefore, the phenol-sulfuric acid method has proved useful in manual analysis of fractions eluted from gel columns with sodium chloride solutions.[27]

Noncorrosive Reagents

The presence of sulfuric acid in high concentration in the reagents generally used in detection systems for carbohydrate chromatography has the disadvantage that pump tubing deteriorates rapidly owing to the highly corrosive nature of this acid. To obviate this difficulty, Mopper and co-workers have developed automated analyzers for sugar chromatography in which noncorrosive reagents, depending on the reducing power of sugars are used.

Tetrazolium blue (*p*-anisyltetrazolium chloride) — In alkaline solution, this is reduced by sugars to diformazan, which has a strong absorbance at 520 nm. This reaction has been utilized by Mopper and Degens[28] in the automation of partition chromatography of sugars on an anion-exchange resin in the sulfate form, with aqueous ethanol (89% w/v) as eluent. The reagent, a solution of tetrazolium blue (2.0 g) in 0.18 M sodium hydroxide (1 l), is pumped at 12 ml/hr and dynamically mixed with the column eluate, pumped at 30 ml/hr; after passing through a heating coil (80°C; residence time 1 min), the reaction mixture flows through a 5-mm cell in a spectrophotometer, which measures the absorbance at 520 nm. Beer's Law holds for sugar concentrations below 50 μg/ml. The method is highly sensitive (detection limit is of the order of 0.1 nmol sugar). The one disadvantage is that its use is confined to systems in which ethanol is present in high proportion in the eluent; the ethanol acts as a solvent for the diformazan produced, which tends to precipitate and clog the pump tubing in aqueous media. For this reason the tetrazolium blue reagent cannot be employed in anion-exchange chromatography using borate buffers.

Copper-bicinchoninate — The colorimetric method subsequently developed by Mopper and Gindler[29] has proved more versatile. The method is based upon the formation of a deep lavender complex between Cu(I) and 2,2'-bicinchoninate, the Cu(I) being produced on reduction of Cu(II) by sugar. The reagent, first applied in automated partition chromatography of sugars on anion-exchange resin in aqueous ethanol,[29-30] is composed as follows:

Solution a	$CuSO_4 \cdot 5H_2O$ (1 g) and aspartic acid (3.7 g) dissolved in 1 l double-distilled water.
Solution b	Disodium 2,2'-bicinchoninate (2 g) and $Na_2CO_3 \cdot 10H_2O$ (38 g) dissolved in 1 l double-distilled water.
Color reagent	Mixture (1:1) of *a* and *b*, degassed before use.

The complex formed in the presence of reducing sugar shows maximum absorbance at 562 nm. The color is stable for at least a week, in contrast to that produced by the tetrazolium blue reagent, which remains stable for only approximately 10 min.

The presence of ethanol increases the color intensity of the Cu(I)-bicinchoninate complex, and, therefore, the reagent is particularly suitable for use in a detection system for partition chromatography of sugars in ethanol. In the automated system developed by Mopper,[30] the sugars are eluted from the anion-exchange column (sulfate form) with 86.7% (w/w) ethanol, flow rate 34 ml/hr, and the reagent is pumped at 18 ml/hr with pulse damping; after mixing, the solution passes through a heating coil at 100°C (5 min residence time), boiling being prevented by the insertion of a 2-atm back-pressure coil, and the absorbance is measured at 562 nm in a colorimeter fitted with a 2-cm flowcell (volume 0.12 ml). Under these conditions the detection limit for most sugars is below 0.5 nmol.

The use of the copper-bicinchoninate method is not confined to sugar chromatography systems involving elution with aqueous ethanol. Mopper has reported the successful application of this method of detection in the anion-exchange separation of uronic acids in acetate medium[31] and of neutral sugars in borate,[32] as well as the chromatography of aminodeoxy sugars on a cation-exchange resin with a citrate buffer as eluent,[33] all of these systems being fully automated. The salient features of the systems, where different from those described above, are summarized in the following table:

AUTOMATED CHROMATOGRAPHIC SYSTEMS WITH COPPER-BICINCHONINATE DETECTION SYSTEM

Compounds	Uronic acids[a]	Neutral sugars	Aminodeoxy sugars
Resin	DA-X8, acetate form	DA-X4, borate form	Locarte, Na form
Eluent	0.08 M sodium acetate, pH 8.5	0.5 $M H_3BO_3$, pH to 8.63 with NaOH	0.35 M sodium citrate, pH 6.72
Flow rates			
Eluent	30 ml/hr	24 ml/hr	36 ml/hr
Reagent	30 ml/hr	36 ml/hr	36 ml/hr
Heating	100°C, 5 min	124°C, 2 min	100°C, 4 min
Sensitivity	∼0.2 nmol	∼0.1 nmol	∼0.1 nmol
Reference	31	32	33

[a]   For detection of uronic acids, reagent is mixed (1:1) with 95% ethanol.

In addition to its versatility and sensitivity, the copper-bicinchoninate detection reagent has the advantage of stability. The mixed reagent is stable for at least a month;[30] in the presence of ethanol it can be used for several months.[31] This reagent is the best yet developed for use as a detector in liquid chromatography of reducing sugars.

Specific Methods
For Ketoses

The *resorcinol-hydrochloric acid* colorimetric method, specific for ketoses, has been automated[34] and is potentially useful in chromatographic analysis of commercial syrups containing D-fructose in the presence of D-glucose and other aldoses. The reagent composition is:

Solution a   Resorcinol (0.5 g) in concentrated hydrochloric acid (1 l) plus water (100 ml).

Solution b   1,1-Diethoxyethane (0.25 ml) in water (500 ml).

The sample stream (0.1 ml/min) is dynamically mixed with *a* (0.53 ml/min) and *b* (0.03 ml/min), with air segmentation (0.42 ml/min), and the reaction mixture is passed through a heating coil (95°C, residence time 3 min), a cooling coil, and a debubbler before the absorbance is measured at 550 nm. Interference by D-glucose and D-mannose present in amounts equimolar to the D-fructose is approximately 2% of the total absorbance.

Hydrochloric acid is the only reagent in an alternative procedure for automated analysis of ketoses in the presence of other sugars.[35] The chromophore produced on dehydration of D-fructose in hydrochloric acid has absorption maxima at 470 (major), 540 nm (minor), and a minimum at 415 nm. In the automated technique, sample solution (0.015 ml/min) is mixed with concentrated HCl (0.53 ml/min), air (3.27 ml/min), and water (0.81 ml/min), and the reaction flow stream is heated at 85°C for 3 min and then cooled and debubbled before the absorbance is measured at 415, 470, and 540 nm in a colorimeter fitted with 10-mm flowcells. The extent of interference from D-glucose present in amount equimolar to D-fructose is only 0.05% of total absorbance at 415 nm, 0.23% at 470 nm, and 0.09% at 540 nm.

For Uronic Acids

The *carbazole-sulfuric acid* method is widely used in the determination of hexuronic acids and polysaccharides (such as acidic glycosaminoglycans) containing uronic acid residues in high proportion. In the procedure originally proposed for manual analysis by this method,[10] 6 ml concentrated sulfuric acid is added to 1 ml sample solution (containing 5 to 100 μg hexuronic acid) with cooling in tap water, and the mixture is heated in a boiling-water bath for 20 min and then cooled to room temperature before addition of 0.2 ml of a freshly prepared solution of carbazole (0.1%) in 95% ethanol. After thorough mixing, the solution is allowed to stand at room temperature for 2 hr. The purple reaction product formed by hexuronic acids under these conditions is stable for about 1 hr after full color development. The absorption maximum is at 535 nm; absorbance is usually measured at 530 nm. Addition of 2 ml water to the reaction mixture after the color has developed fully results in the immediate disappearance of the chromophore due to hexuronic acid; in contrast, with hexoses and pentoses a purple color appears and rapidly increases in intensity. Under the conditions described above, hexoses normally produce a brown-red color (absorbance approximately 5 to 7% of that due to hexuronic acid in equimolar amount) and pentoses a yellow chromophore with a completely different absorption spectrum.

The extinction coefficients of the chromophores produced by different uronic acids in the carbazole reaction differ: those for D-galacturonic, D-mannuronic, L-iduronic, and L-guluronic acids are 120, 17, 29, and 32%, respectively, of that for D-glucuronic acid. Simple hexuronides and some glycosaminoglycans (e.g., dermatan sulfate, chondroitin sulfates) react as equivalent amounts of the hexuronic acids involved; heparin, in contrast, gives product showing an extinction coefficient 50% higher than that for an equivalent amount of D-glucuronic acid. The presence of protein in concentrations above 200 μg/ml depresses the color in the carbazole reaction, while sulfhydryl compounds increase the color intensity.

The introduction of borate ions into the reaction mixture reduces interference by proteins and salts and increases color yield; the time required for color development is also decreased. In this modified procedure,[36] the sample (1 ml, containing 0.02 to 0.2 μmol uronic acid) is treated with 5 ml of a solution of sodium tetraborate (0.025 M) in concentrated sulfuric acid. After thorough mixing, the solution is heated in a boiling-water bath for 10 min and then cooled to room temperature before addition of 0.2 ml of a 0.125% solution of carbazole in ethanol. The mixture is heated in a boiling-water bath for 15 min and cooled; the color is then fully developed and remains stable for 16 hr. Under these conditions L-iduronic acid and derivatives give approximately 80% of the color intensity given by D-glucuronic acid in equivalent amount.

The carbazole method is much used in manual analysis of uronic acids and polysaccharides such as glycosaminoglycans following elution from liquid chromatography columns. Automated procedures have been described.[37] The reagent used consists of carbazole (160 mg/l) dissolved in a 0.025 M solution of $Na_2B_4O_7 \cdot 10H_2O$ in concentrated H_2SO_4; the carbazole is added immediately prior to the analysis. With a sample flow rate of 0.60 ml/min, reagent is pumped at 0.92 ml/min if the sample is free of salts, 1.44 ml/min if salts are present. The reaction mixture is passed through a heating coil at 104°C before its absorbance is measured at 530 nm. The lowest concentration of uronic acid detectable by the automated carbazole method is approximately 0.5 μg/ml.

For Aminodeoxy Sugars

The *Elson-Morgan reaction* has been much used in manual analysis of fractions containing 2-amino-2-deoxyhexoses. The amino sugar is condensed with acetylacetone (pentane-2,4-dione) in alkaline solution and the product subsequently combined with N,N-dimethyl-*p*-aminobenzaldehyde in acid solution to form a colored compound (red; absorption maximum 530 nm). A recommended procedure[10] is as follows:

Solution a   Redistilled pentane-2,4-dione (1 ml) in 50 ml 0.5 *M* sodium carbonate.
Solution b   N,N-Dimethyl-*p*-aminobenzaldehyde (0.8 g) in 30 ml ethanol, to which
                 is added 30 ml concentrated HCl.

Both *a* and *b* should be freshly prepared. Solution *a* (1 ml) is added to sample solution (1 ml, containing 20 to 60 μg aminodeoxy sugar), and the mixture is diluted with water to 3 ml. The test tube, covered with a small condenser, is heated in a boiling-water bath for 20 min and then, after cooling to room temperature, 95% ethanol (5 ml) is added, followed by solution *b* (1 ml). The reaction mixture is diluted to 10 ml with ethanol and again heated, at 65 to 70°C, for 10 min. The solution is allowed to cool to room temperature before the absorbance at 530 nm is measured. The extinction coefficient of the chromophore produced by 2-amino-2-deoxy-D-galactose is approximately 90% of that given by 2-amino-2-deoxy-D-glucose.

Ninhydrin, the standard reagent for amino acids, reacts also with amino sugars, and therefore, amino acid analyzers employing this reagent may be used for automated chromatography of these sugars.[36,38] Borate ion, added to the citrate eluents to improve resolution,[38] decreases the color yield in the ninhydrin reaction to some extent, but the sugars are nevertheless detectable at levels of approximately 5 μg under these conditions.

For Alditols

Alditols are easily oxidized by periodate to formaldehyde, which can be determined by several colorimetric methods. In manual analysis by the *chromotropic acid* method,[39,40] the sample solution, containing alditol that will liberate 0.12 to 0.55 mg formaldehyde upon oxidation, is pipetted into a 100-ml volumetric flask, the volume is adjusted to 20 ml with distilled water, and the solution is acidified by addition of 5 *M* sulfuric acid (1 ml). To this solution 0.1 *M* sodium metaperiodate (5 ml) is added with mixing, and exactly 5 min later, excess periodate is destroyed with *M* sodium arsenite (5 ml). After 10 min the volume of the solution is adjusted to 100 ml with distilled water. Aliquots (1 ml) are pipetted into tubes and to each is added 10 ml chromotropic acid reagent [prepared by dissolving 1 g chromotropic acid in 100 ml water, filtering, and adding sulfuric acid solution (67% v/v) to the filtrate to give a final volume of 500 ml]. The tubes are then heated in a boiling-water bath for 30 min; the contents should not be exposed to direct light during this period or while cooling to room temperature. The absorbance of the purple chromophore produced is measured at 570 nm.

The necessity to exclude light in this reaction is a disadvantage insofar as possible automation is concerned. For this purpose, an alternative procedure, involving determination of the formaldehyde by measuring the absorbance (maximum at 412 nm) of the yellow compound (3,5-diacetyl-1,4-dihydro-2,6-lutidine) produced on reaction with *pentane-2,4-dione* in the presence of ammonia, is preferable. In the manual method,[39] the solution resulting from periodate oxidation of the alditol, followed by arsenite reduction of the excess periodate, is diluted with water so that the concentration of formaldehyde in the sample taken for analysis will not exceed 0.25 μmol/ml. A measured volume of this diluted solution is mixed in a test tube with an equal volume of the pentane-2,4-dione reagent (2 ml redistilled pentane-2,4-dione, 150 g ammonium acetate, and 3 ml glacial acetic acid dissolved in water, final volume 1 *l*), and the tube is heated at 60°C for 10 min. After cooling, the absorbance at 412 nm is measured.

The pentane-2,4-dione method just described, which is less subject to interference from other carbonyl compounds than is the chromotropic acid procedure, has been successfully automated for use as a detection system in liquid column chromatography of alditols.[16,20,41] For partition chromatography in aqueous ethanol, a higher reaction temperature (80°C) is necessary in order to accelerate color development, which is slower in the presence of ethanol.[41] In this automated procedure, the reagents are

Solution a   Sodium metaperiodate, 0.015 M, in 0.12 M HCl.
Solution b   Sodium arsenite, 0.5 M, pH adjusted to 7.0 with HCl.
Solution c   Pentane-2,4-dione, 0.02 M, in aqueous solution containing ammonium acetate, 2 M, and acetic acid, 0.05 M.

The column eluate, flow rate 0.4 m*l*/min, is mixed with periodate (*a*), pumped at 0.6 m*l*/min, with air segmentation (0.6 m*l*/min). After passage through a long mixing coil, the reaction stream is joined by the line containing the arsenite solution (*b*), flowing at 0.6 m*l*/min, and the mixture flows through another long mixing coil before the pentane-2,4-dione reagent (*c*) is introduced, at a flow rate of 1.2 m*l*/min. The stream passes through a short mixing coil and then a heating coil (80°C, residence time 10 min). After cooling and venting of the air bubbles, the solution passes through a colorimeter fitted with a 15-mm flow cell, and the absorbance is measured at 420 nm.

Under these conditions, the yield of formaldehyde from most alditols is high, whereas aldoses produce negligible, or very small, amounts. Interference from D-fructose, however, is appreciable.

This automated procedure for the detection of alditols, which has been applied also to anion-exchange chromatography in a borate-chloride gradient,[40] can be used in combination with automated orcinol[20] or cysteine-sulfuric acid[16] sugar analyses, in a second channel, to facilitate chromatography of the complex mixtures of sugars and alditols frequently encountered in structural studies of polysaccharides and glycoproteins. In anion-exchange chromatography, the color development in pentane-2,4-dione assay depends upon the concentration of salt in the eluate; provided that this is constant, the area of each peak has been found[42] to be proportional to the amount of alditol present over the range from 0.1 to 1 μmol.

For Polysaccharides

Colorimetric methods that involve heating the carbohydrate in concentrated acid are applicable to polysaccharides as well as to sugars, since degradation of the polysaccharides to sugars, and thence to furfural derivatives, occurs under these conditions. A colorimetric procedure that has been suggested by Anderson and co-workers[43] specifically for detecting polysaccharides in gel chromatography is the dyeing of the samples prior to chromatography. This obviates the necessity for the addition of color-forming reagent to the eluate and permits direct monitoring of the emergence of the colored substances from the column. The choice of Procion dyes for this purpose was based on the successful application of these dyestuffs in similar detection of polysaccharides in electrophoresis on cellulose acetate.[44] Procion dyes react appreciably only with molecules having a large number of primary hydroxyl groups, and, therefore, their use is confined to polysaccharides, reactivity decreasing with increasing uronic acid content. The dyestuff recommended by Anderson and co-workers[43] for coloring polysaccharides was Procion Brilliant Red M2B (absorption maximum at 530 nm). This technique was successfully applied by these workers to the detection of dextrans and a wide variety of plant-gum polysaccharides in gel chromatography, but it has not been generally adopted, as its sensitivity is much inferior to that of other detection methods.

Radioassay

The radiochemical methods that have been extensively used by Conrad[45,46] and Sanford[47] and co-workers in the detection of carbohydrates of various types on paper chromatogram are applicable also to liquid chromatography. For example, ^{35}S labeling has been used to follow gel chromatography of chondroitin 4-sulfate,[48] and ^{14}C labeling in the detection of 2-amino-2-deoxy-D-glucose and 2-amino-2-deoxy-D-galactose eluted with a phosphate buffer from a cation-exchange resin column.[49] The results obtained in these two radioassays were in good agreement with those given by simultaneous colorimetric analyses (carbazole-sulfuric acid and Elson-Morgan method, respectively). Reductive labeling, by reaction of the carbohydrates with NaB^{3}H$_4$ prior to chromatography, has also been applied in liquid chromatography, e.g., in the analysis of glycosaminoglycan fractions eluted from a gel column.[50] Radioactivity of the labeled material is measured by liquid scintillation counting. This detection method is sensitive (e.g., for the aminodeoxy sugars detection of amounts as low as 10 nmol has been reported,[49] and improved equipment has increased sensitivity about fivefold[47]) and versatile, having been applied to oligosaccharides,[45] glycosaminoglycans, glycolipids, N-acetylneuraminic acid,[46] and a wide variety of other carbohydrates.

REFERENCES

1. Hettinger, J. and Majors, R. E., Separation of carbohydrates by high performance liquid chromatography, *Varian Instrum. Appl.,* 10, 6, 1976.
2. Palmer, J. K., A versatile system for sugar analysis via liquid chromatography, *Anal. Lett.* 8, 215, 1975.
3. Lehrfeld, J., Separation of some perbenzoylated carbohydrates by high-performance liquid chromatography, *J. Chromatogr.,* 120, 141, 1976.
4. Nachtmann, F. and Budna, K. W., Sensitive determination of derivatized carbohydrates by high-performance liquid chromatography, *J. Chromatogr.,* 136, 279, 1977.
5. Liljamaa, J. J. and Hallén, A. A., Sugar borate chromatography using refractometry for monitoring, *J. Chromatogr.,* 57, 153, 1971.
6. Verhaar, L. A. T. and Dirkx, J. M. H., Ion-exchange chromatography of sugars, sugar alcohols and sugar acids using U.V. spectrometry for direct detection, *Carbohydr. Res.,* 62, 197, 1978.
7. Katz, S. and Thacker, L. H., A new sensitive ultraviolet detection system for carbohydrates eluted during column chromatography, *J. Chromatogr.,* 64, 247, 1972.
8. Katz, S., Pitt, W. W., Mrochek, J. E., and Dinsmore, S., Sensitive fluorescence monitoring of carbohydrates eluted by a borate mobile phase from an anion-exchange column, *J. Chromatogr.,* 101, 193, 1974.
9. Mrochek, J. E., Dinsmore, S. R., and Waalkes, T. P., Liquid-chromatographic analysis for neutral carbohydrates in serum glycoproteins, *Clin. Chem. Winston-Salem, N.C.,* 21, 1314, 1975.
10. Dische, Z., Color reactions of carbohydrates, in *Methods in Carbohydrate Chemistry,* Vol. 1, Whistler, R. L. and Wolfrom, M. L., Eds., Academic Press, New York, 1962, 477.
11. Hodge, J. E. and Hofreiter, B. T., Determination of reducing sugars and carbohydrates, in *Methods in Carbohydrate Chemistry,* Vol. 1, Whistler, R. L. and Wolfrom, M. L., Eds., Academic Press, New York, 1962, 380.
12. Spiro, R. G., Analysis of sugars found in glycoproteins, in *Methods in Enzymology,* Vol. 8, Neufeld, E. F. and Ginsburg, V., Eds., Academic Press, New York, 1966, 3.
13. Scott, T. A. and Melvin, E. H., Determination of dextran with anthrone, *Anal. Chem.,* 25, 1656, 1953.
14. Jenner, H., Automated determination of the molecular weight distribution of dextran, in *Proc. Eur. Technicon Symp. Automation in Anal. Chem.,* Mediad, New York, 1967, 203.
15. Arturson, G. and Granath, K., Dextrans as test molecules in studies of the functional ultrastructure of biological membranes. Molecular weight distribution analysis by gel chromatography, *Clin. Chim. Acta,* 37, 309, 1972.
16. Barker, S. A., How, M. J., Peplow, P. V., and Somers, P. J., Separations of isomeric polyols, *Anal. Biochem.,* 26, 219, 1968.
17. Hough, L., Jones, J.V.S., and Wusteman, P., On the automated analysis of neutral monosaccharides in glycoproteins and polysaccharides, *Carbohydr. Res.,* 21, 9, 1972.
18. Kesler, R. B., Rapid anion-exchange chromatography of carbohydrates, *Anal. Chem.,* 39, 1416, 1967.

19. Kennedy, J. F. and Fox, J. E., The fully automatic ion-exchange and gel-permeation chromatography of neutral monosaccharides and oligosaccharides with a Jeolco JLC-6AH analyser, *Carbohydr. Res.*, 54, 13, 1977.

20. Samuelson, O., Partition chromatography on ion-exchange resins, in *Methods in Carbohydrate Chemistry*, Vol. 6, Whistler, R. L. and BeMiller, J. N., Eds., Academic Press, New York, 1972, 66.

21. John, M. and Dellweg, H., Gel chromatographic separation of oligosaccharides, *Sep. Purif. Methods*, 2, 231, 1973.

22. Dubois, M., Gilles, K. A., Hamilton, J. K., Rebers, P. A., and Smith, F., Colorimetric method for determination of sugars and related substances, *Anal. Chem.*, 28, 350, 1956.

23. Jolley, R. L., Pitt, W. W., and Scott, C. D., Nonpulsing reagent metering for continuous colorimetric detection systems, *Anal. Biochem.*, 28, 300, 1969.

24. Katz, S., Dinsmore, S. R., and Pitt, W. W., A small automated high-resolution analyzer for determination of carbohydrates in body fluids, *Clin. Chem. Winston-Salem, N.C.,* 17, 731, 1971.

25. Brummel, M. C., Mayer, H. E., and Montgomery, R., Continuous autoanalysis of glycopeptides and glycoproteins, *Anal. Biochem.*, 33, 16, 1970.

26. Brummel, M. C., Gerbeck, C. M., and Montgomery, R., Effect of metals on analytical procedures for amino acids and carbohydrates, *Anal. Biochem.*, 31, 331, 1969.

27. Anderson, D. M. W. and Stoddart, J. F. The use of molecular-sieve chromatography in studies on *Acacia senegal* gum, *Carbohydr. Res.*, 2, 104, 1966.

28. Mopper, K. and Degens, E. T., A new chromatographic sugar autoanalyzer with a sensitivity of 10^{-10} moles, *Anal. Biochem.*, 45, 147, 1972.

29. Mopper, K. and Gindler, E. M., A new noncorrosive dye reagent for automatic sugar chromatography, *Anal. Biochem.*, 56, 440, 1973.

30. Mopper, K., Improved chromatographic separations on anion-exchange resins. I. Partition chromatography of sugars in ethanol, *Anal. Biochem.*, 85, 528, 1978.

31. Mopper, K., Improved chromatographic separations on anion-exchange resins. II. Separation of uronic acids in acetate medium and detection with a noncorrosive reagent, *Anal. Biochem.*, 86, 597, 1978.

32. Mopper, K., Improved chromatographic separations on anion-exchange resins. III. Sugars in borate medium, *Anal. Biochem.*, 87, 162, 1978.

33. Dawson, R. and Mopper, K., An automatic analyzer for the specific determination of amino sugars, *Anal. Biochem.*, 84, 191, 1978.

34. Barker, S. A., Hatt, B. W., and Somers, P. J., The effect of areneboronic acids on the alkaline conversion of D-glucose into D-fructose, *Carbohydr. Res.*, 26, 41, 1973.

35. Kennedy, J. F. and Chaplin, M. F., A new, automated assay for the direct analysis of D-fructose in D-fructose-D-glucose mixtures, *Carbohydr. Res.*, 40, 227, 1975.

36. Davidson, E. A., Analysis of sugars found in mucopolysaccharides, in *Methods in Enzymology*, Vol. 8, Neufeld, E. F. and Ginsburg, V., Eds., Academic Press, New York, 1966, 52.

37. Von Berlepsch, K., Two automated procedures for determination of uronic acid in acid glycosaminoglycans, *Anal. Biochem.*, 27, 424, 1969.

38. Donald, A. S. R., Separation of hexosamines, hexosaminitols and hexosamine-containing di- and trisaccharides on an amino acid analyser, *J. Chromatogr.*, 134, 199, 1977.

39. Speck, J. C., Periodate oxidation. Determination of formaldehyde, in *Methods in Carbohydrate Chemistry*, Vol. 1, Whistler, R. L. and Wolfrom, M. L., Eds., Academic Press, New York, 1962, 441.

40. Hay, G. W., Lewis, B. A., and Smith, F., Determination of the average chain length of polysaccharides by periodate oxidation, reduction and analysis of the derived polyalcohol, in *Methods in Carbohydrate Chemistry*, Vol. 5, Whistler, R. L., BeMiller, J. N., and Wolfrom, M. L., Eds., Academic Press, New York, 1965, 377.

41. Samuelson, O. and Strömberg, H., Separation of alditols and aldoses by partition chromatography on ion-exchange resins, *Carbohydr. Res.*, 3, 89, 1966.

42. Hough, L., Ko, A. M. Y., and Wusteman, P., The simultaneous separation and analysis of aminodeoxyhexitols and alditols by automated ion-exchange chromatography, *Carbohydr. Res.*, 44, 97, 1975.

43. Anderson, D. M. W., Hendrie, A., and Munro, A. C., Automated molecular-sieve chromatography of polysaccharides, *J. Chromatogr.*, 44, 178, 1969.

44. Dudman, W. F. and Bishop, C. T., Electrophoresis of dyed polysaccharides on cellulose acetate, *Can. J. Chem.*, 46, 3079, 1968.

45. Conrad, H. E., Bamburg, J. R., Epley, J. D., and Kindt, T. J., The structure of the *Aerobacter aerogenes* A3(S1) polysaccharide. II. Sequence analysis and hydrolysis studies, *Biochemistry*, 5, 2808, 1966.

46. Roll, D. E. and Conrad, H. E., Quantitative radiochromatographic analysis of the major groups of carbohydrates in cultured animal cells, *Anal. Biochem.*, 77, 397, 1977.
47. Sandford, P. A. and Watson, P. R., Liquid scintillation counting of radioactive monomeric and polymeric carbohydrates on paper chromatograms, *Anal. Biochem.*, 56, 443, 1973.
48. Wasteson Å., A method for the determination of the molecular weight and molecular-weight distribution of chondroitin sulfate, *J. Chromatogr.*, 59, 87, 1971.
49. Lohmander, S., Ion exchange chromatography of glucosamine and galactosamine in microgram amounts with quantitative determination and specific radioactivity assay, *Biochim. Biophys. Acta*, 264, 411, 1972.
50. Hopwood, J. J. and Robinson, H. C., The molecular-weight distribution of glycosaminoglycans, *Biochem. J.*, 135, 631, 1973.

Section II.III

DETECTION REAGENTS FOR PAPER AND/OR THIN-LAYER CHROMATOGRAPHY

LIST OF DETECTION REAGENTS FOR CARBOHYDRATES

Aminobenzoic acid	1	Mixed indicators	29
Aminodiphenyl	2	α-Naphthol-phosphoric acid	30
Aminoguanidine sulfate-chromic acid	3	Naphthoresorcinol	31
Aminohippuric acid	4	Naphthylamine-thymol	32
Aminophenol	5	Ninhydrin	33
Ammonium molybdate	6	Nitroaniline-periodic acid	34
Ammonium sulfate	7	Orcinol-trichloroacetic acid	35
Aniline oxalate/citrate	8	Periodate-acetylacetone	36
Aniline phosphate	9	Periodate-anisidine	37
Aniline phthalate	10	Periodate-benzidine	38
Anisaldehyde-sulfuric acid	11	Periodate-permanganate	39
Anisidine hydrochloride	12	Periodate- silver nitrate	40
Anisidine-phthalic acid	13	Permanganate, alkaline	41
Anthrone	14	Phenol-sulfuric acid	42
Benzidine	15	Phenylenediamine	43
Bromophenol blue	16	Resorcinol-hydrochloric acid	44
Carbazole-sulfuric acid	17	Silver nitrate, alkaline	45
Dichromate-sulfuric acid	18	Silver nitrate, ammoniacal	46
Dimedone-phosphoric acid	19	Sulfuric acid	47
3,4-Dinitrobenzoate alkaline	20	Tetrazole blue	48
2,4-Dinitrophenylhydrazine	21	Thiobarbiturate-periodate	49
3,5-Dinitrosalicylate, alkaline	22	Thiobarbituric acid-phosphoric acid	50
Diphenylamine-aniline-phosphoric acid	23	Thymol-sulfuric acid	51
Elson-Morgan reagent	24	Triphenyltetrazolium chloride	52
Fleury's reagent	25	Urea-acid	53
Hydroxylamine-ferric chloride	26	Vanadium pentoxide-sulfuric acid	54
Iodine vapor	27	Vanillin-perchloric acid	55
Lead tetraacetate	28		

These reagents are described, in the alphabetical order given above, on the following pages.

DETECTION REAGENTS

1. *p*-Aminobenzoic Acid

Preparation: Add 3 g *p*-aminobenzoic acid gradually with stirring to 5.0 ml hot H_3PO_4. To this solution add 300 ml 1-butanol-acetone-water (10:5:2). Stable below 25°C for long periods.

Procedure: Treat chromatogram; heat at 100°C for 15 min.

Results: Pentoses — dark red; other sugars — dark brown.

2. *o*-Aminodiphenyl

Preparation: 0.3 g *o*-aminodiphenyl and 5 ml conc H_3PO_4 in 95 ml ethanol.

Procedure: Spray; heat at 100°C for 15—20 min.

Results: Carbohydrates detected as brown spots. Sensitivity 0.1 μg.

3. Aminoguanidine Sulfate-Chromic Acid[1]

Preparation:

Solution a 2.5 g aminoguanidine sulfate monohydrate in 100 ml water. Stable at room temperature.

Solution b 1 ml 1% solution of $K_2Cr_2O_7$ in water mixed with 100 ml conc H_2SO_4. Stable for about 1 month at room temperature.

Procedure: Spray with *a* then, immediately after, with *b;* heat at 100°C for 10 min.

Results: Fucose and galacturonic acid-lilac; other sugars; intense blue or gray-blue spots. Detection limit 0.1 μg; optimal amount 0.3—0.5 μg.

4. *p*-Aminohippuric Acid

Preparation: 0.3 g *p*-aminohippuric acid in 100 ml ethanol.

Procedure: Spray; heat at 140°C for 8 min. Examine under UV light.

Results: Hexoses and pentoses — orange-red fluorescence; uronic acids — purple. Sensitivity 1 μg.

5. *o*-Aminophenol[2]

Preparation: 1% *o*-aminophenol in methanol + 10 ml 85% H_3PO_4 + 5 ml water.

Procedure: Spray; heat at 100°C for 10 min.

Results: See Table IIB.

6. Ammonium Molybdate

Preparation: 20 ml 10% ammonium molybdate added to 3 ml conc HCl with shaking, followed by addition of 5 g ammonium chloride.

Procedure: Spray; heat at 75°C for a few minutes.

Results: Phosphorylated compounds give immediate yellow color due to ammonium phosphomolybdate. After heating, reducing sugars give blue spots due to molybdenum blue. Sucrose is hydrolyzed; gives positive reaction.

7. Ammonium Sulfate

Preparation: Mixture of ammonium sulfate (20%) and sulfuric acid (4%).
Procedure: Spray; heat at 110—130°C for a few minutes.
Alternative preparation: Gal[3] advocates use of 30% aqueous solution of ammonium bisulfate.
Procedure: Spray; heat at 140°C for 20 min.
Results: Brown to black spots with most organic compounds.

8. Aniline Oxalate/Citrate

Preparation: 0.9 m*l* aniline added to 100 m*l* 0.1 *M* oxalic acid.
Procedure: Spray; heat at 100—105°C for 10—20 min.
Results: Reducing sugars give brown spots.
Modification: 2% ethanolic solution of aniline mixed (1:1) with 0.2 *M* citric acid.
Procedure: Stepwise procedure advocated,[4] in which paper is left at 24—26°C for 2 hr, then heated at 55°C for periods of 10 min, 1 hr and a further 1 hr, spots being examined under both white light and UV after each stage; final stage involves heating at 105—110°C for 10 min.
Results: Different sugars develop optimal color and fluorescence at different stages (see Table IID); finally all spots turn brown.

9. Aniline Phosphate

Preparation:
Solution a	Mix in order 20 m*l* aniline, 200 m*l* water, 180 m*l* acetic acid, and 10 m*l* H_3PO_4. Store at 4°C.
Solvent b	Acetone.
Color reagent	Mix *a* and *b* (2:3).

Procedure: Dip paper, dry and heat at 100°C for 5 min.
Alternative preparation: 2 *N* aniline in water-saturated 1-butanol mixed with 2 volumes of 2 *N* H_3PO_4 in 1-butanol.
Procedure: Spray; heat at 105°C for 10 min.
Results: Aldopentoses — red-brown, hexoses — yellow or yellow-brown. Sorbose also gives yellow-brown spot, other ketoses do not react unless present in large amounts.

10. Aniline Phthalate

Preparation: Mix 0.93 g aniline hydrogen phthalate, 1.66 g phthalic acid, and 100 m*l* water-saturated 1-butanol.
Procedure: Spray; heat at 105—130°C for 5 min. Inspect under UV light.
Results: Detects aldoses; weak reaction with ketoses and oligosaccharides. Aldohexoses and deoxyhexoses — brown spots; aldopentoses — red spots.

11. Anisaldehyde-Sulfuric Acid

Preparation: Mix 9 m*l* 95% ethanol, 0.5 m*l* anisaldehyde, and 0.5 m*l* conc H_2SO_4.
Procedure: Spray; heat at 90—100°C for 5—10 min.
Results: Sugars give green, blue, or violet spots. Steroids, terpenes, and phenols interfere.

12. p-Anisidine Hydrochloride

Preparation: 3% p-anisidine hydrochloride in 1-butanol.
Procedure: Spray; heat at 100—120°C for 5—10 min.
Results: Aldohexoses — green-brown; aldopentoses and uronic acids — red; hexuloses — yellow. Many methylated sugars may be differentiated by colors produced with this reagent (shades of red to red-brown).

13. p-Anisidine-Phthalic Acid

Preparation: 1.23 g p-anisidine and 1.66 g phthalic acid in 100 mℓ methanol or ethanol.
Procedure: Spray; heat at 100°C for 10 min.
Results: Hexoses and deoxyhexoses — yellow to green; pentoses — red; uronic acids — brown. Sensitivity: hexoses, deoxyhexoses 0.5 μg; pentoses, uronic acids 0.1—0.2 μg.

14. Anthrone

Preparation: Dissolve 0.3 g anthrone in 10 mℓ glacial acetic acid and add to solution 20 mℓ 96% ethanol, 3 mℓ H_3PO_4, and 1 mℓ water. Solution is stable for several weeks if stored at 4°C.
Procedure: Spray; heat at 110°C for 5—6 min.
Results: Hexuloses and oligosaccharides containing them give yellow spots; heptuloses orange, pentuloses purple.

15. Benzidine

Preparation: Dissolve 0.5 g benzidine in 20 mℓ glacial acetic acid and 80 mℓ ethanol.
Procedure: Spray; heat at 100—105°C for 15 min.
Results: Most sugars give brown spots on light yellow background. Many inorganic compounds interfere.
Caution: Benzidine is highly carcinogenic.

16. Bromophenol Blue

Preparation: Dissolve 40 mg bromophenol blue (bromocresol green or purple may also be used) in 100 mℓ 95% ethanol containing 100 mg boric acid and 7.5 mℓ 1% aqueous borax solution.
Procedure: Spray; observe at room temperature.
Results: Alditols detected as yellow spots on blue background.
Alternative procedure: Oligogalacturonic acids have been detected on cellulose plates by spraying plates with 10% ammonia solution, then with bromophenol blue reagent.[5]
Results: As for alditols.

17. Carbazole-Sulfuric Acid[6]

Preparation: Dissolve 0.5 g carbazole and 5 mℓ conc H_2SO_4 in 95 mℓ ethanol.
Procedure: Spray; heat at 120° for 10 min.
Results: All sugars give violet spots on blue background.

18. Dichromate-Sulfuric Acid

Preparation: 5% $K_2Cr_2O_7$ in 40% H_2SO_4.

Procedure: Spray; heat at 120—130°C for a few minutes.

Results: Black spots on white background given by most nonvolatile organic compounds.

19. Dimedone-Phosphoric Acid[6]

Preparation: 300 mg dimedone in 90 ml ethanol; 10 ml H_3PO_4 added.

Procedure: Spray; heat at 100°C for 15 min.

Results: Ketose-containing carbohydrates detected as dark green to yellow spots; pink fluorescence under UV. 2-Deoxy sugars give dark pink to lilac spots, with pink fluorescence. Sensitivity 1—2 μg.

20. 3,4-Dinitrobenzoate, Alkaline

Preparation: 1%, 3,4-dinitrobenzoic acid in 1 M Na_2CO_3

Procedure: Spray; heat at 100°C for 5 min.

Results: Reducing sugars give blue spots; ketoses react faster.

21. 2,4-Dinitrophenylhydrazine

Preparation: 0.4% solution of 2,4-dinitrophenylhydrazine in 2 M HCl.

Procedure: Spray; heat at 105°C for a few minutes.

Results: Only ketoses react; orange spots on light yellow background.

22. 3,5-Dinitrosalicylate, Alkaline

Preparation: Dissolve 0.5 g 3,5-dinitrosalicyclic acid and 4 g NaOH in 100 ml water.

Procedure: Spray; heat at 100°C for 5 min.

Results: Reducing sugars give brown spots on yellow background. Sensitivity about 1 μg.

23. Diphenylamine-Aniline-Phosphoric Acid

Preparation: Dissolve 4 g diphenylamine and 4 ml aniline in 200 ml acetone and mix solution with 20 ml 85% H_3PO_4.

Procedure: Spray, air-dry, then heat at 100°C for 10 min.

Results: Wide variation in colors produced by different sugars. See Tables IIA, IIB and IIE.

24. Elson-Morgan Reagent

Preparation:

Solution a	25% KOH in water (20 vol) and 95% ethanol (80 vol).
Solution b	1% redistilled pentane-2,4-dione (acetylacetone) in 95% ethanol; prepared immediately before use.
Solution c	10% *p*-dimethylaminobenzaldehyde in conc HCl.
Solution d	95% ethanol.

Procedure: Dried chromatogram dipped through mixture of *a* and *b* (1:10) and heated at 110°C for 5 min, then dipped through mixture of *c* and *d* (1:1) and dried in stream of cold air.

Results: Transient purple spots indicate presence of aminodeoxyhexoses. Heating to 80°C give permanent red spots for free aminodeoxyhexoses. Purple-violet for *N*-acetyl derivatives.

25. Fleury's Reagent

Preparation:

Solution a	Dissolve 10 g mercuric oxide in 10 g conc HNO_3 and 200 ml water. Dilute 1:1 with water before use.
Solution b	Mix 10% aqueous barium acetate and glacial acetic acid (1:10).

Procedure: Spray with *a*, heat at 90—100°C for 10 min., then spray with *b* and heat at 100°C for 10—30 min.

Results: Specific for polyols; inositols give orange spots (sensitivity 5—10 μg inositol), other polyols gray to black spots.

26. Hydroxylamine — Ferric Chloride

Preparation:

Solution a	*M* solution of hydroxylamine hydrochloride in methanol.
Solution b	1.1 *M* solution of KOH in methanol.
Solution c	2 g $FeCl_3$ in 100 ml 1% aqueous HCl.

Procedure: Spray with 1:1 mixture of *a* and *b*, dry at room temperature for 5 min and at 105°C for 5—10 min, then spray with *c*.

Results: Aldonolactones revealed as purple spots.

27. Iodine Vapor

Procedure: Dried chromatogram is placed in closed vessel containing a few iodine crystals. On warming, iodine vaporizes.

Results: Brown spots on faint yellow background with most organic compounds.

28. Lead Tetraacetate

Preparation: 1% lead tetraacetate in benzene.

Procedure: Spray paper; examine at room temperature.

Results: Reducing sugars detected as white spots on brown background. Sensitivity 50—100 μg.

29. Mixed Indicators[5,7]

Preparation: 5 mg thymol blue, 25 mg methyl red, and 60 mg bromothymol blue in 100 ml 95% ethanol; pH adjusted by addition of *M* NaOH until blue-green color attained.

Procedure: Spray; observe at room temperature.

Results: Uronic acids and oligomers give red spots on green background.

30. α-Naphthol-Phosphoric Acid[2]

Preparation: Mix 0.5% α-naphthol in 50% ethanol with H_3PO_4 (10:1).

Procedure: Spray; heat at 90°C for 10—15 min.

Results: See Table IIB.

31. Naphthoresorcinol (1,3-Dihydroxynaphthalene)

Preparation: 0.2% naphthoresorcinol in ethanol mixed (1:1) with 2% trichloroacetic acid in water.

Procedure: Spray; heat at 105°C for 5 min.

Alternative Preparation: For TLC on silica gel, H_2SO_4 may be substituted for trichloroacetic acid; Lato et al.[2] recommend mixing 0.2% naphthoresorcinol in ethanol with conc H_2SO_4 in 25:1 ratio.

Results: Differentiates ketoses from aldoses; see Table IIB. Uronic acids give characteristic blue spots; see Table IIC.

32. β-Naphthylamine-Thymol

Preparation: Mix 0.1 g β-naphthylamine, 1 g thymol, 150 m*l* ethanol, and 2 m*l* H_3PO_4.

Procedure: Spray; heat at 110°C for a few minutes.

Results: Aldohexoses — brown spots; aldopentoses — pink; ketoses — yellow; uronic acids — red.

33. Ninhydrin

Preparation: 0.1% ninhydrin in 1-butanol.

Procedure: Spray; heat at 105—110°C for 10 min.

Results: Amino sugars give purple spots. Other sugars do not react.

34. *p*-Nitroaniline-Periodic Acid

Preparation:

Solution a	33% aqueous solution of periodic acid.
Solution b	1% ethanolic *p*-nitroaniline mixed with conc HCl (4:1).

Procedure: Spray with *a*, wait for 10 min, then spray with *b*.

Results: Deoxy sugars and glycols give yellow spots with intense fluorescence under UV light. Further spraying with 5% methanolic NaOH turns the spots green.

35. Orcinol-Trichloroacetic Acid

Preparation: Dissolve 0.5 g orcinol and 1.5 g trichloroacetic acid in 100 m*l* water-saturated butanol.

Procedure: Spray; heat at 105°C for 15—20 min.

Results: Specific for ketoses; hexuloses give yellow spots, heptuloses blue.

36. Periodate-Acetylacetone[8]

Preparation:

Solution a	Mix 2 m*l* 40% periodic acid, 2 m*l* pyridine, and 100 m*l* acetone.
Solution b	Dissolve 15 g ammonium acetate, 0.3 m*l* glacial acetic acid, and 1 m*l* acetylacetone (pentane-2,4-dione) in 100 m*l* methanol.

Procedure: Dip chromatogram through *a*, allow to dry, then dip through *b*. For TLC, spray with *a*, dry plate in air, then spray with *b*. Examine under UV.

Results: Most sugars give yellow-green fluorescent spots after 10—15 min. Sensitivity: 1 μg for hexoses, aminodeoxy sugars, and N-acetylneuraminic acid; less sensitive for pentoses, deoxy sugars, and uronic acids (see Table IIC). Any substance capable of yielding formaldehyde after periodate oxidation will give a positive result with this reagent.

37. Periodate-*p*-Anisidine

Preparation:
  Solution a   1 g *p*-anisidine in 100 ml 70% ethanol.
  Solution b   10 ml 0.1 *M* sodium periodate in 100 ml acetone.
Procedure: Spray with *a*, heat at 105°C for 5—10 min, dip in *b*.
Results: Spots detected on brown background: polyols — white; aldonic acids — white; uronic acids — red to brown; 2-deoxy sugars — yellow; aminodeoxy sugars — yellow or brown; pentoses and hexoses — red fluorescence in UV light. Sensitivity 1—20 μg.

38. Periodate-Benzidine

Preparation:
  Solution a   0.1% aqueous solution of sodium metaperiodate.
  Solution b   To a solution of 2.8 g benzidine in 80 ml ethanol (96%) add 70 ml water, 30 ml acetone and 1.5 ml M HCl.
Procedure: Spray with *a*, leave at room temperature for 5 min, then spray with *b*.
Results: All compounds with free α-diol configuration appear as white spots on blue background.
Disadvantages: Spots are transient. Benzidine is highly carcinogenic.

39. Periodate-Permanganate

Preparation: Mix 2% aqueous sodium metaperiodate and 1% solution of potassium permanganate in 2% aqueous sodium carbonate (2:1).
Procedure: Spray; leave at room temperature for 20—30 min.
Results: Yellow spots given by most sugars; sensitive reagent (see Table IIC).

40. Periodate-Silver Nitrate[9]

Preparation:
  Solution a   0.02 *M* aqueous sodium metaperiodate.
  Solution b   Ammoniacal silver nitrate (see reagent 46).
Procedure: Spray lightly with *a*, leave at room temperature for 3—5 min, then spray with *b*.
Results: Sugars and derivatives revealed as white spots on brown background.

41. Permanganate, Alkaline

Preparation: 1% aqueous potassium permanganate, containing 2% sodium carbonate.
Procedure: Spray; heat at 100°C for a few minutes.
Results: Sugars appear as yellow spots on purple background.
Disadvantage: Spots are transient.

42. Phenol-Sulfuric Acid[6]

Preparation: 3 g phenol and 5 ml conc H_2SO_4 in 95 ml ethanol.
Procedure: Spray; heat at 110°C for 10—15 min. Additional heating may intensify spots.
Results: Carbohydrates appear as brown spots.

43. *m*-Phenylenediamine

Preparation: 0.2 *M m*-phenylenediamine in 76% ethanol.
Procedure: Spray; heat at 105°C for 5 min. Examine in UV light.
Results: Pentoses give orange-yellow fluorescent spots, most other sugars yellow fluorescence, weak for uronic acids, aminodeoxy sugars, and disaccharides.

44. Resorcinol-Hydrochloric Acid[9]

Preparation: 0.2% solution of resorcinol in 1-butanol, mixed just before use with equal volume of 0.25 *M* HCl.
Procedure: Spray; heat at 105°C for a few minutes.
Results: Detects ketoses; hexuloses give characteristic red spot, pentuloses blue.

45. Silver Nitrate, Alkaline

Preparation:
Solution a   0.1 m*l* saturated aqueous silver nitrate solution diluted to 20 m*l* with acetone; water added dropwise to dissolve precipitated AgNO$_3$.
Solution b   0.5 *M* NaOH in aqueous ethanol, prepared by diluting 40% aqueous NaOH solution with ethanol.
Procedure: Dried chromatogram is dipped in *a*, allowed to dry, then sprayed with *b*.
Results: Reducing sugars and alditols give brown to black spots. Background can be removed by soaking papers in 6 *M* ammonia solution or dilute sodium thiosulfate, then washing in running water for 15—30 min; this gives permanent chromatogram. Sensitive — see Table IIC.

46. Silver Nitrate, Ammoniacal

Preparation: To 5% (w/v) silver nitrate solution add dropwise aqueous ammonia (s.c. 0.85) until precipitate dissolves.
Procedure: Spray; heat at 105°C for 5—10 min.
Results: Reducing sugars and alditols appear as brown to black spots. Other reducing substances may interfere.
Caution: Reagent should be freshly prepared and discarded after use, because of possible formation of explosive silver azide. Do not expose to direct sunlight.

47. Sulfuric Acid

Preparation: 5—10% sulfuric acid in water or ethanol.
Procedure: Spray plate lightly; heat at 130°C for 5—10 min.
Results: Carbohydrates and other organic compounds give brown to black spots.

48. Tetrazole Blue

Preparation:
Solution a   0.2% tetrazole blue [3,3'-dianisole bis-4,4(3,4-diphenyl)-tetrazolium chloride] in water.
Solution b   6 *M* NaOH.
Solvent c   Methanol.
Color reagent   Mix *a*, *b*, and *c* (3:10:5) just before use.

Procedure: Dip paper, allow to dry in dark; heat at 100°C for 20—30 min. Background color is minimized, however, if heating step is eliminated and paper left for 24 hr in dark.[10]

Alternative procedure: In TLC, use of 1 M NaOH in b has been recommended.[11] Plate is sprayed, then heated at 40°C for 5—10 min.

Results: Most sugars give blue to violet spots. See Tables IIA and IIC.

49. Thiobarbiturate-Periodate

Preparation:

Solution a	0.5 M sodium metaperiodate in 0.025 M sulfuric acid.
Solution b	Ethylene glycol-acetone-conc H_2SO_4 (50:50:0.3).
Solution c	Sodium 2-thiobarbiturate, 6% in water.

Procedure: Spray with *a*, leave at room temperature for 15 min, then dry thoroughly in stream of air before spraying with *b* and drying similarly. Spray again with *b* if odor of formaldehyde persists. Finally spray with *c* and heat at 100°C for 10 min.

Results: Only neuraminic acids give red spots.[3] Sensitivity about 3 μg.

50. Thiobarbituric Acid-Phosphoric Acid

Preparation:

Solution a	0.5% thiobarbituric acid in ethanol.
Solution b	85% H_3PO_4.
Color reagent	Mix *a* and *b* (50:1).

Procedure: Spray; heat at 100°C for 10 min.

Results: Specific for ketoses and oligosaccharides containing ketose residues. See Tables IIA and IIB.

51. Thymol-Sulfuric Acid[6]

Preparation: 0.5 g thymol and 5 ml conc H_2SO_4 in 95 ml ethanol.

Procedure: Spray; heat at 120°C for 15—20 min.

Results: Most sugars give dark pink spots, turning violet on further heating.

52. Triphenyltetrazolium Chloride

Preparation: Mix equal volumes of 4% solution of 2,3,5-triphenyltetrazolium chloride in methanol and 1 M NaOH in methanol.

Procedure: Spray; heat at 100°C for 5—10 min.

Results: Reducing sugars give pink to red spots. Sensitive reagent — see Tables IIA and IIC.

53. Urea-Acid

Preparation: Dissolve 5 g urea in 20 ml 2 M HCl or H_2SO_4. Add 100 ml ethanol.

Procedure: Spray; heat at 100°C for 20—30 min.

Results: Specific for ketoses and oligosaccharides containing ketose residues. See Tables IIA and IIB.

54. Vanadium Pentoxide-Sulfuric Acid[12]

Preparation: Dissolve 18.2 g vanadium pentoxide in 300 ml M Na$_2$CO$_3$ solution by heating. After cooling, add 400 ml 2.5 MH$_2$SO$_4$ and dilute mixture to 1 1 with water.
Procedure: Spray; examine at room temperature.
Results: All reducing sugars give blue spots; some react more slowly, so that spots turn from yellow to blue slowly. See Table IIB.

55. Vanillin-Perchloric Acid

Preparation: Mix in equal volumes a 1% solution of vanillin in ethanol with 3% aqueous perchloric acid.
Procedure: Spray; heat at 85°C for 5—10 min.
Results: Polyols — pale blue to lilac spots; ketoses — gray-green; rhamnose — brick red. Sensitivity; 15 μg for hexitols, others 20—30 μg. Inositol and aldoses do not react.

REFERENCES*

1. **Martins, P. M. and Dick, Y. P.,** A new method for the detection of sugars, specific for fucose, by thin-layer chromatography, *J. Chromatogr.,* 32, 188, 1968.
2. **Ghebregzabher, M., Rufini, S., Monaldi, B., and Lato, M.,** Thin-layer chromatography of carbohydrates, *J. Chromatogr.,* 127, 133, 1976.
3. **Gal, A. E.,** Separation and identification of monosaccharides from biological materials by thin-layer chromatography, *Anal. Biochem.,* 24, 452, 1968.
4. **Vitek, V. and Vitek, K.,** Stepwise detection of sugars with aniline citrate on paper and thin-layer chromatograms, *J. Chromatogr.,* 143, 65, 1977.
5. **Liu, Y. K. and Luh, B. S.,** Preparation and thin-layer chromatography of oligogalacturonic acids, *J. Chromatogr.,* 151, 39, 1978.
6. **Adachi, S.,** Thin-layer chromatography of carbohydrates in the presence of bisulfite, *J. Chromatogr.,* 17, 295, 1965.
7. **Pressey, R. and Allen, R. S.,** Paper chromatography of oligogalacturonides, *J. Chromatogr.,* 16, 248, 1964.
8. **Weiss, J. B. and Smith, I.,** Sensitive location reagent for the simultaneous detection of sugars, amino sugars and sialic acids, *Nature (London),* 215, 638, 1967.
9. **Hough, L. and Jones, J. K. N.,** Chromatography on paper, in *Methods in Carbohydrate Chemistry,* Vol. 1, Whistler, R. L. and Wolfrom, M. L., Eds., Academic Press, New York, 1962, 21.
10. **Mes, J. and Kamm, L.,** The relative sensitivity of various reagents for the detection and differentiation of sugars and sugar derivatives in glycoproteins, *J. Chromatogr.,* 38, 120, 1968.
11. **Damonte, A., Lombard, A., Tourn, M. L., and Cassone, M. C.,** A modified solvent system and multiple detection technique for the separation and identification of mono- and oligosaccharides on cellulose thin layers, *J. Chromatogr.,* 60, 203, 1971.
12. **Haldorsen, K. M.,** Vanadium pentoxide in sulfuric acid, a general chromogenic spray reagent for carbohydrates, *J. Chromatogr.,* 134, 467, 1977.

---

* Journal references cited for some of the reagents described are listed.

GENERAL REFERENCES*

13. Lederer, E. and Lederer, M., *Chromatography*, 2nd ed., Elsevier, Amsterdam, 1957, 253.
14. Sherma, J. and Zweig, G., *Paper Chromatography and Paper Electrophoresis*, Vol. 2, Academic Press, New York, 1971, 152.
15. Waldi, D., Spray reagents for thin-layer chromatography in *Thin-Layer Chromatography*, Stahl, E., Ed., Academic Press, New York, 1965, 485.
16. Wing, R. E. and BeMiller, J. N., Qualitative thin-layer chromatography in *Methods in Carbohydrate Chemistry*, Vol. 6, Whistler, R. L. and BeMiller, J. N., Eds., Academic Press, New York, 1972, 46.

TABLES

The following tables give details of color reactions and sensitivity of reagents most recommended for use in detection of sugars and derivatives on paper and/or thin-layer chromatography.

Tables IIA and IIB have been compiled by Professor M. Lato, Dr. M. Ghebregza-bher and Dr. S. Rufini, Instituto di Clinica Pediatrica, Università di Perugia, Italy.

Tables IIC to IIE are reproduced, by permission of the publishers, from data first published in the *Journal of Chromatography*.

---

* In general, the list of detection reagents suitable for paper and thin-layer chromatography has been compiled from the sources above.

Table IIA
DETECTION OF SUGARS IN CELLULOSE THIN-LAYER CHROMATOGRAPHY

Layer	L1	L1	L1	L1	L1	L1	L1	L1	L1	L1	L1	L1	L1
Reagent	R1	R2	R2 + R1	R3	R3 + R1	R4	R4 + R1	R5	R5 + R1	R6	R6 + R1	R7	R7 + R1
Technique	T1	T1	T2	T3	T3	T1	T2	T4	T2	T5	T2	T6	T2
Compound						Colors							
L-Arabinose	ol	pk-br r	bn	br r	r bn	—	gy gr	in pk	in pk	ft l	gy ol	—	o pk
D-Xylose	ol	pk-br r	bn gy	br r	r bn	—	gr	in pk	in pk	ft l	ol	—	o pk
D-Lyxose	ol	pk-br r	bn gy	br r	r bn	—	gr	in pk	in pk	ft l	gy ol	—	o pk
D-Ribose	ol	pk-br r	bn y	br r	r bn	—	gy bl	in pk	in pk	—	gy	—	pk
D-Glucose	bl gy	y	gy	bn y	bn	—	gy bn	pl	gy pk	—	gy	—	—
D-Galactose	gy	y	ol	bn y	bn	—	gy y	pk	pk gy	—	gy	—	—
D-Mannose	bl gy	y	gy	y bn	by gy	y-o	gy bn	pk	gy pk	—	gy p	gy bl	gy bl
D-Fructose	y-o	—	y	y	y bn	y-o	in y-bn	in pk	br r	in l	y	—	gy bn
L-Rhamnose	y gr	y	y gr	y	y gr	—	y gr	pk	y bn	—	pk	—	bn
2-Deoxy-D-ribose	pk	y	p	—	pk bn	y-o	l	pk	p	—	y	—	gy bl
Sucrose	bn y	—	y gy	pl y	bn gy	y-o	in y-bn	—	y bn	in l	y	gy bl	gy bl
Turanose	bn y	—	bn gy	—	gy vt	y	in y-bn	in pk	pk gy	—	gy vt	gy bl	ft gy-bl
Maltose	bl	—	bl	—	bl gy	—	y bn	pk	l gy	—	bl	—	vt
Cellobiose	bl	y bn	gy bl	y	gy gr	y	y	pk	gy	—	gy y	ft gy-bl	gy
Melezitose	y gy	—	gy y	—	gy gr	y	in y-bn	—	y gy	—	bl	ft gy-bl	vt
Lactose	gy t l	ft y	gy bl	y	gy gr	—	y	pk	pk gy	—	gy bl	—	—
Gentiobiose	gy bl	y	gy	y bn	bn gy	—	y gr	pk	pk gy	—	gy bl	—	—
Melibiose	gy bl	y	gy	y bn	bn	—	y gr	pk	pk gy	—	gy bl	—	—
Raffinose	y gr	—	y gy	—	gy y	y	in y-bn	—	y	—	y	gy bl	gy
Maltotriose	gy bl	—	bl	—	bl gy	—	y bn	pk	l gy	—	bl	—	ft gy-bl
Stachyose	y gr	—	y gy	—	gy y	y	in y-bn	y	y	—	y	gy bl	gy
Background	w	w	ft gy	w	ft gy	w	ft y-gr	ft pk	ft pk-gy	ft l	ft gy	w	w
L-Arabinose	—	o bn	y bn	o bn	bn gy	gr y	gy bl	r bn	r bn	pk-br r	pk o	pk o	bn o
D-Xylose	—	o bn	y bn	o bn	bn gy	gr	gy gr	r bn	r bn	pk-br r	pk bn	pk o	bn o
D-Lyxose	—	o bn	y bn	o bn	bn gy	gr	gy gr	r bn	r bn	pk-br r	pk bn	pk o	bn o
D-Ribose	—	o bn	y bn	o bn	bn gy	gr y	gy bl	r bn	r bn	pk-br r	pk bn	pk o	r bn
D-Glucose	—	y bn	gy y	p	bn	y-o	ol	in o	bn	y	gy bn	y gy	gy vt
D-Galactose	—	y bn	gy y	p	bn	y-o	y gr	o	y bn	y	gy bn	y gy	gy
D-Mannose	—	y bn	y bn	p	bn	y-o	y gr	o	bn gy	ft y	gy bn	y gy	gy vt
D-Fructose	gr y	y gr	y gr	y	y gy	o	o	in y	y bn	ol	ol	bl gr	bl gy
L-Rhamnose	—	y	y	y	y gy	y	y	in y	y bn	y	y gy	y	bn y
2-Deoxy-D-ribose	l	y	y	gy vt	gy vt	o	l	—	pk	—	vt gy	ol	bn gy
Sucrose	gr y	y gr	y gr	bn y	bn y	o	y bn	y bn	pk gy	gy y	ol	ol	gy bl
Turanose	y gr	y gr	y gr	bn y	bn y	y-o	y bn	ft y	gy bl	y gy	gy	y gy	gy
Maltose	—	ft y	bl	—	bl	y-o	gy y	—	gy	—	ft bl-gy	—	ft gy
Cellobiose	—	ft y	gy bl	p	gy	ft y	y gy	ft y	y pk	ft y	y gy	ft y-gy	gy
Melezitose	y	y	y gr	bn y	bn	y-o	y bn	y	gy	y gy	gy	y gy	gy

Table IIA (continued)
DETECTION OF SUGARS IN CELLULOSE THIN-LAYER CHROMATOGRAPHY

Layer	L1	L1	L1	L1	L1	L1	L1	L1	L1	L1	L1	L1	L1
Reagent	R8	R8 + R2	R8 + R2 + R1	R8 + R3	R8 + R3 + R1	R4 + R2	R4 + R2 + R1	R4 + R3	R4 + R3 + R1	R7 + R2	R7 + R2 + R1	R7 + R3	R7 + R3 + R1
Technique	T3	T2	T2	T2	T2	T2	T2	T2	T2	T2	T2	T2	T2
Compound							Colors						
Lactose	—	ft y	gy bl	p	gy	ft y	y gy	—	bn gy	ft y	gy bn	ft y-gy	ft gy
Gentiobiose	—	y bn	gy y	p y	gy bn	ft y	ol	ft o	gy bn	ft y	l gy	ft y	gy vt
Melibiose	—	y bn	gy y	o bn	gy bn	ft y	gr y	ft o	bn	ft y	l gy	ft y	gy vt
Raffinose	y	y	y gr	y bn	y bn	o	y bn	y	gy bl	ol	ol	ol	gy bl
Maltotriose	—	ft y	bl	—	bl	—	gy y	—	bn	—	ft bl-gy	—	ft gy
Stachyose	y	y	y gr	y bn	y bn	o	y bn	y	bn	ol	ol	gy y	gy bl
Background	w	lt y	lt y	pk y	lt y	lt y	y gr	lt y	lt gy	w	pl gy	w	w

Colors bl = blue; bn = brown; br = brick; ft = faint; gr = green; gy = grey; in = intense; l = lilac; lt = light; o = orange; ol = olive; p = purple; pk = pink; pl = pale; r = red; vt = violet; w = white; y = yellow. Hyphen between symbols indicates change of color with time.

Layer
L1 = 0.25 mm Cellulose MN 300 (Macherey-Nagel).

Reagent
R1 = modified diphenylamine-aniline-phosphoric acid reagent (2% diphenylamine in acetone/2% aniline in acetone/85% H_3PO_4; 5:5:2).
R2 = p-anisidine-phthalic acid, 0.1 M, in ethanol.
R3 = p-aminobenzoic acid-phosphoric acid.
R4 = thiobarbituric acid-phosphoric acid.
R5 = triphenyltetrazolium chloride-sodium hydroxide.
R6 = tetrazole blue-sodium hydroxide, modified: 1 M NaOH instead of 6 M as in original reagent.
R7 = urea-hydrochloric acid.
R8 = dimedone-phosphoric acid.

Technique
T1 = spray chromatoplate with reagent; heat at 100°C for 10 min.
T2 = spray successively with two or three reagents; colors are observed and recorded after each detection.
T3 = spray; heat at 100°C for 15 min.
T4 = spray; heat at 100°C for 5 min.
T5 = spray; heat at 45°C for 5-10 min.
T6 = spray; heat at 100°C for 40-45 min.

REFERENCE

1. Damonte, A., Lombard, A., Tourn, M. L., and Cassone, M. C., A modified solvent system and multiple detection technique for the separation and identification of mono- and oligosaccharides on cellulose thin layers, *J. Chromatogr.*, 60, 203, 1971.

Table IIB
COLOR AND SENSITIVITY OF DETECTION REAGENTS FOR SUGARS AND DERIVATIVES IN THIN-LAYER CHROMATOGRAPHY

Layer → Reagent → Technique → Reference	L1 R1 T2 (1)		L1 R2 T2 (1)	L2 R3 T3 (2)	L1 R4 T2 (1)		L1 R4+R1 T4 (1)		L1 R5 T2 (1)		L3 R6 T5 (3)		L1 R5+R1 T6 (1)		L1 R7 T7 (1)		L1 R8 T2 (1)		L1 R9 T8 (1)	
Compounds	col	sens	col	col	col	sens	col	sens	col	sens	col	sens	col	sens	col	sens	col	sens	col	sens
Glyoxal											bl	0.5								
Glycolaldehyde											bl	0.5								
Glyceraldehyde											bl	0.25								
Erythrose											bl	0.25								
L-Arabinose	lt bl	1.5	ft gy-bl	vt	br r	0.5	bn y	0.5	—		y-bl	0.25	lt bl	1.0			—		p	0.5
D-Xylose	bn	1.5	ft gy-bl	vt	br r	0.5	bn y	0.5	—		y-bl	0.25	lt bl	1.0			—		p	0.5
D-Lyxose	lt bl	1.5	ft gy-bl	vt	br r	0.5	bn y	0.5	—		y-bl	0.25	lt bl	1.0			—		p	0.5
D-Ribose	lt bl	1.5	ft gy-bl	vt	br r	0.5	bn y	0.5	—		y-bl	0.25	lt bl	1.0			—		p	0.5
D-Glucose	vt	4.0	bl gy	bl	y	0.5	y bn	0.5	—		y-bl	0.25	lt bl	2.0			—		—	
D-Galactose	bl	1.5	bl gy	bl	y	0.5	y bn	0.5	lt bl		y-bl	0.25	lt bl	2.0			—		—	
D-Mannose	bl	1.5	bl gy	bl	y	0.5	y bn	0.5	—		y-bl	0.25	lt bl	2.0			—		—	
L-Fucose	pk	0.5	ol-in bl	gr	y	1.0	bn y	1.0	—		y-bl	0.50	vt	2.0			—		o	0.5
L-Rhamnose	pk	0.5	ol-in gy	gr	y	3.0	bn y	1.0	—		y-bl	0.50	vt	2.0			—		o	0.5
2-Deoxy-D-glucose	lt bl	0.5	vt-in bn		y	3.0	gr	2.0			y-bl	0.50	bn gr	2.0			bl	0.5	o	2.0
2-Deoxy-D-galactose	lt bl	0.5	vt-in bn		y	3.0	gr	2.0			y-bl	0.50	bn gr	2.0			bl	0.5	o	2.0
6-Deoxy-D-glucose				gr																
2-Deoxy-D-ribose	gy	0.5	bn-in vt	r	y	3.0	bl gr	1.0			y-bl	0.50	bn y	2.0	—		bl	3.0	o	2.0
Dihydroxyacetone				gr							bl	0.25								
D-threo-Pentulose	o	0.2	y	gr	y	3.0	o	3.0	y	4.0			vt	0.5	bn y	0.5	y	2.0	o	2.0
D-Fructose	p-r	0.5	lt r	r	y	4.0	p-r	0.1	y	0.5			bn y	0.5	bl	0.5	y	1.0	bl	2.0
L-Sorbose	p-r	0.1	y	r bn	y	4.0	p-r	0.1	y	0.5			bn y	0.5	bl	0.5	y	1.0	—	
D-Tagatose	r	0.5	lt r	r	y	4.0	p-r	0.1	y	0.5			bn y	0.5	bl	0.5	y	1.0	—	
D-erythro-Pentulose	gr	0.5	y		y	3.0	o	0.1	y	4.0			vt	0.1	bn y	0.5	y	0.5	bl	0.5
D-altro-Heptulose	p-r	4.0	p-r				—		y	2.0			—				gr	1.5	vt	1.0
D-manno-Heptulose	p-r	0.5	p-r				—		y	0.1			—		bn y	1.0	gr	1.5	vt	1.0
Arabinonic acid											bl	0.50								
D-Gluconic acid											bl	0.50								

Table IIB (continued)
COLOR AND SENSITIVITY OF DETECTION REAGENTS FOR SUGARS AND DERIVATIVES IN THIN-LAYER CHROMATOGRAPHY

Layer	L1		L1	L2	L1		L1		L3		L3		L1		L1		L1		L1	
Reagent	R1		R2	R3	R4		R4 + R1		R5		R6		R5 + R1		R7		R8		R9	
Technique	T1		T2	T3	T2		T4		T2		T5		T6		T7		T2		T8	
Reference	1		1	2	1		1		1		3		1		1		1		1	
Compounds	col	sens	col	col	col	sens	col	sens	col	sens	col	sens	col	sens	col	sens	col	sens	col	sens
D-Glucuronic acid	—		—				—				bl	0.50								
D-Galacturonic acid	—		—				—				bl	0.50			bl	0.5	bl vt	1.0		
D-Mannuronic acid	—		—								bl	0.50								
D-Glucuronolactone							—				bl	0.50			bl	0.5	bl vt	1.0		
D-Galactonolactone											bl	0.50								
Xylitol											y-bl	0.50								
Galactitol											y-bl	0.50								
D-Mannitol											y-bl	0.50								
D-Glucitol											y-bl	0.50								
Inositol											y-bl	0.50								
Methyl-α-D-glucopyranoside	bl	0.5	—				r	2.0			bl	0.50	—		—		—		—	
Methyl-α-D-mannopyranoside	pk	0.5	—				bl	1.5					—		—		—		—	
Methyl-α-L-arabinopyranoside	bl	0.5	—				vt	0.5					—		—		—		—	
Methyl-α-D-xylopyranoside	bl	0.5	—				bl	0.5			bl	0.50	—		—		—		—	
Methyl-4,6-O-benzylidene-α-D-glucopyranoside											bl	0.75								
4,6-O-ethylidene-α-D-glucopyranose											bl	0.50								
1,2,3,4,5-penta-O-acetyl-α-D-glucopyranose											bl	1.00								
Sucrose	p-r	0.5	o-vt	vt	y		r	0.5	y		bl	0.50	—		bl	0.5				
Maltose	bl	4.0	bl	bl		2.0	bn y	1.0			y-bl	0.50	—				—			

	col	sens	col	col	sens	col	sens	col	sens	col	sens	col
Trehalose	bl											
Lactose	bl	4.0	gy bl	y	2.0	bn y	1.0			y-bl	0.50	—
Gentiobiose	bl											
Isomaltose	bl											
Panose	bl											
Melibiose	bl	4.0	gy bl	y	2.0	bn y	1.0			y-bl	0.50	—
Cellobiose	bl									y-bl	0.50	
Turanose	p-r	0.2	y bn	—		r	0.5	bl	1.0			—
Melezitose	vt									y-bl	0.50	—
Maltotriose	bl											
Raffinose	p-r	0.5	y gr	—		r	0.5	bl	1.0	y-bl	0.50	—
Isomaltotriose	bl											
Gentianose										y-bl	0.50	—
Isokestose										bl	0.50	
Stachyose										y-bl	0.50	

Note: col = colors; sens = sensitivity in μg; — = no reaction.

Colors: bl = blue; bn = brown; br = brick; ft = faint; gr = green; gy = grey; in = intense; l = lilac; lt = light; o = orange; ol = olive; p = purple; pk = pink; pl = pale; r = red; vt = violet; w = white; y = yellow.

Layer:
L1 = Silica Gel G0 (Merck®); elution with solvent systems not containing organic acids or pyridine.
L2 = precoated Silica Gel 60 (Merck®), impregnated with NaH_2PO_4.
L3 = silica gel, kieselguhr, and cellulose layers (see L6, L7, and L8, Table TLC 1).

Reagent:
R1 = 1,3-dihydroxynaphthalene (naphthoresorcinol) in ethanol – conc H_2SO_4.
R2 = diphenylamine-aniline-phosphoric acid.
R3 = modification of R2; aniline, diphenylamine, acetone and 80% phosphoric acid present in amounts 4 mℓ: 4 g: 200 mℓ: 30 mℓ.
R4 = p-anisidine-phthalic acid, 0.1 M, in ethanol.
R5 = thiobarbituric acid-phosphoric acid.
R6 = 0.1 M vanadium pentoxide in M H_2SO_4.
R7 = urea-2 M H_2SO_4.
R8 = o-aminophenol-phosphoric acid.
R9 = α-naphthol-phosphoric acid.

Table IIB (continued)

COLOR AND SENSITIVITY OF DETECTION REAGENTS FOR SUGARS AND DERIVIATIVES IN THIN-LAYER CHROMATOGRAPHY

Technique: T1 = 1,3-dihydroxynaphthalene solution and sulphuric acid mixed before use (25:1 v/v) and sprayed on chromatoplate; plate heated at 100°C for 5 min.

T2 = heat plate at 100°C for 10 min.

T3 = heat at 105°C for 30 min.

T4 = spray with R4, heat to 100°C for 10 min, record developed colors, then spray with R1 and heat at 100°C for 5 min.

T5 = plates sprayed at rate of about 2 mℓ per 400 cm² and left under ambient conditions or heated in Termaks drying cabinet.

T6 = spray with R5, heat at 100°C for 10 min, then spray with R1.

T7 = heat at 100°C for 20—30 min.

T8 = heat at 90°C for 10—15 min.

References: 1. Ghebregzabher, M., Rufini, S., Monaldi, B., and Lato, M., Thin-layer chromatography of carbohydrates, *J. Chromatogr.*, 127, 133, 1976.

2. Hansen, S. A., Thin-layer chromatographic method for the identification of mono-, di- and trisaccharides, *J. Chromatogr.*, 107, 224, 1975.

3. Haldorsen, K. M., Vanadium pentoxide in sulfuric acid, a general chromogenic spray reagent for carbohydrates, *J. Chromatogr.*, 134, 467, 1977.

Table IIC
COLORS AND SENSITIVITY OF DETECTION REAGENTS FOR SUGARS AND DERIVATIVES IN PAPER CHROMATOGRAPHY

Compound	R1		R2		R3		R4		R5		R6	
	Col	sens	Col	sens	Col	sens	Col	sens	Col	sens	Col	sens
D-Glucose	bn	1	gr-bn	1	bn	1	bn	4	bl-gy	4	bl-gy	1
D-Galactose	bn	1	gr-bn	1	bn	2	bn	6	bl-gy	6	bl-gy	6
D-Mannose	bn	1	gr-bn	2	bn	2	bn	6	bl-gy	6	--	--
D-Ribose	bn	2	r	1	r	2	bn	6	--	--	bl-gy	6
L-Fucose	bn	1	gr-bn	2	bn	2	bn	6	--	--	bl-gy	6
D-Glucuronic acid	bn	2	o-bn	1	o-bn	1	bn	4	gy	1	gr-bl	2
D-Glucuronolactone	bn	5	bn	20	o-bn	20	--	--	bl-gy	20	--	--
2-Amino-2-deoxy-D-glucose	bn	2	gr-bn	6	bn	2	--	--	--	--	r	1
2-Amino-2-deoxy-D-galactose	bn	2	gr-bn	6	bn	2	--	--	--	--	r	1
2-Acetamido-2-deoxy-D-glucose	bn	4	gr-bn	4	bn	4	--	--	--	--	p	1
2-Acetamido-2-deoxy-D-galactose	bn	4	gr-bn	4	bn	4	--	--	--	--	p	1

Compound	R7		R8		R9		R10		R11		R12	
	Col	sens	Col	sens	Col[a]	sens	Col	sens	Col	sens	Col	sens
D-Glucose	bn	1	--	--	y-gr	1	y	1	r	1	bl-vt	1
D-Galactose	bn	2	--	--	y-gr	1	y	2	r	1	bl-vt	1
D-Mannose	bn	1	--	--	y-gr	1	y	1	r	1	bl-vt	1
D-Ribose	bn	2	--	--	y-gr	6	y	4	r	1	bl-vt	1
L-Fucose	--	--	--	--	--	--	y	1	r	1	bl-vt	1
D-Glucuronic acid	gr-bn	1	bl	1	y-gr	4	y	1	r	1	bl-vt	1
D-Glucuronolactone	gr-bn	20	bl	5	--	--	y	5	r	2	bl-vt	2
2-Amino-2-deoxy-D-glucose	lt bn	1	--	--	y-gr	1	y	2	r	1	bl-vt	1
2-Amino-2-deoxy-D-galactose	lt bn	1	--	--	y-gr	1	y	2	r	1	bl-vt	1
2-Acetamido-2-deoxy-D-glucose	--	--	--	--	y-gr	1	y	2	r	1	bl-vt	1
2-Acetamido-2-deoxy-D-galactose	--	--	--	--	y-gr	1	y	2	r	1	bl-vt	1

Note: col = color; sens = sensitivity in μg; -- = not detected.

[a]Color as seen under UV.

Colors: bl = blue; bn = brown; br = brick; ft = faint; gr = green; gy = grey; in = intense; l = lilac; lt = light; o = orange; ol = olive; p = purple; pk = pink; pl = pale; r = red; vt = violet; w = white; y = yellow.

Reagents:

R1 = alkaline AgNO$_3$.	R7 = benzidine.
R2 = *p*-anisidine hydrochloride.	R8 = naphthoresorcinol.
R3 = aniline phthalate.	R9 = periodate-acetylacetone.
R4 = 3,5-dinitrosalicylate.	R10 = periodate-permanganate.
R5 = diphenylamine-aniline-phosphoric acid.	R11 = 2,3,5-triphenyl tetrazolium chloride.
R6 = Elson-Morgan reagent.	R12 = tetrazole blue.

From Mes, J. and Kamm, L., *J. Chromatogr.*, 38, 120, 1968. With permission.

Table IID
FLUORESCENCE AND COLORS OF SUGAR SPOTS ON STEPWISE DETECTION WITH ANILINE CITRATE REAGENT ON PAPER CHROMATOGRAPHY

Compound	Fluorescence stage		Color stage	
	First appears	Peak	First appears	Peak
Glyceraldehyde	I	I(lt y)	I	II(bn)
Erythrose	I	II(y)	I	II(y)
L-Arabinose	I	II(y-gr)	I	II(r)
D-Ribose	I	II(y-gr)	I	II(r)
D-Xylose	I	II(y-gr)	I	II(r)
D-Glucose	I	III(gr-y)	II	III(y-bn)
D-Galactose	I	III(gr-y)	II	III(y-bn)
D-Mannose	I	III(gr-y)	II	III(y-bn)
L-Rhamnose	I	III(gr-y)	II	IV(bn)
2-Deoxy-D-ribose	I	II(y)	I	II(y)
D-Fructose	IV	V(gr-y)	IV	V(lt bn)
D-Glucuronic acid	I	III(p)	I	III(o-r)
D-Galacturonic acid	I	II(pl o)	I	III(o-r)
2-Acetamido-2-deoxy-D-glucose	IV	IV(y)	IV	V(lt bn)
2-Acetamido-2-deoxy-D-Mannose	IV	IV(y)	IV	V(lt bn)
2-Acetamido-2-deoxy-D-galactose	IV	IV(y)	IV	V(lt bn)
2-Amino-2-deoxy-D-galactose	I	II(y)	I	III(y-bn)
Lactose	II	III(gr-y)	II	IV(gr-bn)
Melibiose	II	III(gr-y)	II	IV(gr-bn)
Cellobiose	II	III(gr-y)	II	IV(gr-bn)
Maltose	III	IV(260 nm, bl;) (360 nm, gr-y)	IV	V(lt bn)

Stages   I   24—26°C for 2 hr.
            II   55°C for 10 min.
           III   55°C for 1 hr.
           IV   55°C for 1 hr.
           V   ·105—110°C for 10 min.

Colors   bl = blue; bn = brown; br = brick; ft = faint; gr = green; gy = grey; in = intense; l = lilac; lt = light; o = orange; ol = olive; p = purple; pk = pink; pl = pale; r = red; vt = violet; w = white; y = yellow.

From Vitek, V. and Vitek, K., *J. Chromatogr.*, 143, 65, 1977. With permission.

Table IIE
COLORS AND SENSITIVITY OF OLIGOSACCHARIDES WITH
DIPHENYLAMINE-ANILINE-PHOSPHORIC ACID REAGENT ON PAPER
CHROMATOGRAMS AND ELECTROPHEROGRAMS AND THIN-LAYER
CHROMATOGRAMS

Sugar[a]	Paper chromatography[b]		Paper electrophoresis[c]	Thin-layer chromatography[d]	
	Col	Sens	Col	Col	Sens
Kojibiose (1→2)	o	10	o	bn	1
Sophorose (1→2)	o	5	o	bn	1
Nigerose (1→3)	gr	5	y-gr	pl bl	1
Laminaribiose (1→3)	gr	5	y-gr	pl bl	1
Maltose (1→4)	in bl	5	bl	in bl	1
Cellobiose (1→4)	bl	5	bl	bl	1
Lactose (1→4)	bl	5	bl	bl	1
Gentiobiose (1→6)	gy-gr	5	gr-y	bl	1
Melibiose (1→6)	gy-gr	5	gr-y	bl	1
Turanose (1→3)	vt	10	vt	r-bn	1
Lactulose (1→4)	gy-gr	10	gr-y	bn	1
Sucrose	gr	5	gr	gr-bn	1
Trehalose	w→y	300	—	bl	10

Note: Col = color; Sens = sensitivity in μg.

[a] Type of linkage shown in parenthesis when appropriate.
[b] Solvent system: 1-butanol-pyridine-water (6:4:3).
[c] In 0.05 *M* borate buffer (pH 9.8).
[d] Solvent system: 1-propanol-water (17:3); silica gel.

Colors   bl = blue; bn = brown; br = brick; ft = faint; gr = green; gy = grey; in = intense; l = lilac; lt = light; o = orange; ol = olive; p = purple; pk = pink; pl = pale; r = red; vt = violet; w = white; y = yellow.

From Toba, T. and Adachi, S., *J. Chromatogr.*, 154, 110, 1978. With permission.

Table III.

Section III

Sample Preparation and Derivatization

Section III

SAMPLE PREPARATION AND DERIVATIZATION

The topics and specific methods discussed in this section, which are grouped under three main headings, are listed below, in the order in which they are presented.

Preparation of Samples from Biological Specimens
1. Extraction of sugars and polyols from plant material
2. Preparation of clinical samples for sugar analysis
3. Isolation and purification of polysaccharides

Degradation of Polysaccharides
4. Acetolysis
5. Hydrolysis
6. Hydrolysis, partial
7. Methanolysis
8. Methylation analysis

Derivatization for Gas Chromatography
9. Alditol acetates
10. Alditols, permethylated
11. Aldononitrile acetates
12. Butaneboronates
13. Glycosides, Chiral
14. Isopropylidene acetals
15. Methylated sugars, acetates
16. Methylated sugars, methyl glycosides
17. Trifluoroacetates
18. Trimethylsilyl ethers
19. Trimethylsilylated alditols and oximes
20. Uronic acid derivatives

Derivatization for HPLC
21. Benzoylation and 4-nitrobenzoylation of carbohydrates for HPLC with UV detector

PREPARATION OF SAMPLES FROM BIOLOGICAL SPECIMENS

Methods of extraction of carbohydrate components from material of plant or animal origin vary widely with the nature of both carbohydrate and source, and a comprehensive treatment is clearly beyond the scope of this book. The examples cited below cover a number of carbohydrates of general interest.

1. Extraction of Sugars and Polyols from Plant Material*
The sugar and polyol constituents in plant material from a wide variety of sources, including algae, lichens, and fungi, as well as roots and leaves of various plant species, have been determined by Holligan and Drew.[1] The method of extraction used is described below.

All plant material was collected as near as possible to mid-day and killed immediately by immersion in cold 95% ethanol The soluble compounds were extracted with hot 80% ethanol, extraction being performed three times over a period of 24 hr. The

---

* See also Section A, Volume II, Section II.II, p. 204.

plant material was finally washed twice in hot ethanol, the washings being added to the total extract, which was then made up to a known volume from which suitable aliquots were taken for preparation of volatile derivatives for GC. The plant material remaining after ethanol extraction was dried at 55°C in weighed aluminium foil cups for 12 hr (or to constant weight) and the dry weight recorded. The weight of the ethanol-soluble fraction was found by drying an aliquot of the extract in similar cups.

Preparation of Clinical Samples for Sugar Analysis*

Methods of preparation of samples for chromatographic analysis from clinical specimens of various body fluids have been reviewed by Menzies and Seakins[2] and, with special reference to gas chromatography, by Sweeley et al.[3]

Urine

Sugars in urine should be preserved by addition of merthiolate (minimum final concentration 100 μg/mℓ) or thymol (a small crystal), and the samples should be stored at 4°C (or −20°C if storage is likely to be prolonged). For paper chromatography, samples should be free of salts and protein; otherwise distortion and trailing of sugar spots may result. Salts, as well as interfering substances such as urea, uric acid, creatinine, ascorbic acid, and other organic acids and bases, are removed by treatment with ion-exchange resins. The procedure recommended by Menzies and Seakins[2] is to shake the sample with a mixed resin (Amberlite MB1, Zerolit DM-F, or AG 501-X8) with the anion-exchanger in the acetate form (converted by prior treatment with 2 M acetic acid). The resin is added to the urine sample (3 mℓ) in a plastic tube until it occupies about 70% of the total volume, the tube is shaken for 3 min, and the contents are then centrifuged. Most of the organic and inorganic electrolytes and normal urinary pigment are completely removed by this treatment, but urate and urea are only partly removed; these authors claim, nevertheless, that up to 1 mℓ of urine thus desalted can be applied for PC, and 100 μℓ for TLC, without serious distortion of the sugar spots. For most clinical purposes, applications need not exceed 100 μℓ for PC or 10 μℓ for TLC to enable significant concentrations of sugars to be detected.

If minor sugar components are to be investigated, larger samples (up to 10 mℓ) of urine are required, and, in this case, desalting by the more effective technique proposed by Vitek and Vitek[4] is necessary. In this procedure, columns (100 mℓ each) of a strong-acid cation-exchange resin (e.g., Amberlite IR-120, Zeo-karb 225, Dowex 50) in the H+-form and a strong-base anion-exchange resin (e.g., Amberlite IRA-400, IRA-410, Dowex 1) in the acetate form are used. The cation-exchange column is set up above the anion-exchanger, with the two connected by an air-tight joint so that effluent from the upper column enters directly into the lower. The urine sample (30 to 50 mℓ) is added carefully to the upper resin bed, just covered with water, and after the urine has soaked into the resin, the sample is washed in with 3 or 4 portions of water (20 to 30 mℓ). The column system is then eluted with water at a maximum flow rate of 2.0 mℓ/min. The first 40 mℓ of effluent should be discarded; the remainder of the effluent (approximately 300 mℓ) is collected, together with the solution emerging when the column system is drained at the end of the run, and evaporated on a rotary evaporator at 36 to 38°C The concentrate and flask washings are transferred through a small filter into a 50-mℓ centrifuge tube and then lyophilized; the freeze-dried residue is stored, if necessary, at −15°C.

Urea is not completely removed even by this desalting technique, but the residual amount does not interfere in most PC and TLC systems. For GC analysis of urine, Sweeley et al.[3] recommend treatment of the urine with urease prior to ion-exchange

---

* See also Section A, Volume II, Section II.II, pages 205 to 206.

desalting. In this procedure, each aliquot (generally 1 ml) of urine is placed in a 50-ml centrifuge tube, distilled water (2 to 3 ml) is added, and the tube is kept in a boiling-water bath for 15 min to eliminate nonspecific enzymatic activity. After the tube has been cooled to 45°C, urease powder (5 mg) is added, and the mixture is incubated at 45°C for 30 min. The tube is then heated in a boiling-water bath for a further 5 min before the solution is percolated through a small column packed with mixed-bed ion-exchange resin (Amberlite MB-3, 2 g). The effluent is collected in a 50-ml centrifuge tube and water is removed *in vacuo* at 50°C; the residue may then be derivatized for GC analysis.

If protein is present in the urine, it should be removed prior to desalting. Various methods have been proposed by different authors. Menzies and Seakins[3] recommend the addition of a 5% aqueous solution of sulfosalicylic acid in a 1:1 volume ratio with the sample, followed, after standing for 10 min, by centrifuging. Sweeley et al.[3] advocate the Somogyi method, in which a 1.8% aqueous solution of $Ba(OH)_2 \cdot 8H_2O$, balanced to neutral pH, and a 2.0% solution of $ZnSO_4 \cdot 7H_2O$ (5 ml of each solution) are added to the urine sample (1 ml), with sufficient water to bring the total volume to 20 ml. After thorough mixing, the precipitated proteins and insoluble Zn and Ba salts are removed by filtration or centrifugation. In preparing samples of rat urine for HPLC, Meagher and Furst[5] removed protein by dialysis in a Bio-Fiber® 50 Minibeaker. The urine collected in a 24-hr period (approximately 7 ml) was diluted to 10 ml, and 8 ml of the diluted sample was used to fill the Minibeaker. Distilled water (50 ml) was passed through the fibers, and the dialysate (which contained the sugars) was then concentrated to 10 ml on a rotary evaporator and passed through a Metricel® filter (0.2-μm pore size) to remove all particulate matter.

Where rapid clinical screening of urine samples by PC or TLC is necessary, pretreatment may be omitted if steps are taken to obviate interference of other substances with the sugar spots. For this purpose, Saini[6] has recommended, for PC of untreated urine on Whatman® 3MM paper, a preliminary run in 90% aqueous pyridine, in which solvent sugars (average R_F approximately 0.65) migrate much faster than the other constituents of urine (e.g., urea has R_F 0.45, alanine 0.18, NaCl 0.15). In this procedure, the urine sample (100 $\mu$$l$) is applied at one corner of a sheet (46 × 19 cm) of Whatman® 3MM paper. After two consecutive ascending runs (20 min, 90 min) in 90% aqueous pyridine along the breadth of the paper, a composite sugar spot forms near the solvent front, well separated from the urea spot and the various amino acids and salts. The section of the paper containing these other components is cut off and the sugars can then be separated by PC in an appropriate solvent by development along the length of the paper.

For TLC of untreated urine, Lato et al.[7] have used plates of silica gel impregnated with sodium tetraborate and sodium tungstate (see Table TLC 1 for method of preparation). As these anions complex the sugars selectively, their migration becomes independent of that of the other, non-carbohydrate substances.

Blood, Plasma, or Serum

Plasma or hemolyzed whole blood must be deproteinized, as described above for urine, before being desalted by one of the ion-exchange techniques described.

Fecal Matter

In the technique described by Menzies and Seakins,[2] suitably preserved material (approximately 5 g) is homogenized with an equal amount of acetone, and the suspension is then centrifuged. Any superficial lipid layer is discarded. The acetone extract is evaporated to dryness, and the residue, redissolved in water, is desalted by one of the procedures described for urine. The ion-exchange treatment partly removes the pigment present in such samples.

Milk

As in the case of blood, samples must be deproteinized before being desalted.

3. Isolation and Purification of Polysaccharides

Polysaccharides occur naturally as components of plant or animal tissue and are usually chemically bound or intimately mixed with other components such as proteins, nucleic acids, or other polysaccharides. There is no general procedure for purification of all polysaccharides, the method used in a particular case depending upon the possible contaminants and the properties of the polysaccharide. The details of a number of purification procedures for various types of polysaccharide will be found in several excellent books dealing with this subject,[8-11] and only a brief review is appropriate here.

In many cases, for example dextrans and plant-gum polysaccharides, isolation by fractional precipitation with ethanol or acetone from an aqueous solution of the crude material is possible. Dialysis is usually necessary to free the precipitated material of coprecipitating inorganic salts. A polysaccharide having high average molecular weight may readily be desalted by passage through a column packed with a gel of low porosity (e.g., Sephadex G-25, Bio-Gel P-2) from which the large molecules are totally excluded, so that the polysaccharide is eluted in maximum concentration at the void volume for the column, the salts emerging much later.

Acidic polysaccharides, such as pectin, alginic acid, glycosaminoglycans, and many bacterial polysaccharides, can be precipitated with quaternary ammonium salts; cetyltrimethylammonium bromide and cetylpyridinium chloride are extensively used for this purpose. By judicious variation of the salt concentration or pH, separation of acidic and neutral polysaccharides and the fractionation of mixtures of acidic polysaccharides are possible by this method. Neutral polysaccharides may be precipitated as quaternary ammonium compounds if they are first converted to borate complexes.

Other methods used in the purification of acidic polysaccharides include fractionation on ion-exchangers such as DEAE-cellulose (e.g., see Table LC 27). Electrophoretic techniques are also widely used in the isolation and purification of charged polysaccharides (e.g., see Tables EL 4 to 6). Gel permeation chromatography is useful for fractionation and purification of polysaccharides of all types (see Tables LC 39 to 45). In certain specific cases, affinity chromatography affords an excellent method of purification (see review at end of Section I.II). This technique is being used to an increasing extent in the isolation of glycoproteins.

Where polysaccharides occur closely associated with protein (e.g., bacterial polysaccharides, glycosaminoglycans), a deproteinization step is necessary before further purification is attempted. This involves digestion with proteolytic enzymes, such as pronase, pepsin, or trypsin. The resulting mixture of peptides and amino acids is separated from the polysaccharide by dialysis or gel permeation chromatography. Alternatively, if the protein is present in comparatively small proportion, as in the case of bacterial polysaccharides, the Sevag method may be used. In this procedure, an aqueous solution of the crude preparation is shaken for 30 min with chloroform and a small amount (20% of the chloroform volume) of 1-butanol or 1-pentanol; the proteins are thereby denatured, so that they become insoluble and accumulate at the chloroform-water interface. The aqueous phase is separated, and the chloroform phase is washed with the aqueous medium used (water or buffer), the washings being added to the aqueous phase. This operation is repeated until no further precipitation occurs. A modification of the Sevag method involves the use of trichlorotrifluoroethane instead of chloroform; this has been advocated[12] for deproteinization of pneumococcal polysaccharides.

Isolation of polysaccharides that are not extractable with water, such as cellulose and the so-called ''hemicelluloses'' of plant cell walls, presents special difficulty. Extraction processes for wood cellulose include acid pulping, in which the wood is digested in nitric acid or a solution of SO_2 in aqueous calcium bisulfite, or alkaline pulping, where wood chips are digested at 160 to 180°C in 5% aqueous NaOH. Hemicelluloses are extracted from plant cell walls by treatment with alkaline solutions,[13,14] but seldom without some degradation.

DEGRADATION OF POLYSACCHARIDES

4. Acetolysis*

The procedure recommended by Wolfrom and Thompson[15] for controlled acetolysis of cellulose is as follows:

The cellulose sample (45 g; air-dried cotton linters used in original reference) is stirred into a mixture of glacial acetic acid (170 ml), acetic anhydride (170 ml), and conc H_2SO_4 (18 ml), the temperature being held at 40°C by external cooling with iced water during this addition. The mixture is allowed to react at 30°C for at least 48 hr (maximum yields of peracetylated biose and triose require 72 to 100 hr), and then neutralized with $NaHCO_3$ until just acid to Congo Red paper. After standing for 18 hr, the white precipitate formed is collected by filtration, washed on the filter with water until the filtrate is neutral to litmus, and dried to constant weight over solid NaOH *in vacuo.*

5. Hydrolysis**

The quantitative liberation of the constituent monosaccharides from a polysaccharide or glycoprotein presents a major problem, since different types of glycosidic linkage are hydrolyzed at different rates and the free monosaccharides vary considerably in their stability to hot acid. The difficulties involved have been discussed in a review article by Dutton.[16] This problem has also been thoroughly investigated by Hough and co-workers,[17] who determined the recovery of the various neutral sugars commonly encountered in polysaccharides and glycoproteins (viz., L-arabinose, D-xylose, D-ribose, D-glucose, D-galactose, D-mannose, L-rhamnose and L-fucose) after treatment with sulfuric, hydrochloric, and trifluoroacetic acids at 100°C for 6 hr. Losses accompanying various neutralization procedures were also investigated. On the basis of the results of these experiments, Hough et al.[17] recommended the following procedure for optimal yield of the neutral constituents of polysaccharides and glycoproteins, prior to analysis by GC or LC.

To the polysaccharide solution (0.5 ml, containing 50 to 1000 nmol of neutral sugars) in a test tube (1 × 15 cm) is added 2 M H_2SO_4 (0.5 ml). The tube is then drawn out and sealed only after flushing with oxygen-free nitrogen for 20 min. The sealed ampule is heated at 100°C for 6 hr. The neutralization procedure recommended by these authors is to transfer the hydrolysate, after cooling, to a stoppered test tube containing N,N-dioctylmethylamine (20% v/v in $CHCl_3$; 5 ml), shake vigorously, and clarify by centrifugation; the $CHCl_3$ layer is drawn off by pipette, and residual amounts of the amine in the aqueous layer are extracted with $CHCl_3$ (5 ml).

Albersheim and co-workers[18] recommended the use of trifluoroacetic acid for the hydrolysis of the polysaccharides of plant cell walls, and this acid has subsequently been widely used for the hydrolysis of various polysaccharides, as well as glycoproteins and glycosaminoglycans. Trifluoroacetic acid has the considerable advantage over mineral acids of being sufficiently volatile to permit its removal simply by freeze-drying

---

* See also Section A, Volume II, Section II.II, page 207.
** See also Section A, Volume II, Section II.II, pages 206 to 207.

the hydrolysate. The work of Hough et al.,[17] mentioned above, has demonstrated that hydrolysis in 2 *M* trifluoroacetic acid at 100°C under nitrogen is a suitable alternative to hydrolysis in 1 *M* H₂SO₄ under the same conditions.

A hydrolysis time of 6 to 8 hr suffices for degradation of polysaccharides composed of neutral sugars, but the presence of uronic acid residues in appreciable proportion introduces the further difficulty that glycosiduronic acid linkages are, in general, much more resistant to acid hydrolysis than other glycosidic linkages. For polysaccharides (such as the capsular polysaccharides of bacteria) containing uronic acid to the extent of approximately 16 to 30 mol%, hydrolysis in 2 *M* trifluoroacetic acid at 100°C under nitrogen for 18 hr has proved satisfactory in some cases.[19] Where sugar residues particularly susceptible to degradation by acid (such as D-ribose, D-xylose, or L-rhamnose[17]) are present, it is advisable to limit the time of hydrolysis to 8 hr and subsequently correct analytical results for sugar remaining linked to uronic acid, the proportion of aldobiuronic acid in the hydrolysate being found by gel chromatography on a tightly cross-linked gel.[20]

Hydrolysis of polysaccharides containing aminodeoxy sugar residues also presents difficulties arising from the resistance to hydrolysis of linkages involving such residues. These sugars are more stable under strongly acidic conditions than are neutral sugars, however, so that analysis for aminodeoxy sugars is possible under conditions that cause destruction of the other sugar residues present. For this purpose, hydrolysis of a glycoprotein with 4 *M* HCl at 100°C for 4 to 8 hr has been recommended;[21] this should be carried out in sealed tubes under N₂. After cooling, the hydrolysate is evaporated to dryness *in vacuo* over pellets of NaOH or KOH.

Polysaccharides composed entirely of acetamidodeoxyhexose residues (chitin) or of such residues alternating with uronic acid residues (glycosaminoglycans) can be hydrolyzed completely only by the use of specific enzymes (e.g., hyaluronidase). Details of such procedures are beyond the scope of this book; the reader is referred to standard biochemical texts (e.g., cited in References 3 and 21).

6. Hydrolysis, Partial

Series of oligosaccharides are produced from polysaccharides by mild acid hydrolysis or enzymic hydrolysis under carefully controlled conditions, which vary, according to the nature of the sugar residues and linkages involved, too widely to permit any generalization here. Methods of improving yields of higher oligosaccharides by artificially protecting them from further hydrolysis have been developed by Painter.[22]

7. Methanolysis

Cleavage of polysaccharides by methanolysis has some advantages over hydrolysis, which have been discussed by Chambers and Clamp[23,24] in presenting the results of a thorough investigation of the application of this method to GC analysis of oligosaccharides and glycopeptides. Higher recoveries of neutral sugars are possible, the liberated reducing group is protected by the formation of the methyl glycoside, and the carboxyl groups of uronic acids and neuraminic acid derivatives are converted to the methyl esters, which are sufficiently volatile to permit direct analysis by GC. The use of methanolysis prior to GC analysis, therefore, makes possible the simultaneous analysis of all sugars — neutral, acidic, and basic — commonly encountered in polysaccharides and glycoproteins.

Chambers and Clamp[24] established the optimal conditions for the release and stability of neutral sugars, hexuronic acids, acetamidodeoxyhexoses, and *N*-acetylneuraminic acid on methanolysis of glycoproteins, by investigating both the release of these sugars (as their methyl glycosides) on methanolysis of such materials under various conditions and the recovery of samples of the monosaccharides treated with methanolic HCl under the conditions used in the methanolysis experiments. All the monosaccharides were found to be stable for 24 hr in methanolic 1 *M* HCl at 85°C; at 100°C slight loss (approximately 9%) of hexuronic acid occurred, but other sugars remained stable. In 2 *M* HCl at 85°C loss of hexuronic acid was approximately 3%, with the other sugars stable for 24 hr; at 100°C hexose, deoxyhexose, and acetamidodeoxyhexose remained stable, but appreciable loss of pentose and hexuronic acid and slight loss of *N*-acetylneuraminic acid occurred. At higher concentration of HCl, all monosaccharides were destroyed to various extents. Analysis of glycopeptides and oligosaccharides of known composition showed that release of carbohydrate was complete within 3 hr in methanolic 1 *M* HCl at 85°C.

Removal of methanolic HCl by rotary evaporation was found by these authors to result in extensive losses of all monosaccharides except acetamidodeoxyhexoses. These losses could be prevented by neutralization prior to evaporation.

Where acetamidodeoxy sugars are present, methanolysis should be followed by a re-*N*-acetylation step, involving treatment of the methanolysate with acetic anhydride. The procedure recommended by Clamp et al.[23-24] for methanolysis of glycoproteins is as follows.

Dry methanol, prepared by adding Mg turnings (2.5 g) and iodine (0.1 g) to methanol (500 mℓ) and heating the mixture under reflux for 1 hr, is distilled into a clean, dry container, and dry HCl gas is slowly bubbled in until the solution is 1.5 *M*. The glycoprotein sample (approximately 0.25 μmol total carbohydrate) is added to a 2 ml ampule together with internal standard (D-mannitol; 0.025 — 0.10 μmol) and dried over P_2O_5 *in vacuo*. Methanolic HCl (0.5 mℓ) is added, and a steady stream of dry nitrogen is bubbled in for 30 sec before the ampule is sealed and then heated at 90°C for 24 hr. After neutralization of the acid (addition of Ag_2CO_3 was recommended in the original paper), acetic anhydride (0.05 mℓ) is added to the ampule, which is then left at room temperature for at least 6 hr (covered with Parafilm). After thorough trituration, the sample is centrifuged and the supernatant transferred to a fresh tube. The trituration and centrifugation steps are repeated with three additions of methanol, and the pooled supernatants are then evaporated under reduced pressure at 35°C. The sample is finally dried over P_2O_5 *in vacuo* for at least 12 hr prior to trimethylsilylation (see Section 18 below) for GC analysis.

Various modifications have been introduced into this procedure by other workers. Reinhold[25] advocates methanolysis in methanolic 0.5 *M* HCl at 65°C for 16 hr; the sample is then dried under a stream of nitrogen and re-*N*-acetylation of amino groups is achieved by suspending the residue in pyridine (0.1 mℓ) and adding acetic anhydride (0.1 mℓ). The sample is immediately evaporated to dryness, and the partial *O*-acetylation occurring during the *N*-acetylation step is reversed by another short methanolysis step, the sample being refluxed in the methanolic HCl (1 mℓ) for 1 hr before derivatization for GC. More recently, Pritchard and Todd,[26] while retaining the conditions recommended by Clamp et al.[23,24] for the actual methanolysis step (except for the addition of internal standard; see below), have recommended the addition of more pyridine (0.15 mℓ) to neutralize the HCl as well as serve as a catalyst for the *N*-acetylation reaction with acetic anhydride (0.1 mℓ). After addition of pyridine and acetic anhydride, the methanolysate is allowed to stand at room temperature for 30 min and then dried over P_2O_5 *in vacuo*. These authors claim that less *O*-acetylation of the monosaccharides accompanies *N*-acetylation under these conditions, rendering a second methanolysis step unnecessary. It has been shown, however, that *O*-acetylation of alditol internal standards does occur, and, therefore, it is recommended that these be added only after both methanolysis and re-*N*-acetylation steps have been completed.

8. Methylation Analysis

Methylation analysis, which is of paramount importance in determining the nature and proportions of various types of glycosidic linkage in polysaccharide structures, has been comprehensively reviewed by Lindberg et al.[27] and by Dutton.[28] Of the various methods available for the permethylation of poly- and oligosaccharides, that developed by Hakomori[29] is the procedure of choice and has now virtually superseded the older techniques. The Hakomori method, which involves conversion of the hydroxyl groups in the sugar residues to the alkoxide form by the action of the dimethylsulfinyl ("dimsyl") carbanion in dimethyl sulfoxide followed by reaction with methyl iodide, is described below in the adaptation recommended by Sandford and Conrad.[30]

For the preparation of the dimethylsulfinyl anion, 1.5 g sodium hydride (55%, coated with mineral oil) is weighed into a dry, 300-mℓ three-necked, round-bottomed flask fitted at one neck with a rubber serum cap and containing a magnetic stirring bar. The sodium hydride is washed three times with 30-mℓ portions of *n*-pentane, and then the flask is fitted with a thermometer and a stoppered condenser. The residual *n*-pentane is removed by successive evacuation through an 18-gauge hypodermic needle inserted into the serum cap, the flask being regassed with nitrogen after each evacuation. The stopper is then removed from the condenser, and nitrogen is passed continuously through the flask via the needle. By means of a hypodermic syringe, 15 mℓ dimethyl sulfoxide, distilled from calcium hydride and stored over dried molecular sieves (Linde, type 4A), is transferred into the flask, which is then heated, with stirring, at 50°C until the solution becomes green and evolution of hydrogen ceases (approximately 45 min). The concentration of the basic "dimsyl" anion can be determined by withdrawing 1-mℓ aliquots for titration with 0.1 M HCl in aqueous solution.

For generation of the polysaccharide alkoxide, the polysaccharide sample (1 g in original Reference), which must be completely deionized and (where applicable) dephosphorylated to be soluble in dimethyl sulfoxide,[28] is, after drying overnight at 50°C *in vacuo,* added to 50 mℓ dry dimethyl sulfoxide in a three-necked flask fitted with a thermometer, condenser, and serum cap as described above. With nitrogen passing continuously through the flask, the suspension is stirred at 60°C until all the polysaccharide has dissolved (approximately 1 hr) and then, after cooling to 25°C, "dimsyl" solution (sufficient to give a 35% excess of the base over the number of equivalents of hydroxyl plus carboxyl present in the polysaccharide) is added. A gel forms immediately upon addition of the "dimsyl" ion; this gel gradually liquefies to a homogeneous viscous solution as the reaction proceeds. The reaction mixture is stirred at 25°C until alkoxide formation is complete (approximately 4 hr for most polysaccharides). An excess of "dimsyl" ion must be maintained throughout the reaction; this may be checked by using triphenylmethane as an external indicator.[31]

For the actual methylation reaction, the solution of the polysaccharide alkoxide is cooled to 20°C, and methyl iodide (3 mℓ) is added gradually, over a period of 6 to 8 min, so that the temperature does not rise above 25°C. Within a few minutes a clear solution, having markedly reduced viscosity, is formed; the reaction is then complete.

The dimethyl sulfoxide and other reagents are removed from the methylated carbohydrate by dialysis (if the molecular size of the latter permits) and repeated extraction with chloroform; the chloroform extract is then evaporated to dryness at 40°C *in vacuo.* Methylated polysaccharides should be purified before analysis by gel permeation chromatography on Sephadex LH-20 with chloroform-methanol (1:1) as eluent.[32] For methylated oligosaccharides, Merckogel OR-PVA 500, with methanol as eluent, affords a means of purification;[19] dialysis is not possible in this case.

The methylated carbohydrate must be hydrolyzed or methanolyzed under conditions that minimize demethylation. Hydrolysis with 2 M trifluoroacetic acid at 100°C for 18 hr has proved satisfactory;[19] for methanolysis, 2.4% methanolic HCl at 100°C for 18 hr is recommended.[33]

Albersheim et al.[34] have used *ethylation analysis* as a complement to methylation analysis, since some polysaccharide components that are not separable on GC as their partially methylated alditol acetates are resolved as the partially ethylated derivatives, and vice versa (see Tables GC 2 and GC 3). Ethylation is performed by the procedure described above for the Hakomori methylation, but with ethyl iodide replacing the methyl iodide in the final reaction. The whole reaction must be repeated at least once to ensure complete ethylation. Hydrolysis of the ethylated product with 2 M trifluoroacetic acid at 121°C for 2 hr, the acid and water being subsequently removed by evaporation at 50°C with a stream of filtered air, is the procedure recommended by Albersheim et al.[34] for the preparation of partially ethylated sugars.

DERIVATIZATION FOR GAS CHROMATOGRAPHY

Since carbohydrates are not sufficiently volatile for gas chromatography, conversion to volatile derivatives is necessary. The preparation of the various types of derivatives used for this purpose is briefly described below. Aspects of the most frequently used derivatization processes, such as trimethylsilylation, are discussed in depth in an excellent review article by Dutton.[16]

9. Alditol Acetates

Aldoses are reduced to alditols by sodium borohydride. The borate formed in the process must be removed before the subsequent acetylation step, as borate complexes involving the polyhydroxy compounds are liable to interfere with acetylation. This may readily be achieved by distillation with methanol after addition of acid or treatment with acidic cation-exchange resin; the trimethyl borate ester is very volatile.

The procedure recommended by Sloneker[35] for the determination of aldoses in poly- and oligosaccharides (1 to 5 mg samples) by the alditol acetate method is given below.

After hydrolysis, and neutralization or removal of the acid, sodium borohydride (1 to 5 mg depending upon the quantity of aldose) is added, and reduction is allowed to proceed for 3 hr. Excess sodium borohydride is then destroyed by acidification of the reaction mixture with acetic acid, and the sample is passed through a small column containing 1 to 2 mℓ of a cation-exchange resin in the H+-form. The column is washed several times with water (1-mℓ portions), and the washings are combined with the first effluent from the column, after which the resulting solution is evaporated to dryness on a rotary evaporator at 40°C. Methanol (1 mℓ) is added and the evaporation is repeated; this is done three times to remove borate quantitatively. After removal of the last traces of methanol, 0.1 to 0.2 mℓ of the acetylating agent is added; this consists of a 1:1 (v/v) mixture of acetic anhydride and dry pyridine. The reaction mixture, in a tightly capped tube, is shaken vigorously to dissolve the alditol residue in the acetylating agent, and the solution is heated at 100°C for 3 hr. After cooling, this solution may be injected directly into the gas chromatograph.

An alternative method of acetylation is to add anhydrous sodium acetate (10 to 15 mg) and acetic anhydride (0.2 mℓ) and heat as described. This is preferable when alditols of low molecular weight, such as ethylene glycol and glycerol, are present in a sample. Albersheim et al.[18] use the sodium acetate remaining after removal of boric acid by evaporation with methanol to catalyze the subsequent acetylation with acetic anhydride. In the procedure recommended by these authors, the deionization step described above is omitted and, after destruction of excess borohydride with glacial acetic acid, addition of methanol (1 mℓ) and evaporation to dryness is repeated five times before acetic anhydride (1 mℓ) is added and the reaction mixture heated at 121°C for 3 hr.

The mixtures of *partially methylated sugars* obtained on hydrolysis of permethylated poly- and oligosaccharides are readily derivatized to alditol acetates for GC analyses (see Table GC 2) in a manner similar to that described for the free sugars, except that the time required in the acetylation step is shorter. In the procedure recommended by Lindberg et al.,[36] the methylated sugars (1 mg) are reduced in water (5 mℓ) with sodium borohydride (10 mg) for 2 hr and, after treatment with cation-exchange resin (H⁺-form) and evaporation of the resulting solution, boric acid is removed by codistillation with methanol and the product is treated with a 1:1 (v/v) mixture of acetic anhydride and pyridine (2 mℓ) at 100°C for 10 min. Jones,[37] however, prefers acetylation to be performed at 60°C for 1 hr. If the solvent "tail" produced by pyridine interferes with GC peaks associated with methylated sugars having short retention times, the reaction mixture is not injected directly into the gas chromatograph, but is diluted with water, evaporated to dryness, and redissolved in chloroform.[36-37]

For analysis of mixtures of partially ethylated sugars (see Table GC 3). Albersheim et al.[34] form the alditol acetate derivatives as follows: The mixture of partially ethylated aldoses released on hydrolysis of the ethylated poly- or oligosaccharide (approximately 6 mg) is reduced to the alditols by addition of sodium borodeuteride (10 mg) in 1 *M* ammonia solution (1 mℓ); this reaction mixture is stirred for 2 hr at room temperature, and the excess reductant is then decomposed by dropwise addition of acetic acid until evolution of hydrogen ceases. After the reduction step, the solution is evaporated at 50°C under a stream of filtered air to remove ammonia and water, and the borate is removed by evaporating from the sample 10% glacial acetic acid in methanol (twice) and then methanol (eight times). The partially ethylated alditols are then acetylated by heating them in a sealed tube with acetic anhydride (0.5 mℓ) at 121°C for 3 hr.

10. Alditols, Permethylated*

Alditols and reduced oligosaccharides are prepared by reduction of the parent sugars with sodium borohydride or borodeuteride (as described in Section 9) and permethylated by the Hakomori method (see Section 8).

11. Aldononitrile Acetates

In the procedure recommended by Varma and Varma[38] for GC analysis of mixtures of sugars in glycoprotein hydrolysates by the aldononitrile acetate method (see Table GC 5), the hydrolysate sample (2 to 3 mg) is mixed with 6 to 7 mg dried hydroxylamine hydrochloride and dry pyridine (8 to 10 drops) in an ampule which is then sealed and heated at 90°C for 45 min. After cooling to room temperature, the ampule is opened, and dry acetic anhydride (25 to 30 drops) is added. The ampule is then resealed and heating at 90°C is continued for an additional 45 min. The cooled solution is evaporated to dryness under a stream of nitrogen and the residue is dissolved in dry chloroform (25 µℓ), from which solution aliquots are injected into the gas chromatograph.

For analysis of mixtures of partially methylated sugars by the aldononitrile acetate method (see Table GC 6), Seymour et al.[39] have described a procedure in which the sample (3 mg) is mixed with 15 mg hydroxylamine hydrochloride and 12 drops of pyridine, and the stirred mixture is heated at 60°C for 20 min. Acetic anhydride (12 drops) is added, and heating and stirring are continued for 20 min. The solution is cooled and then transferred to chloroform-water (1:2) and vigorously shaken. The chloroform layer, after re-extraction with water (2 mℓ) and drying with a molecular sieve, is used in GC analysis.

---

*   See Table GC 4 for references.

12. Butaneboronates

The cyclic esters formed when sugars or polyols react with butaneboronic acid in pyridine have been found suitable as volatile derivatives for GC (see Table GC 7). Eisenberg[40] formed the boronate by simply combining sugar and butaneboronic acid in a 1:5 ratio in pyridine (final concentration of sugar was 1 mg/mℓ). Siddiqui et al.[41] have reported that this reaction takes 4 hr or longer to reach completion at 30°C; heating the reaction mixture at 100°C for 10 min was recommended, as was subsequent trimethylsilylation (see Section 18 below).

13. Glycosides, Chiral

Glycosides formed by the D- and L-enantiomers of a sugar with a chiral alcohol are diastereoisomeric and can be separated by GC of suitable derivatives. Vliegenthart and co-workers[42] have resolved enantiomers by capillary GC of the trimethylsilylated derivatives of the glycosides formed with (−)-2-butanol. Butanolysis is performed in a manner analogous to methanolysis (see Section 7 above): the dry monosaccharide (0.5 mg) is dissolved in (−)-2-butanolic M HCl (0.5 mℓ) and heated under nitrogen in a sealed ampule at 80°C for 8 hr. The solution is then neutralized with Ag_2CO_3 the precipitate triturated thoroughly and centrifuged (2000 g for 10 min), and the supernatant solution concentrated at 45°C under reduced pressure. The residue is dried for 12 hr over P_2O_5 *in vacuo* and then trimethylsilylated (see Section 18 below) for GC analysis.

Lindberg and co-workers[43] have achieved resolution of enantiomers by capillary GC of the acetylated derivatives of glycosides formed with (+)-2-octanol. In this procedure, the sugar (0.3 to 3 mg) is placed in an ampule with (+)-2-octanol (0.5 mℓ) containing trifluoroacetic acid (1 drop) as a catalyst, a small, glass-coated magnetic rod is inserted, and the ampule is sealed and heated, with stirring, in an oil bath at 130°C overnight. The solution is then concentrated to dryness on a rotary evaporator, connected to an oil pump, at a bath temperature of 55°C, and the product is acetylated by treatment with acetic anhydride-pyridine (1:1, 1 mℓ) at 100°C for 20 min. After concentration, excess of this reagent is removed by codistillation with ethanol.

Results obtained by both of these methods are summarized in Table GC 8.

14. Isopropylidene Acetals

In the procedure recommended by Morgenlie[44] (see Table GC 9), the syrup (1 to 2 mg) obtained on concentration of an aqueous solution of the sugar is shaken with acetone (1.5 mℓ) containing 1% (v/v) of H_2SO_4. After 2 hr, the solution is neutralized with solid $NaHCO_3$ and injected into the gas chromatograph. This author claims that complete acetonation of aldoses in syrupy mixtures occurs within much shorter times of reaction than are usually considered necessary.

For analysis of an acid hydrolysate by this method,[44] the acidic residue was acetonated, and the acetone was then removed under reduced pressure and the acetal mixture dissolved in chloroform prior to injection.

15. Methylated Sugars, Acetates

Dutton and co-workers[45] have reported satisfactory resolution of some methyl ethers, notably tetra-*O*-methyl-aldohexoses, by GC as their acetates (see Table GC 10). In the procedure recommended by these authors, the methylated sugars are acetylated in acetic anhydride-pyridine (1:1), the mixtures being heated under anhydrous conditions on a steam bath for 15 min. The acetates are then extracted with chloroform, and the solution is washed with 1 M HCl (twice), followed by aqueous $NaHCO_3$ and water. The extract, dried with $CaCl_2$, is concentrated to small volume for injection. Omission of the washing procedure may lead to the production of extraneous peaks.

16. Methylated Sugars, Methyl Glycosides

The preparation of these derivatives (see Table GC 11), which involves methanolysis of the permethylated derivative of a poly- or oligosaccharide, has been described in the articles dealing with methanolysis (Section 7) and methylation analysis (Section 8).

17. Trifluoroacetates

Trifluoroacetylation of carbohydrates produces particularly volatile derivatives, well suited to use in GC analysis. Several different methods of trifluoroacetylation have been reported. Treatment with trifluoroacetic anhydride in the presence of pyridine appears satisfactory for alditols (see Table GC 15), which give single peaks on GC analysis. Shapira[46] added 2 mℓ trifluoroacetic anhydride containing approximately 10 $\mu\ell$ pyridine to an alditol mixture (160 mg) and heated at 35°C for at least 2 hr to dissolve hexitols. For mono-*O*-methyl-D-glucitols, Anderle and Kováč[47] obtained complete conversion by treating each sample (5 mg) with 0.3 mℓ trifluoroacetic anhydride in the presence of 3 $\mu\ell$ pyridine and, after vigorous shaking for 30 sec, leaving the reaction mixture to stand at room temperature for 1 hr. As in the preparation of alditol acetates (see Section 9), borate formed when sodium borohydride is used to reduce sugars to alditols must be removed before derivatization.

This procedure is unsatisfactory for trifluoroacetylation of free sugars, since extraneous peaks are obtained on GC. Zanetta et al.[48] have recommended use of a 1:1 mixture of trifluoroacetic anhydride and dichloromethane, with heating at 150°C for two periods of 5 min, for trifluoroacetylation of both free sugars (see Table GC 12) and the mixtures of *O*-methyl glycosides produced on methanolysis of polysaccharides and glycoproteins (Table GC 14). These authors found 200 $\mu\ell$ of the mixed reagent sufficient to derivatize 1 to 200 $\mu\ell$ of free carbohydrate or methyl glycoside.

If an electron detector is to be used in GC, the trifluoroacetylating reagent must be completely removed from the reaction mixture before analysis; otherwise the detector may be overloaded. For this purpose, Eklund et al.[49] recommended the following procedure:

After addition of 100 $\mu\ell$ trifluoroacetic anhydride and 100 $\mu\ell$ dichloromethane to the sugar sample under a stream of dry nitrogen, the reaction mixture, in a tightly stoppered vial, is heated in an oil bath at 130°C for 2 hr. The mixture is then cooled and evaporated under dry nitrogen, with gentle heating from an IR lamp, to a volume of approximately 10 $\mu\ell$. Dry *n*-hexane (100 $\mu\ell$) containing 1% of trifluoroacetic anhydride is added, and the mixture is again evaporated to approximately 10 $\mu\ell$. A further 100 $\mu\ell$ *n*-hexane is added to dissolve the residue prior to injection into the gas chromatograph.

Sullivan and Schewe[50] have recently reported the successful application of a newer trifluoroacetylating agent, *N*-methylbis (trifluoroacetamide), to the derivatization of free sugars. The procedure recommended by these authors is as follows:

To approximately 5 to 10 mg sugar mixture in a 5-mℓ vial, 0.5 mℓ *N*-methylbis(trifluoroacetamide) is added, followed by 0.5 mℓ pyridine. The vial is capped and heated at 65°C, with occasional shaking, for 1 hr. Small aliquots (1 $\mu\ell$) are then injected directly into the gas chromatograph.

It is claimed that this procedure results in more reproducible derivatization of free sugars than does the use of trifluoroacetic anhydride.

18. Trimethylsilyl Ethers*

Trimethylsilylation is the most frequently used method of derivatization for GC of carbohydrates, as of many other classes of organic compound, owing to the rapidity

---

*   See also Section A, Volume II, Section II.11, pages 214 to 226.

and relative simplicity of the technique. The method originally proposed by Sweeley et al.[51] in 1963 is still widely used; this is described below.

The carbohydrate sample (10 mg or less) is dissolved in 1.0 mℓ of anhydrous pyridine (dried over KOH pellets) and treated successively with 0.2 mℓ of hexamethyldisilazane (HMDS) and 0.1 mℓ of trimethylchlorosilane (TMCS). The mixture, in a stoppered vial, is shaken vigorously for about 30 sec and then allowed to stand for at least 5 min (reaction is normally complete within this time). If the carbohydrate is sparingly soluble in the mixture, the vial may be warmed at 75 to 85°C for 2 to 3 min. In cases where no molecular rearrangement is likely to occur, the sugar may be dissolved by warming in pyridine prior to addition of the silylating reagents. This method may be applied to proportionately smaller amounts of sugar and reagents.

In the original procedure, aliquots of the reaction mixture were injected directly into the gas chromatograph, but the solvent "tailing" arising from the presence of pyridine is a severe disadvantage if components having short retention times are present. For this reason, it is advisable to replace the pyridine, after the trimethylsilylation reaction, by another solvent (*n*-hexane is recommended[52]), the pyridine being removed by evaporation at 40°C under reduced pressure. An alternative procedure is extraction of the trimethylsilyl ethers into chloroform followed by washing of the chloroform layer with hydrochloric acid. Details of this method, originally proposed by Partridge and Weiss,[53] are as follows.

After trimethylsilylation of the carbohydrate sample by the method of Sweeley, chloroform (1 mℓ) is added to the reaction mixture, which is then well shaken. The mixture is cooled in ice, treated with 3 *M* HCl (3 mℓ), and again shaken vigorously. The biphasic system formed is allowed to settle, still cooled in ice, and the upper aqueous layer is pipetted off. The chloroform layer is washed twice with 1 mℓ of water, the aqueous layer being discarded each time, and the remaining chloroform extract is then used in GC analysis.

The use of dimethylsulfoxide or *N,N*-dimethylformamide as the reaction solvent instead of pyridine has been recommended by Ellis,[54] following a thorough investigation of the trimethylsilylation of monosaccharides in these solvents. In the procedure suggested by this author, the monosaccharides were equilibrated in water at 20°C for 24 hr, or by refluxing for 1 hr, and then the water was removed by evaporation at 40°C *in vacuo,* and the sugar mixture was dissolved in sufficient solvent (1 to 3 mℓ) to give a concentration of 0.2 to 6.0 mg of each individual sugar per milliliter. The solvents were dried over KOH pellets before use. The trimethylsilylation reaction was carried out in an 8-mℓ tube fitted with a plastic-lined cap. HMDS (0.3 volumes) and TMCS (0.2 volumes) were added successively to the solution, with intermediate mixing and final thorough equilibration by vigorous shaking, with inversion, of the stoppered tube. The mixture was then allowed to stand for a time dependent upon the solvent used — with dimethylsulfoxide separation into two phases occurred rapidly (after 1 to 3 min), but with dimethylformamide 16 to 20 hr was necessary. The trimethylsilyl ethers of the sugars were found to be present in very high concentration in the upper layer, which resembled a silicone and was assumed to be hexamethyldisiloxane, in which the trimethylsilylated derivatives are highly soluble. This obviated the necessity for any further concentration prior to GC analysis, for which aliquots of this hexamethyldisiloxane layer were used.

This method has since been used fairly extensively for trimethylsilylation of sugars; it is also applicable to sugar phosphates.[55] Dimethylsulfoxide is clearly the solvent of choice in view of the greater speed of formation of the hexamethyldisiloxane layer. To dissolve sparingly soluble carbohydrates, sonication should be used,[55] as heating is to be avoided.

Another variant in the HMDS procedure for trimethylsilylation is the use of trifluoroacetic acid as a catalyst instead of TMCS, which has been recommended by Brobst and Lott[56-57] for the analysis of corn syrup. The Sweeley procedure[51] demands anhydrous conditions and, therefore, where hydrolysates such as corn syrup are to be analyzed, problem arise owing to the presence of traces of water that are difficult to remove. Moderate amounts of water may be tolerated if the procedure of Brobst and Lott is used. Details are given below.

Syrups containing less than 25% of water may be sampled directly, without further drying. The sample (60 to 70 mg) is transferred to a vial fitted with a stopper and dissolved in 1.0 mℓ pyridine; this usually requires allowing the mixture to stand overnight at approximately 25°C or warming it in a bath at 60°C for 15 min. The tube must be kept tightly capped throughout this time. When the sample has dissolved, HMDS (0.9 mℓ) and 99% trifluoroacetic acid (0.1 mℓ) are added successively, and the tube is capped, shaken vigorously for 30 sec, and allowed to stand for 30 min with occasional shaking and intermittent release of gas pressure (ammonia is evolved). Properly prepared derivatives should give solutions clear to the point of brilliance. If the solution is not clear, it should be warmed at 60°C for 5 to 10 min. No precipitate is formed during this derivatization process, in contrast to the Sweeley method, which results in the formation of a fine precipitate of ammonium chloride; ammonium trifluoroacetate is soluble in pyridine.

An alternative method of trimethylsilylating sugars in the presence of water is the use of the reagent *N*-(trimethylsilyl)imidazole (TSIM), which can be added to sugars in aqueous solution;[58] it has also been successfully applied to silylation of an aqueous polyhydric alcohol mixture.[59] The reagent can be added directly to dry residues as well. For example, Cahour and Hartmann,[60] in trimethylsilylating the residues obtained after evaporation of the methanolysates from glycoproteins, added TSIM (20 μℓ) to each dry residue, dissolved the solid by grinding, and then added 80 μℓ of water (doubly distilled) and 50 μℓ of carbon disulfide with thorough mixing. This resulted in extraction of the trimethylsilylated derivatives into the organic phase, the excess TSIM remaining in the aqueous phase.

Another alternative trimethylsilylating reagent that has been applied to carbohydrates, though with somewhat erratic results, is *N,O*-bis(trimethylsilyl)acetamide (BSA). Janauer and Englmaier[61] recently reported the application of this reagent in GC analysis of carbohydrates ranging from tetroses to tetrasaccharides, with a derivatization procedure in which BSA (130 μℓ) was added to a solution of the sugar mixture (0.5 to 1.0 mg) in pyridine (70 μℓ) in a stoppered vial; the vial was capped immediately and kept at room temperature for 30 min before aliquots were taken for GC analysis. The derivatives obtained were described as "only moderately volatile", and it is possible, therefore, that the sugars were not completely trimethylsilylated by this method. For the derivatization of aminodeoxy sugars, BSA in pyridine proved unsuccessful unless it was used together with other silylating reagents. Complete conversion was effected within 30 min with a mixture of BSA, TMCA, HMDS, and pyridine (1:0.5:1:10) at room temperature.[62] This procedure gave both *N*- and *O*-trimethylsilylation.

The fluorinated analogue of BSA, *N,O*-bis(trimethylsilyl)-trifluoroacetamide (BSTFA), has approximately the same donor strength as BSA but gives products of greater volatility. This reagent has been applied to derivatization of carbohydrates in several instances, an example being its use in GC analysis of heptuloses from the avocado.[63]

The application of yet another trimethylsilylating agent, dimethyl-*N*-(trimethylsilyl)amine (TMSDMA), to the derivatization of carbohydrates was recently reported by Yokota and Mori,[64] who used this reagent, with TMCS as catalyst, in the trimethylsilylation of the methanolysates from glycosaminoglycans. The dried residue obtained after methanolysis (10 to 1000 μg of carbohydrate) was dissolved in a mixture of pyridine (0.6 mℓ), TMSDMA (0.3 mℓ), and TMCS (0.1 mℓ) in an ampule that was then sealed and heated at 120°C for 24 hr. After cooling, the trimethylsilylated derivatives were extracted with chloroform. These authors have commented, however, that TMSDMA was not found more effective than HMDS in trimethylsilylating amino-deoxy sugars, despite a reported greater effectiveness in derivatizing amino groups in amino acids, and, therefore, this reagent has no special advantages.

For most purposes, the reagents originally proposed by Sweeley are perfectly satisfactory. The diversity of the classes of carbohydrate that can be analyzed by GC of their trimethylsily derivatives is exemplified by the data presented in Tables GC 16 to 25.

19. Trimethylsilylated Alditols and Oximes

One disadvantage of GC analysis of reducing sugars as their trimethylsilyl ethers is the production of multiple peaks, corresponding to the various isomeric forms present at equilibrium. This may be a useful diagnostic feature, but for many practical purposes a simple chromatogram is desirable. One method of achieving this is by reduction of the sugars to *alditols* by treatment with NaBH₄ or NaBD₄ (procedure described in Section 9) prior to trimethylsilylation. Simple alditols are generally rather poorly resolved on GC as their trimethylsilyl ethers, but the method has proved more successful when applied to reduced disaccharides[65] (Table GC 21). For analysis of the mixtures of methylated sugars obtained on hydrolysis of permethylated poly- and oligosaccharides, this technique is particularly useful, since the simpler chromatograms obtained greatly facilitate peak assignment. In the procedure described by Stephen and co-workers,[66] the partially methylated sugars are reduced in water with sodium borohydride, and the resulting mixture of partially methylated alditols is extracted into chloroform and dried *in vacuo* before trimethylsilylation by the method of Sweeley.[51] The mixtures of trimethylsilylated, partially methylated alditols are allowed to stand in sealed vials for 1 hr before GC analysis (see Table GC 22).

Another method of simplifying GC of trimethylsilylated sugars is by prior conversion of the sugars to their *oximes*. Trimethylsilylated oximes may give two peaks on GC, owing to the presence of *syn* and *anti* forms (of which one usually preponderates), but the chromatograms are still far less complex than those given by the trimethylsilyl ethers of sugars as such. The O-methyl oximes are also frequently used. Oximes are prepared by treatment of the sugars with hydroxylamine hydrochloride in pyridine. In the procedure used by Sweeley et al.,[51] the sugar (10 mg) and hydroxylamine hydrochloride (6 mg) were dissolved in pyridine (1 mℓ), and the mixture was heated at 70 to 80°C for 30 min, after which the solution was cooled and the silylating reagents were added. Petersson[67] advocates sonication at 30°C for 2 hr rather than heating. The O-methyl oximes are prepared in analogous manner by substitution of methoxyamine hydrochloride for hydroxylamine hydrochloride. In the procedure proposed by Laine and Sweeley,[68] the sugar (1 mg) was mixed with methoxyamine hydrochloride (1 mg) and pyridine (50 μℓ) in a Teflon-capped vial, and the stoppered vial was heated at 80°C for 2 hr, after which the trimethylsilyl ether was prepared by addition of BSTFA and heating at 80°C for a further 15 min. Satisfactory results have been obtained with both derivatives (see Table GC 20).

20. Uronic Acid Derivatives

Uronic acids can be analyzed as their methyl ester methyl glycosides (see Table GC 11) or the trimethylsilylated derivatives of these (see Tables GC 19 and GC 24). The acids themselves, usually as salts,[69,70] can be trimethylsilylated and have been examined by GC in this form (Table GC 24), but the chromatograms obtained are very complicated owing to the production of multiple peaks. It has been noted[70] that dimethyl sulfoxide is the best solvent for trimethylsilylation of hexuronic acids and lactones.

The problem of multiple peaks can be eliminated by reduction of the uronic acids to aldonic acids, followed by lactonization and trimethylsilylation of the lactones. The procedure recommended by Perry and Hulyalkar[71] is as follows.

The mixture of hexuronic acids (approximately 1 to 10 mg) is dissolved in water (2 m*l*), and the acids are converted to salts by stirring the solution with barium carbonate (150 mg) at 65°C for 10 min. After insoluble matter has been removed by filtration, the neutralized hexuronic acids in the solution are reduced to their corresponding aldonic acids by treatment with sodium borohydride (50 mg). The reaction mixture is allowed to stand at room temperature for 1 hr, and then excess borohydride is destroyed by addition of cation-exchange resin (H[+]-form) and the solution is passed through a small column containing the same resin (approximately 6 m*l*). The eluate from the column, combined with the washings, is evaporated to dryness under reduced pressure, and the boric acid is removed from the residue as volatile methyl borate by several distillations with dry methanol (5 times, 20 m*l* each). The residue finally obtained is dissolved in concentrated hydrochloric acid (0.5 m*l*) and the solution is evaporated to dryness *in vacuo*. The aldono-1,4-lactones thereby produced are then trimethylsilylated by the method of Sweeley,[51] the trimethylsilylated derivatives being extracted into chloroform for GC analysis.

Petersson[67] has recommended analysis of the aldonic acids themselves as trimethylsilyl esters, without prior lactonization. This requires initial saponification of lactones existing in solution in equilibrium with the acids. The sample (0.1 to 10 mg) containing aldonic acids is dissolved in water (2 to 5 m*l*), and the pH is adjusted to 8.5 by addition of 0.05 *M* NaOH. The solution is stirred for 4 hr at room temperature, the pH being maintained at 8.5 by automatic titration with 0.05 *M* NaOH. If the sample is not known to be alkali-stable, the pH is adjusted to 7 with 0.01 *M* HCl before evaporation, which is carried out at 35°C on a rotary evaporator. The sample is further dried by two successive evaporations with dichloromethane (1 to 2 m*l*) at 35°C, or by vacuum desiccation over P_2O_5 for at least 24 hr. Dry pyridine (1 to 2 m*l*) is then added as a solvent for trimethylsilylation with BSTFA + TMCS (2:1), with sonication for 2 hr. The resulting solutions may be injected directly into the gas chromatograph, but removal of pyridine by evaporation at 30°C under reduced pressure and subsequent extraction of the residue into anhydrous diethyl ether (0.5 to 1 m*l*) containing 0.5% of BSTFA was considered preferable by Petersson,[67] who has also applied this method of derivatization to other carbohydrate related acids, such as aldaric acids[72] (see Table GC 24).

DERIVATIZATION FOR HPLC

21. Benzoylation and 4-Nitrobenzoylation of Carbohydrates for HPLC with UV Detection

The introduction of a UV-absorbing chromophore by prior derivatization makes possible the use of the more sensitive UV detector in HPLC of carbohydrates (see Section II.II). This can be achieved by *benzoylation,* for which Lehrfeld[73] recommends the following procedure.

The sample (1 to 10 mg) is dissolved in pyridine (2 m*l*) in a stoppered vial, and the solution is heated at 62°C for 1 hr to promote anomeric equilibration. After cooling to room temperature, the solution is treated with benzoyl chloride (0.2 m*l*) and heated at 62°C for a further 45 min. The reaction mixture is cooled, methanol (0.5 m*l*) is added to decompose residual benzoyl chloride, and the solution is then evaporated to dryness under a stream of dry nitrogen. The residue is redissolved in chloroform (2.5 ml), and the solution is washed with 1 M HCl (1 ml), neutralized with solid NaHCO$_3$ and dried with anhydrous Na$_2$SO$_4$. The resulting suspension is filtered, the residue is twice washed with chloroform, and the filtrate and washings are combined and evaporated to approximately 1 ml.

More recently, *nitrobenzoylation* has been recommended as a means of derivatization of sugars for HPLC with a UV detector; it is claimed[74] that the incorporation of the additional chromophoric group into the aryl portion of the derivative lowers the detection limit by a factor of 10. The successful separation of a number of carbohydrates as their 4-nitrobenzoates by HPLC has been reported (see Table LC 7). The method of 4-nitrobenzoylation used by Nachtmann et al.[74,75] is as follows:

The sample is dissolved in pyridine, as is the reagent 4-nitrobenzoyl chloride (100 mg/ml pyridine). To 50 μl of the sample solution (containing not more than 0.5 mg) 150 μl of the reagent solution is added, and the mixture is well shaken and then allowed to stand at room temperature for 10 min. The pyridine is removed by evaporation *in vacuo* at 50°C (unless fructose or any other sugar likely to be affected by warming in pyridine is present, in which case room temperature is not exceeded), and the sample tube is then flushed with air or nitrogen, after which 2 ml of a 5% aqueous solution of NaHCO$_3$, containing 5 mg of 4-dimethylaminopyridine, is added. The residual 4-nitrobenzoyl chloride is hydrolyzed after 5 min sonication or vigorous shaking of this mixture. The derivatized carbohydrates are then extracted with chloroform (2 ml), and the extracts are washed with 5% NaHCO$_3$ (2 ml) and twice with 0.05 M HCl containing 5% NaCl (3 ml each time).

Derivatization has so far not been much used in HPLC of carbohydrates, but it has obvious advantages where greater sensitivity of detection compared to the RI detector is necessary.

REFERENCES

1. **Holligan, P. M. and Drew, E. A.,** Routine analysis by gas-liquid chromatography of soluble carbohydrates in extracts of plant tissues. II. Quantitative analysis of standard carbohydrates, and the separation and estimation of soluble sugars and polyols from a variety of plant tissues, *New Phytol.,* 70, 271, 1971.
2. **Menzies, I. S. and Seakins, J. W. T.,** Sugars, in *Chromatographic and Electrophoretic Techniques,* Vol. 1, 4th ed., Smith, I. and Seakins, J. W. T., Eds., Heinemann Medical Books, London, 1976, 183.
3. **Sweeley, C. C., Wells, W. W., and Bentley, R.,** Gas chromatography of carbohydrates, in *Methods in Enzymology,* Vol. 8, Neufeld, E. F. and Ginsburg, V., Eds., Academic Press, New York, 1966, 95.
4. **Vitek, V. and Vitek, K.,** Chromatography of sugars in body fluids. I. Preparation of urine for paper chromatographic analysis, *J. Chromatogr.,* 60, 381, 1971.
5. **Meagher, R. B. and Furst, A.,** Reversed phase high-pressure liquid chromatography of normal rat urinary carbohydrates, *J. Chromatogr.,* 117, 211, 1976.
6. **Saini, A. S.,** Separation of sugars from untreated urine, *J. Chromatogr.,* 61, 378, 1971.
7. **Ghebregzabher, M., Rufini, S., Ciuffini, G., and Lato, M.,** A two-dimensional thin-layer chromatographic method for screening carbohydrate anomalies, *J. Chromatogr.,* 95, 51, 1974.
8. **Whistler, R. L., BeMiller, J. N., and Wolfrom, M. L., Eds.,** *Methods in Carbohydrate Chemistry,* Vol. 5, Academic Press, New York, 1965.
9. **Pigman, W. and Horton, D., Eds.,** *The Carbohydrates: Chemistry and Biochemistry,* Vols. 2A and 2B, 2nd ed., Academic Press, New York, 1970.
10. **Whistler, R. L., Ed.,** *Industrial Gums: Polysaccharides and Their Derivatives,* 2nd ed., Academic Press, New York, 1973.
11. **Brimacombe, J. S. and Webber, J. M.,** *Mucopolysaccharides: Chemical Structure, Distribution and Isolation,* Elsevier, Amsterdam, 1964.

12. Markowitz, A. S. and Lange, C. F., Removal of proteins with trifluorotrichloroethane, in *Methods in Carbohydrate Chemistry*, Vol. 5, Whistler, R. L. and Wolfrom, M. L., Eds., Academic Press, New York, 1965.

13. Whistler, R. L. and Feather, M. S., Hemicellulose: Extraction from annual plants with alkaline solutions, in *Methods in Carbohydrate Chemistry*, Vol. 5, Whistler, R. L., BeMiller, J. N., and Wolfrom, M. L., Eds., Academic Press, New York, 1965.

14. Blake, J. D., Murphy, P. T., and Richards, G. N., Isolation and A/ B classification of hemicelluloses, *Carbohydr. Res.*, 16, 49, 1971.

15. Wolfrom, M. L. and Thompson, A., Acetolysis: polymer-homologous series of oligosaccharides from cellulose; their alditols and anomeric acetates, in *Methods in Carbohydrate Chemistry*, Vol. 3, Whistler, R. L., Green, J. W., BeMiller, J. N., and Wolfrom, M. L., Eds., Academic Press, New York, 1963, 143.

16. Dutton, G. G. S., Applications of gas-liquid chromatography to carbohydrates. I, *Adv. Carbohydr. Chem. Biochem.*, 28, 11, 1973.

17. Hough, L., Jones, J. V. S., and Wusteman, P., On the automated analysis of neutral monosaccharides in glycoproteins and polysaccharides, *Carbohydr. Res.*, 21, 9, 1972.

18. Albersheim, P., Nevins, D. J., English, P. D., and Karr, A., A method for the analysis of sugars in plant cell-wall polysaccharides by gas-liquid chromatography, *Carbohydr. Res.*, 5, 340, 1967.

19. Merrifield, E. H., and Stephen, A. M., Structural studies on the capsular polysaccharide from *Klebsiella* serotype K64, *Carbohydr. Res.*, 74, 241, 1979.

20. Dutton, G. G. S., Stephen, A. M., and Churms, S. C., Structural investigation of *Klebsiella* K7 polysaccharide, *Carbohydr. Res.*, 38, 225, 1974.

21. Spiro, R. G., Analysis of sugars found in glycoproteins, in *Methods in Enzymology*, Vol. 8, Neufeld, E. F. and Ginsburg, V., Eds., Academic Press, New York, 1966, 3.

22. Painter, T. J., Partial acidic and enzymic hydrolysis: some methods for improving the yields of oligosaccharides, in *Methods in Carbohydrate Chemistry*, Vol. 5, Whistler, R. L., BeMiller, J. N., and Wolfrom, M. L., Eds., Academic Press, New York, 1965.

23. Bhatti, T., Chambers, R. E., and Clamp, J. R., The gas chromatographic properties of biologically important N-acetylglucosamine derivatives, monosaccharides, disaccharides, trisaccharides, tetrasaccharides and pentasaccharides, *Biochim. Biophys. Acta*, 222, 339, 1970.

24. Chambers, R. E. and Clamp, J. R., An assessment of methanolysis and other factors used in the analysis of carbohydrate-containing materials, *Biochem. J.*, 125, 1009, 1971.

25. Reinhold, V. N., Gas-liquid chromatographic analysis of constituent carbohydrates in glycoproteins, in *Methods in Enzymology*, Vol. 25 (Part B9), Hirs, C. H. W. and Timasheff, S. N., Eds., Academic Press, New York, 1972, 244.

26. Pritchard, D. G. and Todd, C. W., Gas chromatography of methyl glycosides as their trimethylsilyl ethers. The methanolysis and re-N-acetylation steps, *J. Chromatogr.*, 133, 133, 1977.

27. Björnal, H., Hellerqvist, C. G., Lindberg, B., and Svensson, S., Gas-liquid chromatography and mass spectrometry in methylation analysis of polysaccharides, *Angew. Chem. Int. Ed. Engl.*, 9, 610, 1970.

28. Dutton, G. G. S., Applications of gas-liquid chromatography to carbohydrates. II, *Adv. Carbohydr. Chem. Biochem.*, 30, 9, 1974.

29. Hakomori, S., Rapid permethylation of glycolipids and polysaccharides, catalyzed by methylsulfinyl carbanion in dimethyl sulfoxide, *J. Biochem. (Tokyo)*, 55, 205, 1964.

30. Sandford, P. A. and Conrad, E. E. The structure of the *Aerobacter aerogenes* A3 (S1) polysaccharide. I. A re-examination using improved procedures for methylation analysis, *Biochemistry*, 5, 1508, 1966.

31. Aspinall, G. O. and Stephen, A. M., Polysaccharide methodology and plant polysaccharides, in *Carbohydrates (MTP International Review of Science: Organic Chemistry)*, Ser. 1, Vol. 7, Aspinall, G. O., Ed., Butterworth, London, 1973, 285.

32. Talmadge, K. W., Keegstra, K., Bauer, W. D., and Albersheim, P., The structure of plant cell walls. I. The macromolecular components of the walls of suspension-cultured sycamore cells with a detailed analysis of the pectic polysaccharides, *Plant Physiol.*, 51, 158, 1973.

33. Stephen, A. M., Kaplan, M., Taylor, G. L., and Leisegang, E. C., Application of gas-liquid chromatography to the structural investigation of polysaccharides. I. The structure of the gums of *Virgilia oroboides* and *Agave americana*, *Tetrahedron*, 7, 233, 1966.

34. Sweet, D. P., Albersheim, P., and Shapiro, R. H., Partially ethylated alditol acetates as derivatives for elucidation of the glycosyl linkage-composition of polysaccharides, *Carbohydr. Res.*, 40, 199, 1975.

35. Sloneker, J. H., Gas liquid chromatography of alditol acetates, in *Methods in Carbohydrate Chemistry*, Vol. 6, Whistler, R. L., and BeMiller, J. N., Eds., Academic Press, New York, 1972, 20.

36. Björndal, H., Lindberg, B., and Svensson, S., Gas-liquid chromatography of partially methylated alditols as their acetates, *Acta Chem. Scand.*, 21, 1801, 1967.

37. Jones, H. G., Gas-liquid chromatography of methylated sugars, in *Methods in Carbohydrate Chemistry*, Vol. 6, Whistler, R. L. and BeMiller, J. N., Eds., Academic Press, New York, 1972, 25.

38. Varma, R. and Varma, R. S., Simultaneous determination of neutral sugars and hexosamines in glycoproteins and acid mucopolysaccharides (glycosaminoglycans) by gas liquid chromatography, *J. Chromatogr.*, 128, 45, 1976.

39. Seymour, F. R., Plattner, R. D., and Slodki, M. E., Gas-liquid chromatography-mass spectrometry of methylated and deuteriomethylated per-*O*-acetyl-aldononitriles from D-mannose, *Carbohydr. Res.*, 44, 181, 1975.

40. Eisenberg, F., Jr., Cyclic butaneboronic acid esters: novel derivatives for the rapid separation of carbohydrates by gas-liquid chromatography, *Carbohydr. Res.*, 19, 135, 1971.

41. Wood, P. J., Siddiqui, I. R. and Weisz, J., The use of butaneburonic acid esters in the gas-liquid chromatography of some carbohydrates, *Carbohydr. Res.*, 42, 1, 1975.

42. Gerwig, G. J., Kamerling, J. P., and Vliegenthart, J. F. G., Determination of the D and L configuration of neutral monosaccharides by high-resolution capillary g.l.c., *Carbohydr. Res.*, 62, 349, 1978.

43. Leontein, K., Lindberg, B., and Lonngren, J., Assignment of absolute configuration of sugars by g.l.c. of their acetylated glycosides formed from chiral alcohols, *Carbohydr. Res.*, 62, 359, 1978.

44. Morgenlie, S., Analysis of mixtures of the common aldoses by gas chromatography-mass spectrometry of their *O*-isopropylidene derivatives, *Carbohydr. Res.*, 41, 285, 1975.

45. Bebault, G. M., Dutton, G. G. S., and Walker, R. H., Separation by gas-liquid chromatography of tetra-*O*-methylaldohexoses and other sugars as acetates, *Carbohydr. Res.*, 23, 430, 1972.

46. Shapira, J., Identification of sugars as their trifluoroacetyl polyol derivatives, *Nature (London)*, 222, 792, 1969.

47. Anderle, D. and Kovač, P., Gas chromatography of partially methylated alditols as trifluoroacetyl derivatives. I. Separation of mono-*O*-methyl-per-*O*-TFA-D-glucitols, *J. Chromatogr.*, 49, 419, 1970.

48. Zanetta, J. P., Breckenridge, W. C., and Vincendon, G., Analysis of monosaccharides by gas-liquid chromatography of the *O*-methyl glycosides as trifluoroacetate derivatives. Application to glycoproteins and glycolipids, *J. Chromatogr.*, 69, 291, 1972.

49. Eklund, G., Josefsson, B., and Roos, C., Gas-liquid chromatography of monosaccharides at the picogram level using glass capillary columns, trifluoroacetyl derivatization, and electron-capture detection, *J. Chromatogr.*, 142, 575, 1977.

50. Sullivan, J. E. and Schewe, L. R., Preparation and gas chromatography of highly volatile trifluoroacetylated carbohydrates using *N*-methylbis (trifluoroacetamide), *J. Chromatogr. Sci.*, 15, 196, 1977.

51. Sweeley, C. C., Bentley, R., Makita, M., and Wells, W. W., Gas-liquid chromatography of trimethylsilyl derivatives of sugars and related substances, *J. Am. Chem. Soc.*, 85, 2497, 1963.

52. Bethge, P. O., Holmström, C., and Juslin, S., Quantitative gas chromatography of mixtures of simple sugars, *Sven. Papperstidn.*, 69, 60, 1966.

53. Partridge, R. D. and Weiss, A. H., Extraction procedure for TMS carbohydrate ethers, *J. Chromatogr. Sci.*, 8, 553, 1970.

54. Ellis, W. C., Solvents for the formation and quantitative chromatography of trimethylsilyl derivatives of monosaccharides, *J. Chromatogr.*, 41, 325, 1969.

55. Leblanc, D. J. and Ball, A. J. S., Fast one-step method for the silylation of sugars and sugar phosphates, *Anal. Biochem.*, 84, 574, 1978.

56. Brobst, K. M. and Lott, C. E., Jr., Determination of some components in corn syrup by gas-liquid chromatography of the trimethylsilyl derivatives, *Cereal Chem.*, 43, 35, 1966.

57. Brobst, K. M., Gas-liquid chromatography of trimethylsilyl derivatives: analysis of corn syrup, in *Methods in Carbohydrate Chemistry*, Vol. 6, Whistler, R. L. and BeMiller, J. N., Eds., Academic Press, New York, 1972, 3.

58. Brittain, G., Development of a reagent for gas chromatography, *Am. Lab.*, 57, 1969.

59. van Ling, G., Silylation and g.l.c. analysis of an aqueous polyhydric alcohol mixture, *J. Chromatogr.*, 44, 175, 1969.

60. Cahour, A. and Hartmann, L., Study of natural and aminomonosaccharides by gas-liquid differential chromatography: application to three reference glycoproteins, *J. Chromatogr.*, 152, 475, 1978.

61. Janauer, G. A. and Englmaier, P., Multistep time program for the rapid gas-liquid chromatography of carbohydrates, *J. Chromatogr.*, 153, 539, 1978.

62. Kärkkäinen, J. and Vihko, R., Characterisation of 2-amino-2-deoxy-D-glucose, 2-amino-2-deoxy-D-galactose and related compounds, as their trimethylsilyl derivatives by gas-liquid chromatography-mass spectrometry, *Carbohydr. Res.*, 10, 113, 1969.

63. Johansson, I. and Richtmyer, N. K., The isolation of both a *talo-* heptulose and an *allo*-heptulose from the avocado *Carbohydr. Res.*, 13, 461, 1970.

64. **Yokota, M. and Mori, T.**, Gas-liquid chromatographic analysis of the monosaccharide composition of acid glycosaminoglycans (mucopolysaccharides) derived from animal tissues, *Carbohydr. Res.*, 59, 289, 1977.

65. **Kärkkäinen, J.**, Determination of the structure of disaccharides as *O*-trimethylsilyl derivatives of disaccharide alditols by gas-liquid chromatography-mass spectrometry, *Carbohydr. Res.*, 11, 247, 1969.

66. **Freeman, B. H., Stephen, A. M., and van der Bijl, P.**, Gas-liquid chromatography of some partially methylated alditols as their trimethylsilyl ethers, *J. Chromatogr.*, 73, 29, 1972.

67. **Petersson, G.**, Gas-chromatographic analysis of sugars and related hydroxy acids as acyclic oxime and ester trimethylsilyl derivatives, *Carbohydr. Res.*, 33, 47, 1974.

68. **Laine, R. A. and Sweeley, C. C.**, *O*-methyl oximes of sugars. Analysis as *O*-trimethylsilyl derivatives by gas-liquid chromatography and mass spectrometry, *Carbohydr. Res.*, 27, 199, 1973.

69. **Raunhardt, O., Schmidt, H. W. H., and Neukom, H.**, Gas chromatographic investigations of uronic acids and uronic acid derivatives, *Helv. Chim. Acta*, 50, 1267, 1967.

70. **Kennedy J. F., Robertson, S. M. and Stacey, M.**, GLC of the *O*-trimethylsilyl derivatives of hexuronic acids, *Carbohydr. Res.*, 49, 243, 1976.

71. **Perry, M. B. and Hulyalkar, R. K.**, The analysis of hexuronic acids in biological materials by gas-liquid partition chromatography, *Can. J. Biochem.*, 43, 573, 1965.

72. **Petersson, G.**, Retention data in GLC analysis: carbohydrate-related hydroxy carboxylic and dicarboxylic acids as trimethylsilyl derivatives, *J. Chromatogr. Sci.*, 15, 245, 1977.

73. **Lehrfeld, J.**, Separation of some perbenzoylated carbohydrates by high-performance liquid chromatography, *J. Chromatogr.*, 120, 141, 1976.

74. **Nachtmann, F. and Budna K. W.**, Sensitive determination of derivatized carbohydrates by high performance liquid chromatography, *J. Chromatogr.*, 136, 279, 1977.

75. **Nachtmann, F., Spitzy, H., and Frei, R. W.**, Ultraviolet derivatization of digitalis glycosides as 4-nitrobenzoates for liquid chromatographic trace analysis, *Anal. Chem.*, 48, 1576, 1976.

Section IV

Products and Sources of Chromatographic

Materials

PRODUCTS AND SOURCES OF CHROMATOGRAPHIC MATERIALS

This section contains descriptions and sources of various widely used chromatographic materials, with emphasis on those used in the examples cited in the tables (Section I). It should be noted that:

1. Inclusion of specific materials and sources should not be regarded as an endorsement of these by the Editors or CRC Press, Inc.
2. Materials not mentioned may serve equally well in many cases.
3. Materials may be available from other sources in addition to those given.

Readers interested in comprehensive listings of chromatographic instruments, accessories, supplies, services, manufacturers and suppliers are referred to the current issues of the following guides:

1. *Analytical Chemistry* Lab Guide
2. *Journal of Chromatographic Science* International Chromatography Guide
3. *Science* Guide to Scientific Instruments
4. *American Laboratory* Laboratory Buyer's Guide Edition

Table 1
LIQUID PHASES FOR GAS CHROMATOGRAPHY

Phase	Type	Max temp (°C)	Solvent	McReynolds constants					
				X′	Y′	Z′	U′	S′	H
Apiezon L	Hydrocarbon	300	B	32	22	15	32	42	13
Apiezon M	Hydrocarbon	275	B	31	22	15	30	40	12
Apiezon N	Hydrocarbon	325	B	38	40	28	52	58	25
Butanediol succinate	Polyester	225	C	370	571	448	657	611	457
Carbowax 6000 }	Polyethylene	200	C	322	540	369	577	512	390
Carbowax 20M }	glycols	225	C	322	536	368	572	510	387
Dexsil 300GC	*meta*-Carborane	500	C	47	80	103	148	96	55
ECNSS-M	Silicone polyester	200	C	421	690	581	803	732	548
Ethylene glycol adipate	Polyester	200	A	372	576	453	655	617	462
Ethylene glycol succinate	Polyester	200	C	537	787	643	903	889	633
Neopentyl glycol sebacate	Polyester	240	C	172	327	225	344	326	257
Neopentyl glycol succinate	Polyester	240	A	272	469	336	539	474	371
Polyphenyl ether (6-ring)	Ether	350	A, C, L	182	233	228	313	293	181
Silicones									
JXR }	100% methyl	300	C, T	15	53	45	64	41	31
OV-1 }	silicone gums	350	C, T	16	55	44	65	42	32
OV-17	50% phenyl	300	C, T	119	158	162	243	202	112
OV-22	65% phenyl	300	C	160	188	191	283	253	133
OV-101	100% methyl, fluid	350	C	17	57	45	67	43	33
OV-210	50% trifluoropropyl	275	E	146	238	358	468	310	206
OV-225	25% cyanopropyl, 25% phenyl	275	A, C	228	369	338	492	386	282
OV-275	Cyano silicone	250	A	629	872	763	1106	849	—
QF-1	50% trifluoropropyl	250	A	144	233	355	463	305	203
SE-30	100% methyl, gum	300	T	15	53	44	64	41	31
SE-52	5% phenyl, gum rubber	300	T	32	72	65	98	67	44
SE-54	5% phenyl, 1% vinyl, gum rubber	300	T	33	72	66	99	67	46
XE-60	25% cyanoethyl	250	A	204	381	340	493	367	289
XF-1150	50% cyanoethyl	200	A	308	520	470	669	528	401

Solvent Code   A = acetone.   E = ethyl acetate.
                 B = benzene.   L = dichloromethane.
                 C = chloroform.   T = toluene.
McReynolds Constants   X′ = ΔI benzene.
                       Y′ = ΔI 1-butanol.
                       Z′ = ΔI 2-Pentanone.
                       U′ = ΔI 1-nitropropane.
                       S′ = ΔI pyridine.
                       H = ΔI 2-methyl-2-pentanol.

Compiled from data in catalogues issued by Analabs, Inc., North Haven, Connecticut, and Varian Associates, Palo Alto, California. Other sources of GC phases include Alltech Associates, Deerfield, Ill., Applied Science Laboratories, Inc., State College, Pennsylvania, Hewlett-Packard Co., Palo Alto, Calif., E. Merck, Darmstadt, West Germany, and Supelco, Inc., Bellefonte, Pa.

Table 2
SUPPORT MATERIALS FOR GAS CHROMATOGRAPHY

The preparations of diatomaceous earths that are most often used as solid supports in GC of carbohydrates are Chromosorb W AW DMCS (Johns-Manville) and Gas-Chrom Q (Applied Science Laboratories). Information on these and other suitable materials from these sources is given below; similar materials from other sources are also listed.

Support	Pretreatment and properties	Special uses	Mesh range	Similar materials
Chromosorb A	High capacity for liquid phase; not highly adsorptive	Preparative chromatography	20—30 45—60 60—80	Anaprep
Chromosorb G	High density; low surface area	Polar compounds	60—80 80—100	
Chromosorb G-HP	Special acid and DMCS treatment		80—100 100—120	
Chromosorb G AW DMCS	Acid-washed, silanized		60—80 80—100 100—120	
Chromosorb W	Light, friable; surface not highly adsorptive	Polar compounds	45—60 60—80 80—100 100—120	Anakrom U Celite 545 Gas-Chrom S
Chromosorb W-HP	Special acid and DMCS treatment		80—100 100—120	Anakrom Q
Chromosorb W AW	Acid-washed		45—60 60—80 80—100 100—120	Anakrom A Gas-Chrom A Supelcon AW
Chromosorb W AW DMCS	Acid-washed, silanized		60—80 80—100 100—120	Anakrom AS Gas-Chrom Z Supelcoport Varaport 30
Gas-Chrom P	Washed with acid and alcoholic base	General-purpose support	60—80 80—100 100—120	Anakrom AB
Gas-Chrom Q	Washed with acid and alcoholic base, silanized	Excellent general-purpose support	60—80 80—100 100—120	Anakrom ABS

Code AW   = acid-washed.
      DMCS = treated with dimethylchlorosilane.
      HP   = high performance.

Chromosorb products from Johns-Manville, Denver, Colo. Gas-Chrom products from Applied Science Laboratories, Inc., State College, Pennsylvania. Anakrom and Anaprep from Analabs, Inc., North Haven, Connecticut. Supelcon and Supelcoport from Supelco, Inc., Bellefonte, Pa., Varaport from Varian Associates, Palo Alto, California.

Table 3

MICROPARTICULATE SILICA AND BONDED-PHASE PACKINGS SUITABLE FOR HPLC OF CARBOHYDRATES

Source	Packing[a]	Description
Waters Associates	μ-Porasil	Microparticulate silica
	μ-Bondapak Carbohydrate	Polar organic phase chemically bonded to microparticulate silica support (8—12 μm). Packing inherently reactive lifetime ca. 4—6 months with continuous use.
Whatman Inc. (Reeve Angel division)	Partisil 5, 10, or 20	Microparticulate silica, available in particle size 5, 10, or 20 μm.
	Partisil 10-PAC	Polar amino-cyano phase chemically bonded to Partisil 10.
Varian Associates	MicroPak Si 5 or 10	Microparticulate silica, available in particle size 5 or 10 μm.
	MicroPak NH$_2$	Alkylamine chemically bonded to MicroPak Si 10.
E. Merck	Lichrosorb Si 60	Microparticulate silica, available in mean particle size 5, 10, and 30 μm.
	Lichrosorb NH$_2$	Amino phase, chemically bonded to silica, 5 or 10 μm.
Johns-Manville	Chromosorb LC 9	Amino phase, bonded to Chromosorb LC-6 silica (10μm).

[a] These packings are a available in pre-packed columns: 30 cm length, 4 mm I.D. from Waters and Varian; 25 cm length, 4.6 mm I.D. from Whatman; 25 cm length, 4 mm I.D. from Merck.

Table 4
ION-EXCHANGE RESINS

Product description	Column size (cm)	Resin required (g)	Resin type	Cross-linkage %	Particle size (µm)
Aminex A-4	0.9 × 50	36	Sulfonic acid	8	20 ± 4
Aminex A-5	0.9 × 7 0.9 × 22 0.9 × 60	6 16 g 45	Sulfonic acid	8	13 ± 2
Aminex A-6	0.6 × 19 0.9 × 50	6 36	Sulfonic acid	8	17.5 ± 2
Aminex A-7	0.6 × 19	6	Sulfonic acid	8	7—11
Aminex A-8	—	—	Sulfonic acid	—	5—8
Aminex A-9	0.9 × 34.5	37	Sulfonic acid	—	11.5 ± 5
Aminex Q-150S	0.9 × 50 0.9 × 150 2.0 × 150	36 108 450	Sulfonic acid	8	28 ± 7
Aminex Q-15S	0.9 × 15	11	Sulfonic acid	8	22 ± 3
Aminex 50W-X2 200—325 mesh	0.9 × 150 1.8 × 150	100 375	Sulfonic acid	2	44—74
Aminex 50W-X4 20—30 µ(H +)	0.9 × 150	110	Sulfonic acid	4	25 ± 5
Aminex 50W-X4 30—35 µ(H +)	0.9 × 150	110	Sulfonic acid	4	32.5 ± 2.5

Table 4 (continued)
ION-EXCHANGE RESINS

Product description	Column size (cm)	Resin required (g)	Resin type	Cross-linkage (%)	Particle size (µm)
Aminex A-14	0.9 × 60	30	Quaternary amine	4	20 ± 3
Aminex A-25	0.9 × 20 0.9 × 60	13 40	Quaternary amine	8	17.5 ± 2
Aminex A-27	0.22 × 150	5	Quaternary amine	8	13.5 ± 1.5
Aminex A-28	0.22 × 150	5	Quaternary amine	8	9 ± 2
Aminex A-29	—	—	Quaternary amine	8	5—8

ION EXCHANGE RESIN TRADE NAME INDEX[a]

Type and exchange group	Bio-Rad analytical grade ion exchange resins	Dow Chemical Company "Dowex"	Diamond-Shamrock "Duolite"	Rohm & Haas "Amberlite"	Permutit Company (England)	Ionac Chemical Company
Anion exchange resins Strongly basic polystyrene ϕ-CH$_2$N$^+$(CH$_3$)$_3$Cl$^-$						
	AG 1-X1	1-X1[b]			Zerolit FF (lightly cross-linked)	A-540
	AG 1-X2	1-X2[b]				
	AG 1-X4	1-X4[b,c]	A-101D	IRA-401	Zerolit FF	ASB-1
	AG 1-X8	1-X8[b,c](SBR)		IRA-400 CG-400		
	AG 1-X10	1-X10[b]		IRA-430		
	AG 21K	21K[b](SBR-P)		IRA-402		ASB-1-P

Anion exchange resins (continued)

$\phi\text{-}CH_2N^+(CH_3)_2(C_2H_4OH)\,Cl^-$

(structure: ring with $N^+\!\!-\!CH_3\;Cl^-$ and CH_3)

Resin type	Bio-Rad (AG)	Dowex	Duolite (C-20)	Amberlite	Zeo-Karb / Permutit	Other
$\phi\text{-}CH_2N^+(CH_3)_2(C_2H_4OH)\,Cl^-$	AG 2-X4; AG 2-X8; AG 2-X10; Bio-Rex 9	2X4[b]; 2-X8[e](SAR); A-102D		IRA-410		A-550; ASB-2; A-580
Intermediate Base, epoxy/polyamine $R\text{-}N^+(CH_3)_2Cl^-$ and $R\text{-}N^+(CH_3)_2(C_2H_4OH)Cl^-$	Bio-Rex 5	A-30B		IRA-47	F	A-305
Weakly basic polystyrene or phenolic polyamine $\phi\text{-}CH_2N^+(R)_2Cl^-$	AG 3-X4 A	WGR[b]; A-6; A-7[c]; A-4F		IR-45; IR-48; IRA-68	G	AFP-329
Cation exchange resins						
Strong acidic, polystyrene $\phi\text{-}SO_3H^+$	AG 50W-X1; AG 50W-X2; AG 50W-X2; AG-50W-X4; AG 50W-X8; AG 50W-X10; AG 50W-X12; AG 50W-X16	50W-X1[b]; 50W-X2[b]; 50W-X2[b]; 50W-X4[b]; 50W-X8[b,c] (HCRW); 50W-X10; 50W-X12[b,c]; 50W-X16[b]	C-20; C-20X10; C-20X12	IR-116; IR-116; IR-118; IR-120; CG-120; IR-122; IR-124	Zeocarb 225(X4); Zeocarb 225; C-249	C-240; C-250; C-255
Weakly acidic, acrylic $R\text{-}COO^-Na^+$	Bio-Rex 70	CC-3		IRC-84; IRC-50; CG-50	Zeocarb 226	CC; CNN
Weakly acidic, chelating resin, polystyrene $\phi\text{-}CH_2N(CH_2COO^-H^+)(CH_2COO^-H^+)$	Chelex 100	A-1			XE-318	

Table 4 (continued)
ION EXCHANGE RESIN TRADE NAME INDEX[a]

Type and exchange group	Bio-Rad analytical grade ion exchange resins	Dow Chemical Company "Dowex"	Diamond-Shamrock "Duolite"	Rohm & Haas "Amberlite"	Permutit Company (England)	Ionac Chemical Company
Macroporous resins						
Strong base, polystyrene ϕ-CH$_2$N$^+$(CH$_3$)$_3$Cl$^-$	AG MP-1	MSA-1	A-161	IRA-900		A-641
Strong acid, polystyrene ϕ-SO$_3$H$^+$	AG MP-50	MSC-1	C-250	200		CFP-110
No exchange group, styrene/divinylbenzene copolymer bead	SM-2	FSP-4022		XAD-2		
Mixed bed resins						
ϕ-SO$_3$H$^+$ & ϕCH$_2$N$^+$(CH$_3$)$_3$OH$^-$	AG 501-X8		GPM-331 G	IRN-150	Bio-Demineralit	NM-60
ϕ-SO$_3$H$^+$ & ϕ·CH$_2$N$^+$(CH$_3$)$_3$OH$^-$ indicator dye	AG 501-X8D				Indicator Bio-Demineralit	NM-65

Note: Other sources of ion-exchange resin suitable for chromatography of carbohydrates include Durrum Chemical Corporation(e.g., Durrum DA-X4, DA-X8, both available as fine particles, approximately 11 μm) and Technicon Corporation (e.g., Type S Chromobeads, developed specifically for use in carbohydrate analyzer). The Aminex resins manufactured by Bio-Rad are especially suitable for use in high-resolution liquid chromatography. Those having 8% cross-linking can be used at high pressure in HPLC. ϕ = polystyrene matrix.

[a] Manufacturers and distributors throughout the world are listed with the trade names and numerical designations of their commercial grade ion exchange resins. Listed with these are the Analytical Grade Ion Exchange Resins which Bio-Rad Laboratories supplies. Resins with the same active groups made by different manufacturers may in many cases be used interchangeably although they are not necessarily identical in composition or physical properties. Some resins have been discontinued by a particular manufacturer and are listed here for reference purposes.

[b] Denotes the commercial resin selected by Bio-Rad for further sizing and purification to an Analytical Grade Ion Exchange Resin in our own laboratories.

[c] Denotes a resin also supplied by Bio-Rad in bulk commercial grade at manufacturers list price.

From Catalogue D, Bio-Rad Laboratories, Richmond, California, 1978. With permission.

Table 5
ION-EXCHANGE GELS

From Bio-Rad Laboratories

Type	Description	Suspension medium	Capacity (μmol/ml)	Hemoglobin capacity (mg/ml)	Mesh range
DEAE Bio-Gel A	Weakly basic anion exchanger Agarose matrix, functional groups diethylaminoethyl	0.01 M Tris-HCl pH 8.0, with 0.02% sodium azide	15	100[a]	100—200
CM Bio-Gel A	Weakly acidic cation exchanger. Agarose matrix, functional groups carboxymethyl	Aqueous 0.02% sodium azide	20	100[b]	100—200

[a]   At pH 8 in Tris-HCl buffer.
[b]   At pH 5 in 0.01 M sodium acetate buffer.

From Pharmacia Fine Chemicals

The information on Sephadex and Sepharose ion exchangers given here has been contributed by Dr. John Brewer, Pharmacia Fine Chemicals, Uppsala, Sweden.

Pharmacia Fine Chemicals has developed a range of ion exchangers from Sephadex G-types especially suitable for the separation of labile substances, such as enzymes, hormones, and other charged compounds of biological origin. Both strong and weak anion and cation exchanger types are available in two porosities. The use of a porous matrix significantly increases the capacity for macromolecules and the bead form gives outstanding flow properties. Non-specific adsorption effects are minimal.

Sephadex Ion Exchangers

Sephadex ion exchangers are derivatives of Sephadex G-25 and G 50 containing anion or cation exchange groups. Anion and cation exchangers are denoted by the designations A and C respectively; the number following this designation refers to the Sephadex G-type used as starting material. A total of eight types is available according to the table below. The dry bead diameter is 40 to 120 μ.

The total capacities and haemoglobin capacities are given in the table below. Sephadex ion exchangers are stable in the pH range 2 to 10. Strong oxidizing agents should be avoided. Sephadex ion exchangers may be sterilized by autoclaving.

Sephadex ion exchangers swell in all aqueous electrolyte solutions.

Types[a]		Description	Counter ion	Total capacity (meq/g)	Hemoglobin capacity (g/g)
DEAE-	A-25	Weakly basic	Chloride	3.5 ± 0.5	0.5
Sephadex	A-50	anion exchanger			5
QAE	A-25	Strongly basic	Chloride	3.0 ± 0.4	0.3
Sephadex	A-50	anion exchanger			6
CM-	C-25	Weakly acidic	Sodium	4.5 ± 0.5	0.4
Sephadex	C-50	cation exchanger			9
SP-	C-25	Strongly acidic	Sodium	2.3 ± 0.3	0.2
Sephadex	C-50	cation exchanger			7

Note: The hemoglobin capacity is the amount of hemoglobin (MW 69,000) reversibly bound by 1g (dry weight) of ion exchanger. This value was measured for DEAE- and QAE-Sephadex in tris-HCl buffer pH 8.0, 0.01 M, and CM- and SP-Sephadex in acetate buffer pH 5.0, 0.01 M.

Table 5 (continued)
ION-EXCHANGE GELS

From Pharmacia Fine Chemicals

" Functional groups: DEAE = diethylaminoethyl, QAE = diethyl-(2-hydroxypropyl) aminoethyl, CME = carboxymethyl, and SP = sulphopropyl.

A detailed discussion of ion exchange chromatography, applications and experimental techniques is to be found in the handbook *Sephadex Ion Exchangers — A Guide to Ion Exchange Chromatography*, published by Pharmacia Fine Chemicals.

DEAE-Sepharose CL-6B and CM-Sepharose CL-6B

Sepharose ion exchangers are based on the cross-linked agarose gel, Sepharose CL-6B. They are supplied preswollen and ready for use; precycling and the removal of fines are not required. Sepharose ion exchangers may be regenerated without removing them from the column.

Type	DEAE-Sepharose CL-6B	CM-Sepharose CL-6B
Description	Weakly basic anion-exchanger	Weakly acidic cation exchanger
Functional groups	Diethylaminoethyl	Carboxymethyl
Ionic form	Cl^-	Na^+
Capacity (meq/100 ml gel)	12 ± 2	12 ± 2
Approximate capacity for hemoglobin	100 mg/ml at pH 8.0	100 mg/ml at pH 5.0
Bead size (μm)	45—165	45—165

Table 6
ION-EXCHANGE CELLULOSES FOR COLUMN CHROMATOGRAPHY[a]

A. Whatman[b]

Grade	Type	Ionic form	Small ion capacity (meq/g)	Protein capacity (mg/g) (pH)		Water regain (g/g dry exchanger)
DE 22	Fibrous cellulose anion exchanger DEAE-diethyl-aminoethyl functional group	Free base	1.0 ± 0.1	750 insulin	8.5	1.7—2.2
DE 23	Fibrous cellulose anion exchanger (fines reduced) DEAE-diethyl-aminoethyl functional group	Free base	1.0 ± 0.1	750 insulin	8.5	1.7—2.2
CM 22	Fibrous cellulose cation exchanger CM-car-boxy-methyl functional group	$Na^{+\ c}$	0.6 ± 0.06	600 lysozyme	5.0	2.0—2.8

Table 6 (continued)
ION-EXCHANGE CELLULOSES FOR COLUMN CHROMATOGRAPHY[a]

A. Whatman[b]

Grade	Type	Ionic form	Small ion capacity (meq/g)	Protein capacity (mg/g) (pH)		Water regain (g/g dry exchanger)
CM 23	Fibrous cellulose cation exchanger (fines reduced) CM-carboxy-methyl cation exchanger	Na$^+$ [c]	0.6 ± 0.06	600 lysozyme	5.0	2.0—2.8
DE 32	Microgranular cellulose anion exchanger, DEAE-diethyl-aminoethyl functional group.	Free base	1.0 ± 0.1	850 insulin	8.5	2.3—2.8
DE 52	Microgranular cellulose anion exchanger (preswollen). DEAE-diethyl-aminoethyl functional group.	Free base	1.0 ± 0.1	850 insulin	8.5	2.3—2.8
CM 32	Microgranular cellulose cation exchanger. CM-carboxymethyl functional group.	Na$^+$ [c]	1.0 ± 0.1	1260 lysozyme	5.0	2.3—2.7
CM 52	Microgranular cellulose cation exchanger (preswollen). CM-carboxymethyl functional group.	Na$^+$ [c]	1.0 ± 0.1	1260 lysozyme	5.0	2.3—2.7

B. Bio-Rad Laboratories[d]

Product description	Exchange capacity (meq/g)	Literature designation	Exchange group and ionic form
Cellex T	0.5 ± 0.1	TEAE Triethylaminoethyl	Intermediate base $\int -O-C_2H_4N^+-(C_2H_5)_3Br^-$
Cellex QAE	0.8 ± 0.1	DIETHYL 2-Hydroxypropylamino	Strong base $\int -O-C_2H_4-N^+(C_2H_5)_2$ $CH_2-CHOH-CH_3Br^-$
Cellex D	0.4 ± 0.1 (low) 0.7 ± 0.1 (std) 0.9 ± 0.05 (high)	DEAE Diethylaminoethyl	Weak base $\int -O-C_2H_4N-N+H-(C_2H_5)_2OH-$
Cellex BD[e]	0.38 ± 0.08	B-DEAE Benzoylated DEAE	Weak base
Cellex E	0.3 ± 0.05	ECTEOLA (mixed amines)	Intermediate base (undefined) + OH$^-$

Table 6 (continued)
ION-EXCHANGE CELLULOSES FOR COLUMN CHROMATOGRAPHY[a]

B. Bio-Rad Laboratories[d]

Product description	Exchange capacity (meq/g)	Literature designation	Exchange group and ionic form
Cellex CM	0.7 ± 0.1	CM Carboxymethyl	Weak acid $\int -O-CH_2-COO^-H^+$
Cellex P	0.85 ± 0.1	P Phosphate	Intermediate acid $\int -O-P(O) \begin{matrix} O^-H^+ \\ O^-H^+ \end{matrix}$
Cellex N-1	—	Cellulose powder	Normal cellulose
Cellex 410	—	Ethanolyzed cellulose powder	Normal cellulose

C. Pharmacia Fine Chemicals[f]

DEAE-Sephacel is a bead-formed DEAE-cellulose ion exchanger supplied pre-swollen and ready for use. Precycling and the removal of fines are not required. The bead-form minimizes fines formation during use, and columns packed with DEAE-Sephacel can usually be regenerated without repacking.

Type	DEAE-Sephacel
Description	Weakly basic anion exchanger; functional groups: diethylaminoethyl
Ionic form	Cl⁻
Capacity (meq/g)	1.4 ± 0.1
Approximate capacity for human serum albumin	150 mg/ml at pH 8.0
Bead size (μm)	40—160

[a] For data on products from Schleicher and Schuell, Inc., Machery-Nagel & Co., and Applied Science Laboratories, Inc., see Section A, Volume II, Section II. III, pages 300 to 301.

[b] Data and comments from Technical Bulletins 1032, 2000, and 15M, H. Reeve Angel and Co., Clifton, New Jersey, from which these products are available.

[c] Small proportion in the H⁺ form due to water washing.

[d] Data from Catalogue D, Bio-Rad Laboratories, Richmond, California, 1978. With permission.

[e] Cellex BD is supplied swollen in ethanol-salt solution. Nominal volume is 80 ml for 5g (dry). This material is 100 to 200 mesh.

[f] Data contributed by Dr. John Brewer, Pharmacia Fine Chemicals, Uppsala, Sweden.

Table 7
PACKINGS FOR GEL-PERMEATION (FILTRATION) CHROMATOGRAPHY*

From Pharmacia Fine Chemicals[b]

1. Sephadex G Cross-Linked Dextran

Some important physical properties of Sephadex of the G-series are given in the table below. These types differ in degree of cross-linkage and thus in swelling properties. The type numbers refer to the water regain, i.e., the amount of water imbibed per gram of dry Sephadex. Sephadex G-10 and G-200 are thus gels with water regains of about 1 and 20 ml/g dry material respectively.

Sephadex is stable in the pH range 2 to 10 and to dissociating agents such as urea, guanidine hydrochloride, and sodium dodecyl sulfate. Strong oxidizing agents should be avoided. Sephadex may be sterilized by autoclaving.

Sephadex G-types swell in all aqueous solutions and also in dimethyl sulfoxide, formamide, glycol, and mixtures of water and the lower alcohols. Sephadex G-10 and G-15 may also be used in dimethylformamide.

The table below shows the fractionation ranges for the different Sephadex G-types

Choice of gel type for gel filtration (gel-permeation) depends on the shape and size of the molecules to be separated. For other applications the chemical properties of the substances must be taken into account. A detailed discussion of gel selection is to be found in the handbook *Sephadex — Gel Filtration in Theory and Practice*, published by Pharmacia Fine Chemicals.

Sephadex		Dry bead diameter (μm)	Fractionation range (mol wt)		Bed volume (ml/g dry Sephadex)
Type	Grade		Peptides and globular proteins	Dextrans	
G-10		40—120	—700	—700	2—3
G-15		40—120	—1,500	—1,500	2.5—3.5
G-25	Coarse	100—300	1,000—5,000	10—5,000	4—6
	Medium	50—150			
	Fine	20—80			
	Superfine	10—40			
G-50	Coarse	100—300	1,500—30,000	500—10,000	9—11
	Medium	50—150			
	Fine	20—80			
	Superfine	10—40			
G-75		40—120	3,000—80,000	1,000—50,000	12—15
	Superfine	10—40	3,000—70,000		
G-100		40—120	4,000—150,000	1,000—100,000	15—20
	Superfine	10—40	4,000—100,000		
G-150		40—120	5,000—300,000	1,000—150,000	20—30
	Superfine	10—40	5,000—150,000		18—22
G-200		40—120	5,000—600,000	1,000—200,000	30—40
	Superfine	10—40	5,000—250,000		20—25

2. Sephadex LH-20 and LH-60

For separation of lipids, fatty acids, glycerides, steroids, vitamins, hormones, etc. by gel filtration, adsorption chromatography and partition chromatography.

Properties

Sephadex LH-20 is a hydroxypropyl derivative of Sephadex G-25 having both hydrophilic and lipophilic properties. The dry bead diameter is 25 to 100 μm. In gel filtration, the exclusion limit of Sephadex LH-20 is at about molecular weight 4,000 but various somewhat with the solvent system used.

Sephadex LH-20 is stable in the pH range 2 to 10 and in all organic solvents. Strong oxidizing agents should be avoided. Sephadex LH-20 may be sterilized by autoclaving.

Sephadex LH-20 swells in water, polar organic solvents, aqueous solvent mixtures, and mixtures of polar organic solvents.

In mixed solvent systems, Sephadex LH-20 preferentially takes up the more polar component and may thus be used for straight phase partition chromatography. Choice of the appropriate solvent accentuates the adsorption properties of Sephadex LH-20 and provides a basis for adsorption chromatography.

Table 7 (continued)
PACKINGS FOR GEL-PERMEATION (FILTRATION) CHROMATOGRAPHY[a]

From Pharmacia Fine Chemicals[b]

Sephadex LH-60 is a hydroxypropyl derivative of Sephadex G-50, with properties simlar to LH-20 but capable of fractionating larger molecules.

Swelling characteristics of Sephadex LH-20 and LH-60 are given below:

Solvent	Ca. bed volume (ml/g dry gel) LH-20	LH-60	Solvent	Ca. bed volume (ml/g dry gel) LH-20	LH-60
Dimethyl sulphoxide	4.5	13.6	Methylene dichloride	3.8	11.2
Pyridine	4.3	13.6	Butanol	3.7	11.2
Dimethylformamide	4.2	12.6	Isopropanol	3.5	10.2
Methanol	4.2	12.1	Dioxane	3.4	10.0
Ethylene dichloride	4.0	11.2	Acetone	2.5	5.7
Chloroform[c]	4.0	12.5	Acetonitrile	2.3	3.2
Propanol	3.9	11.2	Carbon tetrachloride	2.0	2.0
Ethanol[d]	3.8	12.2	Benzene	1.8	2.5
Isobutanol	3.8	11.0	Ethyl acetate	1.7	3.1
Formamide	3.8	8.8	Toluene	1.6	2.0

3. Sepharose Agarose Gel[e]

For separation of DNA, RNA, viruses, high molecular weight proteins, and polysaccharides by gel filtration. Now the widely accepted matrix for immobilization of ligands in affinity chromatography.

Properties

Sepharose is a bead-formed agarose gel. Some of its physical properties are given in the table below. Three types of Sepharose of different agarose concentration are supplied as sterilized suspensions of swollen beads in distilled water containing sodium azide. The agarose concentration refers to the concentration in the gel particles and not to that of the suspension.

Sepharose is stable in aqueous suspension in the pH range 4—9. Sepharose can be used in solutions containing high concentrations of salt, urea, guanidine hydrochloride, and detergents. Exposure to temperatures above 40°C or below 0°C should be avoided. Diethylpyrocarbonate may be used for sterilization.

Sepharose is used in aqueous solutions. The swollen beads should not be dried since dehydration leads to breakdown of the gel structure.

Choice of Gel Type

Each type of Sepharose fractionates over a different molecular weight range as given in the table below. Sepharose is used for separations of substances totally excluded by Sephadex G-200.

Sepharose is now established as the matrix of choice for the immobilization of enzymes, antigens, antibodies, and other biospecific substances for affinity chromatography.

Sepharose type	Agarose conc (%)	Wet bead diameter (µm)	Fractionation range (mol/wt) Polysaccharides	Proteins
Sepharose 2B	2	60—250	10^5—20×10^6	7×10^4—40×10^6
Sepharose 4B	4	40—190	3×10^4—5×10^6	6×10^4—20×10^6
Sepharose 6B	6	40—210	10^4—1×10^6	10^4—4×10^6

4. Sepharose CL Gels, Cross-Linked Agarose

Sepharose CL-2B, CL-4B, and CL-6B are available. These gels are prepared by cross-linking Sepharose 2B, Sepharose 4B, and Sepharose 6B, respectively, with 2,3-dibromopropanol. The resulting gels have prac-

Table 7 (continued)
PACKINGS FOR GEL-PERMEATION (FILTRATION) CHROMATOGRAPHY[a]

From Pharmacia Fine Chemicals[b]

tically the same exclusion properties as the uncross-linked gels, but have greatly improved chemical and physical stability. They are unaffected by strong solutions of urea or guanidine hydrochloride, can be used in organic solvents, and can be autoclaved.

	Ca. exclusion limits (mol wt)	
	Polysaccharides	Globular proteins
Sepharose CL-6B	1×10^6	4×10^6
Sepharose CL-4B	5×10^6	20×10^6
Sepharose CL-2B	20×10^6	40×10^6

5. Sephacryl S Gels

New products for fast gel filtration: Sephacryl S-200 and S-300 Superfine are copolymers of N,N'-methylene bisacrylamide and allyl dextran. Data on these products are given below:

	Sephacryl S-200	Sephacryl S-300
Wet particle diameter, (μm)	40—50	40—105
Molecular weight fractionation range for globular proteins	5,000—250,000	10,000—1.5×10^6
Molecular weight fractionation range for dextrans	5,000—80,000	5,000—400,000

From Waters Associates

1. μ-Bondagel

Microparticulate (10 μm) silica with bonded ether, developed for high-speed exclusion chromatography of water-soluble polymers. Available, in prepacked columns only, with fractionation ranges as follows:

Type	Mol wt range[f]
E-125	2,000—50,000
E-300	3,000—100,000
E-500	5,000—500,000
E-1000	50,000—2×10^6
E-Linear (blend of pore diameters)	2,000—2×10^6

2. Porasil Porous Silica Beads

Type[g]	Average pore diameter (A)	Surface area (m^2/g)	Particle size range (μm)
A (60)	<100	350—500	37—75
A (60)			75—125
B (250)	100—200	125—250	37—75
B (250)			75—125

Table 7 (continued)
PACKINGS FOR GEL-PERMEATION (FILTRATION) CHROMATOGRAPHY[a]

From Waters Associates

Type[e]	Average pore diameter (A)	Surface area (m²/g)	Particle size range (μm)
C (400)	200—400	50—100	37—75
C (400)			75—125
D (1000)	400—800	25—45	37—75
D (1000)			75—125
E (1500)	800—1500	10—20	37—75
E (1500)			75—125
F (2000 +)	>1500	2—6	37—75
F (2000 +)			75—175
Porasil T	150	300	15—25
Porasil T	150	300	25—37
Porasil T	150	300	37—50

From E. Merck

1. Merckogel Type SI (Porous Silica)

Type	Mol wt exclusion limit (polystyrene in CHCl₃)	Mean pore diameter (A)	Particle size range (μm)
SI 50	10,000	50	37—75
SI 50	10,000	50	75—125
SI 150	50,000	150	37—75
SI 150	50,000	150	75—125
SI 500	400,000	500	37—75
SI 500	400,000	500	75—125
SI 1,000	1,000,000	1,000	37—75
SI 1,000	1,000,000	1,000	75—125

2. Merckogel Type OR-PVA (Vinyl Acetate Copolymers)

	Exclusion limit[i] (mol wt)		Swelling volumes[h] (approximate)			
			Tetrahy-drofuran	Benzene	Dimethyl-formamide	1,2,4-Tri-chloroben-zene
9377 Merckogel® OR-PVA	500	300	3.2	3.2	3.2	3.0
9375 Merckogel® OR-PVA	2,000	1,000	4.7	4.7	4.7	4.0
9373 Merckogel® OR-PVA	6,000	4,000	7.3	7.8	7.4	6.0
9371 Merckogel® OR-PVA	20,000	14,000	7.2	7.6	5.2	3.2
9369 Merckogel® OR-PVA	30,000	50,000	6.1	7.6	5.5	5.1
9392 Merckogel® OR-PVA	300,000	200,000	7.2	5.6	5.4	9.2
9367 Merckogel® OR-PVA	1,000,000	1,000,000				

[a] For data on products from Bio-Rad Laboratories and Waters Associates see Section A, Volume II, Section II. III, pages 287 and 290 to 291.
[b] Data contributed by Dr. John Brewer, Pharmacia Fine Chemicals, Piscataway, New Jersey.
[c] Containing 1% ethanol.
[d] Containing 1% benzene.
[e] Detailed information about applications and experimental techniques is given in the booklet *Sepharose 2B—4B—6B*, published by Pharmacia Fine Chemicals.
[f] For dextrans and similar polymers in water.
[g] Old designation given in parentheses.
[h] Determined with 1 g gel in 20 mℓ solvent. Values in mℓ/g.
[i] For polystyrenes in tetrahydrofuran.

Table 8
MEDIA FOR AFFINITY CHROMATOGRAPHY

Affinity chromatography is the most important new chromatographic technique to have been developed in the last decade. It is now used in practically every biochemistry laboratory concerned with the preparation of pure protein and other biopolymers.

Pharmacia Fine Chemicals produces a wide range of media for affinity chromatography.

A. Base Media

Type	Description	Use
Sepharose 4B	4% agarose gel	General supports for affinity
Sepharose 6B	6% agarose gel	chromatography media.
Sepharose CL-4B	Cross-linked 4% agarose gel	Supports for media to be used under harsh elution condi-
Sepharose CL-6B	Cross-linked 6% agarose gel	tions or when derivatization requires use of organic solvents
Sephadex G-200	Cross-linked dextran gel	Special applications where a support capable of enzymic degradation is required

B. Derivatives for Preparing Affinity Chromatography Media

Type	Description	Functional group required on ligand	Bond formed
CNBr-activated Sepharose 4B	Sepharose 4B activated with CNBr[a]	$-NH_2$	Isourea
CNBr-activated Sepharose 6MB	Macrobeads of Sepharose 6B activated with CNBr	$-NH_2$	Isourea
Epoxy-activated Sepharose 6B	1,4-Bis-(2,3-epoxy-propoxy-)butane coupled to Sepharose 6B	$-OH$ $-SH$ $-NH_2$	Ether Thioether Secondary amine
AH-Sepharose 4B	1,6-Diamino-hexane coupled to Sepharose 4B	$-COOH$	Peptide
CH-Sepharose 4B	6-Aminohexanoic acid coupled to Sepharose 4B	$-NH_2$	Peptide
Activated CH-Sepharose 4B	N-Hydroxysuccinimide ester of CH-Sepharose	$-NH_2$	Peptide
Activated Thiol-Sepharose 4B	Mixed disulfide between 2,2'-dipyridyl disulfide and glutathione coupled via its primary amino group to Sepharose 4B	$-SH$	Mixed disulfide (reversible)
Thiopropyl-Sepharose 6B	Mixed disulfide between 2,2'-dipyridyl disulfide and 2-hydroxy-propane thiol coupled to Sepharose 6B	$-SH$	Mixed disulfide (reversible)

[a]   CNBr = cyanogen bromide.

C. Pre-Made Media for Affinity Chromatography

Type	Ligand	Specificity
Lysine-Sepharose 4B	L-Lysine	Plasminogen; rRNA
5'-AMP-Sepharose 4B	N^6-(6-aminohexyl) 5'-AMP	Enzymes which have NAD[+] as cofactor and ATP-dependent kinases

Table 8 (continued)
MEDIA FOR AFFINITY CHROMATOGRAPHY

C. Pre-Made Media for Affinity Chromatography

Type	Ligand	Specificity
2′,5′-ADP-Sepharose 4B	N^6-(6-aminohexyl) adenosine 2′,5′-bis-phosphate	Enzymes which have NADP⁺ as cofactor.
Blue Sepharose CL-6B	Cibacron Blue F3G-A	Broad range of enzymes which have dinucleotide cofactors; serum albumin
Poly(U)-Sepharose 4B	Polyuridylic acid	Nucleic acids, especially mRNA, and oligonucleotides which contain poly(A) sequences
Poly(A)-Sepharose 4B	Polyadenylic acid	Nucleic acids and oligonucleotides which contain poly(U) sequences; RNA specific proteins
Con A-Sepharose	Concanavalin A *(Canavalia ensiformis)*	Terminal α-D-glucopyranosyl, α-D-mannopyranosyl or sterically similar residues
Lentil Lectin-Sepharose 4B	Hemagglutinating lectin from the common lentil *(Lens culinaris)*	Similar to that of Con A-Sepharose but lower binding affinity
Wheat germ Lectin-Sepharose 6MB	Wheat germ lectin *(Triticum vulgare)*	2-acetamido-2-deoxy-D-glucose
Helix pomatia Lectin-Sepharose 6MB	Lectin from the albumin gland of the vineyard snail *(Helix pomatia)*	2-acetamido-2-deoxy-D-galactose; blood group substance A
Protein A-Sepharose CL-4B	Protein A *(Staphylococcus aureus)*	F_c region of IgG and related molecules

Data contributed by Dr. John Brewer, Pharmacia Fine Chemicals, Uppsala, Sweden.

Table 9
CHROMATOGRAPHY PAPERS[a]

Whatman Inc., Clifton, New Jersey

1. Unmodified Cellulose Papers Specifically Manufactured for Chromatography

Whatman
1	Standard paper for chromatography
2	Slightly slower than Whatman 1
3MM	Thick, smooth-surface paper with flow rate similar to Whatman 1
3	Similar to 3MM but with a rough surface
4	Fast flow rate
17	Extremely thick with fast flow rate for preparative chromatography
20	Very slow flow rate; for high resolution
31ET	Very rapid flow rate and low ash, useful for electrophoresis

2. Adsorbent-Loaded Paper

Grade	Basis weight (g/m²)[b]	Thickness (mm)[c]	Diffusion rate (min) [d]	Diffusion rate (min) [e]	Loading[f]	pH[g]
SG 81	100	0.25	15	100	22% SiO₂	7

Table 9 (continued)
CHROMATOGRAPHY PAPERS[a]

3. Ion-Exchange Cellulose Papers

Type	Basis weight (g/m²)	Thickness (mm)	Capacity (μequiv/cm²)	Flow rate[d] (min)	Ionic form as supplied	
P81	85	0.17	18.0	10	Mono-ammonium	(cellulose phosphate paper)
DE 81	85	0.17	3.5	20	Free base	(diethylaminoethyl cellulose paper)

Note: Glass fiber papers — Grades GF/A, GF/B, GF/C, GF/D, and GF/F are available. These are manufactured primarily as filters but GF/C has been used in electrophoresis of polysaccharides (see note below Table EL 4).

[a] For data on papers from Schleicher and Schuell, Inc., Keene, New Hampshire and Macherey-Nagel & Co., Düren, Germany see Section A, Volume II, Section II.III, pages 304 to 307.
[b] Mean total basis weight, cellulose + loading.
[c] By deadweight micrometer.
[d] Time taken for water to ascend 7.5 cm up a vertical strip of paper.
[e] Time taken for benzene to descend 15 cm down a vertical strip of paper.
[f] Estimated as ignited oxides by weight.
[g] pH of aqueous suspension.

Table 10
SORBENTS FOR THIN-LAYER CHROMATOGRAPHY[a]

From E. Merck, Darmstadt, Germany

E. Merck Sorbents
1. Analytical Sorbents

Aluminum oxide G, type-E[b] (1090)
Aluminum oxide GF-254, type-E[b] (1092)
Aluminum oxide H, type-E[b] (1085)
Aluminum oxide HF-254, type-E[b] (1094)
Aluminum oxide, acidic, type-T[b] (1106)
Aluminum oxide, basic, type-T[b] (1105)
Aluminum oxide, neutral, type-T[b] (1101)
Kieselguhr G (8129)
Polyamide II (7435)
Silica gel G (7731)

Silica gel GF-254 (7730)
Silica gel H (7736)
Silica gel HR (7744)
Silica gel HF-254 (7739)
Silica gel HF-254, silanized (7750)
Silica gel HF-254 + 366 (7741)
Fluorescent indicator (luminous pigment) F-254 (9182)
Avicel micro crystalline cellulose (2330)
PEI—Cellulose (2336)

2. Preparative Sorbents

Aluminum oxide PF-254, type-E[b] (1103)
Aluminum oxide PF-254, type-T[c] (1064)
Aluminum oxide PF-254 + 366, type-T[c] (1065)
Aluminum oxide PF-254 + 366, type E[b] (1104)

Silica gel PF-254 (7747)
Silica gel PF-254, silanized (7751)
Silica gel PF-254 + 366 (7748)
Silica gel PF-254 with calcium sulfate (7749)

[b] Type-E materials are especially recommended for pesticides.
Type-T materials are especially recommended for steroids and higher peptides.

Suffix
G—indicates CaSO₄ binder.
H—without binder.
S—starch binder.
HR—exceptionally pure.
F—2% inclusion of UV indicator.
P—sorbents for making preparative layers.

Table 10 (continued)
SORBENTS FOR THIN-LAYER CHROMATOGRAPHY[a]

3. Precoated Glass Plates

Analytical (layer thickness 100—250 μm)
    Aluminium oxide F-254 (type E)
    Aluminium oxide F-254 (type T)
    Cellulose (without fluorescent indicator)
    Cellulose F
    PEI-Cellulose (without fluorescent indicator)
    PEI-Cellulose F
    Silica gel 60 (without fluorescent indicator)
    Silica gel F-254
    Silica gel F-254, fast-running (mixed with kieselguhr)
    Silica gel F-254, silanized
    Silica gel 60 silanized (without fluorescent indicator)
    Kieselguhr F-254
    Polyamide 11 F-254
Plates with concentration zones
    (Porous SiO$_2$, layer thickness 150 μm, 25 mm length precedes chromatographic layer, 250 μm thick)
    Silica gel 60 (without fluorescent indicator)
    Silica gel F-254
HPTLC plates, for TLC at nanogram levels
    (Thinner layers, narrower pore-size distribution; rapid separations, with high resolution of small samples)
    Silica gel 60 (without fluorescent indicator)
    Silica gel F-254
    Silica gel F-254, with concentration zone
Preparative plates (layer thickness 1.5—2 mm)
    Aluminium oxide F-254 (type T)
    Silica gel 60 (without fluorescent indicator)
    Silica gel F-254
    Silica gel F-254 + 366

4. Precoated Aluminium Sheets

(Layers 100—250 μm)
Aluminium oxide, neutral, F-254 (type E)
Aluminium oxide, neutral, F-254 (type T)
Cellulose (without fluorescent indicator)
Cellulose F-254
Silica gel 60 (without fluorescent indicator)
Silica gel F-254
Silica gel 60/Kieselguhr F-254
Kieselguhr F-254
Polyamide 11 F-254
HPTLC aluminium sheets
    Silica gel 60 (without fluorescent indicator)
    Silica gel F-254

5. Precoated Plastic Sheets

Aluminium oxide neutral, F-254 (type E)
Cellulose (without fluorescent indicator)
Cellulose F-254
PEI-Cellulose F
Silica gel 60 (without fluorescent indicator)
Silica gel F-254

---

[a]   For data on products from Schleicher & Schuell Inc., Keene, New Hampshire, J. T. Baker Chemical Co., Phillipsburg, New Jersey, Bio-Rad Laboratories, Richmond, California, Camag, Inc., New Berlin, Wisconsin, and Macherey-Nagel & Co., Düren, Germany, see Section A, Volume II, Section II.III, pp. 310—316.

[b]   Type-E materials are especially recommended for pesticides.

[c]   Type-T materials are especially recommended for steroids and higher peptides.

Table 11
MEDIA FOR ELECTROPHORESIS

From Pharmacia Fine Chemicals

Polyacrylamide Gradient Gels PAA 4/30 are carefully standardized precast gel slabs 2.7 mm thick in glass cassettes 82 × 82 × 4.9 mm. The concave monomer concentration gradient 4 to 30% gives excellent resolution of complex mixtures of biopolymers according to molecular size in the range 50,000 to 2,000,000 daltons.

The monomer gradient produces a gradient of pore size in the slabs so that molecules moving through the gradient move into progressively smaller pores. Since the trailing edge of a zone is always moving faster than the leading edge, the zone becomes sharper during its passage through the gradient. This zone sharpening results in distinct narrow bands, which at the completion of the run are practically nondiffusable. About 40 bands can be distinguished in a 5-μl sample of normal human serum.

Polyacrylamide Gradient Gels PAA 4/30 are cast in a solution containing Na_2 EDTA (1 g/l), boric acid (1 g/l), $(NH_4)_2SO_4$ (2.5 g/l), and NaN$_3$ (0.2 g/l).

From Bio-Rad Laboratories

Acrylamide	mol wt 71.08	Bis	mol wt 154.17
Acrylic acid	<0.001%	(N,N'-methylene-bis-acrylamide)	
(conductometric titration of total acrylate)		Acrylic acid	<0.05%
Conductivity	<2.0 μmho	pH, 1% solution	>5.5
(35% solution)		OD$_{290}$, 1% solution	<0.2
pH	6.5 ± 0.5	DATD	mol wt 228.25
(10% solution in 0.1 MNaCl)		(N,N' diallyltartardiamide)	
Insolubles		White, crystalline	
H$_2$O	<0.005%	Melting point	183—185°C
MeOH	<0.005%	Assay	>99%
		Riboflavin-5'-phosphate	mol wt 514.37
Acrylamide/Bis, preweighed		(flavin mononucleotide)	
Mixture ratios	29:1 (3.3% C)	TEMED	mol wt 116.21
	19:1 (5.0% C)	(N,N,N',N'-tetramethylethylenediamine)	
		Boiling range	119—121°C
Total weight	200 ± 0.5 g	Refractive index	1.4180 ± 0.0003
Makes stock solution 400 ml, 50% T		Ammonium sulfate	mol wt 132.14
Shelf life, dry >2 years at 25°C		Pb	<0.0001%
Shelf life, solution >90 days at 4°C		As	<0.00002%

Table 11 (continued)
MEDIA FOR ELECTROPHORESIS

From Bio-Rad Laboratories

Ammonium persulfate	mol wt 228.21
Assay	99%
Meets ACS standards	
Sodium dodecylsulfate	mol wt 288.38
(sodium lauryl sulfate)	
Dithiothreitol	
(DDT, Cleland's Reagent)	
White, crystalline	
Melting point	39—41°C
OD_{283}	<0.04
(0.02 M solution)	
Agarose, standard low (m.)	
White powder	
Relative mobility (m.)	0.13 ± 0.02
Gel strength, 1% gel	>1000 g/cm²
Jerne plaque assay	pass test
Immunoelectrophoresis	pass test
Gelation temperature	34—45°C
Sulfur	<0.1%
Agarose, high (m.)	
Relative mobility (m.)	>0.25
Gelation temperature	35—45°C
Sulfur	<0.1%
Agarose, ultra pure	
Relative mobility (m.)	<0.1
Gel strength, 1%	>1000 g/cm²
Gelation temperature	40—50°C
Sulfur	<0.05%
Agarose, low gel temperature	
Relative mobility (m.)	<0.10
Gelation temperature (1%)	<37°C
Sulfur	<0.05%
Glycine	mol wt 75.07
(aminoacetic acid)	
Heavy metals (as Pb)	<20 ppm
OD_{280}, 1.0 M solution	<0.15
Tris	mol wt 121.14
Tris (hydroxymethyl) aminomethane or tromethamine, tris buffer	

Data from Catalogue D, Bio-Rad Laboratories, Richmond, California, 1978. With permission.

Section V

Literature References

LITERATURE REFERENCES

In this section, an updated bibliography (see earlier International Chromatography Book Directory in Section A, Volume II, Section II.IV) is followed by a list of review articles and selected journal references that are considered to represent definitive treatments of, or important advances in, various aspects of carbohydrate chromatography.

For the bibliography, the format used in Section A, Volume II has been followed. The review articles and journal references are grouped according to the type of chromatography involved; within each group, references are listed in alphabetical order of first authors' surnames.

BIBLIOGRAPHY

Books containing chapters dealing specifically with chromatography of carbohydrates are listed below, together with some that are recommended as general guides to chromatographic principles and techniques.

	Gas	HPLC	Ion exchange	Gel	Affinity	Paper	Thin-layer	Electrophoresis	General
Academic Press, Inc., New York and London									
Block, R. J., Durrum, E. L., and Zweig, G., *A Manual of Paper Chromatography and Paper Electrophoresis*, 2nd ed., 1958	—	—	—	—	—	X	—	X	—
Hais, I. M. and Macek, K., *Paper Chromatography: A Comprehensive Treatise*, 1963	—	—	—	—	—	X	—	—	—
Littlewood, A. B., *Gas Chromatography*, 2nd ed., 1970	X	—	—	—	—	—	—	—	—
Randerath, K., *Thin-Layer Chromatography*, 2nd ed., 1966	—	—	—	—	—	—	X	—	—
Walker, J. Q., Jackson, M. T., Jr., and Maynard, J. B., *Chromatographic Systems: Maintenance and Trouble-Shooting*, 1972	X	X	—	—	—	—	—	—	—
Whistler, R. L. and Wolfrom, M. L., Eds., *Methods in Carbohydrate Chemistry*, Vol. 1, 1962	X	—	—	—	—	X	—	X	X
Whistler, R. L., Green, J. W., BeMiller, J. N., and Wolfrom, M. L., Eds., *Methods in Carbohydrate Chemistry*, Vol. 3, 1963	—	—	—	—	—	X	—	—	X
Whistler, R. L., Smith, R. J., BeMiller, J. N., and Wolfrom, M. L., Eds., *Methods in Carbohydrate Chemistry*, Vol. 4, 1964	X	—	—	—	—	X	X	—	X
Whistler, R. L., BeMiller, J. N., and Wolfrom, M. L., Eds., *Methods in Carbohydrate Chemistry*, Vol. 5, 1965	—	—	—	X	—	—	—	X	X
Whistler, R. L. and BeMiller, J. N., Eds., *Methods in Carbohydrate Chemistry*, Vol. 6, 1972	X	—	—	—	—	—	X	—	X
Whistler, R. L. and BeMiller, J. N., Eds., *Methods in Carbohydrate Chemistry*, Vol. 7, 1976	X	—	—	—	—	X	—	—	X
Zweig, G. and Sherma, J., *Paper Chromatography*, 1971	—	—	—	—	—	X	—	—	—
Ann Arbor Science Publishers, Inc., Ann Arbor, Michigan									
Grushka, E., Ed., *Bonded Stationary Phases in Chromatography*, 1974	—	X	—	—	—	—	—	—	—
Scott, R. M., *Clinical Analysis by Thin-Layer Chromatography Techniques*, 1969	—	—	—	—	—	—	X	—	—
Butterworth and Co., Ltd., London									
Dickes, G. J. and Nicholas, P. V., *Gas Chromatography in Food Analysis*, 1976	X	—	—	—	—	—	—	—	—
Marcel Dekker, Inc., New York									
Keller, R. A., Ed., *Separation Techniques in Chemistry and Biochemistry*, 1967	X	—	—	—	—	—	—	—	X
Rajcsanyi, P. M. and Rajcsanyi, E., *High-Speed Liquid Chromatography*, 1975	—	X	—	—	—	—	—	—	—
Elsevier Publishing Company, Amsterdam									
Deyl, Z., Macek, K., and Janak, J., *Journal of Chromatography Library*, Vol. 3, 1973	—	X	X	X	—	—	—	—	—

	Gas	HPLC	Ion exchange	Gel	Affinity	Paper	Thin-layer	Electrophoresis	General
Parris, N. A., Ed., *Journal of Chromatography Library*, Vol. 5, 1976	—	X	X	X	—	—	—	—	—
Lederer, E. and Lederer, M., *Chromatography*, 2nd ed., 1957	—	—	X	—	—	X	—	—	X
Zlatkis, A. and Kaiser, R. E., Eds., *TLC: High-Performance Thin-Layer Chromatography*, 1976	—	—	—	—	—	—	X	—	—
Heinemann Medical Books Ltd., London									
Smith, I. and Seakins, J. W. T., Eds., *Chromatographic and Electrophoretic Techniques*, Vol. 1, 4th ed., 1976	—	—	—	—	—	X	X	—	—
Macmillan Press, Ltd., London									
Denney, R. C., *A Dictionary of Chromatography*, 1976	—	—	—	—	—	—	—	—	X
McGraw-Hill Book Company, New York									
Helfferich, F., *Ion Exchange*, 1962	—	—	X	—	—	—	—	—	—
North-Holland Publishing Company, Amsterdam									
Work, T. S. and Work, E., Eds., *Laboratory Techniques in Biochemistry and Molecular Biology*, Vol. 1 (Part 2), 1969	—	—	—	X	—	—	—	—	—
Work, T. S. and Work, E., Eds., *Laboratory Techniques in Biochemistry and Molecular Biology*, Vol. 2 (Part 2), 1970	—	—	X	—	—	—	—	—	—
Pierce Chemical Company, Rockford, Illinois									
Brittain, G. D., *Handbook of Silylation with Other Special Techniques for Gas Phase Analysis: Methods, Reagents and Accessories*, 1970	X	—	—	—	—	—	—	—	—
Pierce, A., *Silylation of Organic Compounds: A Technique for Gas Analysis*, 1968	X	—	—	—	—	—	—	—	—
Plenum Press, New York									
Ettre, L. S., *Open Tubular Columns in Gas Chromatography*, 1965	X	—	—	—	—	—	—	—	—
Springer-Verlag, Berlin and New York									
Determann, H., *Gel Chromatography*, 2nd ed., 1968	—	—	—	X	—	—	—	—	—
Stahl, E., Ed., *Thin-Layer Chromatography: A Laboratory Handbook*, 2nd ed., 1969	—	—	—	—	—	—	X	—	—
Van Nostrand Reinhold Company, New York									
Heftmann, E., Ed., *Chromatography; A Laboratory Handbook of Chromatographic and Electrophoretic Methods*, 3rd ed., 1975	X	X	X	X	—	X	X	X	X
John Wiley & Sons, Wiley-Interscience, New York									
Done, J. N., Knox, J. H., and Luheac, J., *Applications of High Speed Liquid Chromatography*, 1974	—	X	—	—	—	—	—	—	—
Ettre, L. S. and Zlatkis, A., *Practice of Gas Chromatography*, 1967	X	—	—	—	—	—	—	—	—
Grob, R. L., Ed., *Modern Practice of Gas Chromatography*, 1977	X	—	—	—	—	—	—	—	—

	Gas	HPLC	Ion exchange	Gel	Affinity	Paper	Thin-layer	Electrophoresis	General
Kirchner, J. G., *Thin-Layer Chromatography*, 1967	—	—	—	—	—	—	X	—	—
Kirkland, J. J., Ed., *Modern Practice of Liquid Chromatography*, 1971	—	X	—	—	—	—	—	—	—
Lowe, C. R. and Dean, P. D. G., *Affinity Chromatography*, 1974	—	—	—	—	X	—	—	—	—
McFadden, W. H., *Techniques of Combined Gas Chromatography/Mass Spectrometry: Applications in Organic Analysis*, 1973	X	—	—	—	—	—	—	—	—
Purnell, H., *New Developmens in Gas Chromatography*, 1973	X	—	—	—	—	—	—	—	—
Snyder, L. R. and Kirkland, J. J., *Introduction to Modern Liquid Chromatography*, 1974	—	X	—	—	—	—	—	—	—
John Wiley & Sons, Ltd., Chichester, U.K.									
Epton, R., Ed., *Chromatography of Synthetic and Biological Polymers*, Vol. 1, 1978	—	—	—	X	—	—	—	—	X
Epton, R., Ed., *Chromatography of Synthetic and Biological Polymers*, Vol. 2, 1978	—	—	X	—	X	—	—	—	X
Gasparaič, J. and Churáček, J., *Laboratory Handbook of Paper and Thin-Layer Chromatography*, 1978	—	—	—	—	—	X	X	—	—
Touchstone, J. C. and Dobbins, M. F., *Practice of Thin-Layer Chromatography*, 1978	—	—	—	—	—	—	X	—	—

REVIEWS AND SELECTED JOURNAL REFERENCES

Review Articles

Gas Chromatography

1. Bishop, C. T., Gas-liquid chromatography of carbohydrate derivatives, in *Advances in Carbohydrate Chemistry*, Vol. 19, Wolfrom, M. L. and Tipson, R. S., Eds., Academic Press, New York, 1964, 95.
2. Björndal, H., Hellerqvist, C. G., Lindberg, B., and Svensson, S., Gas-liquid chromatography and mass spectrometry in methylation analysis of polysaccharides, *Angew. Chem. Int. Ed. Engl.*, 9, 610, 1970.
3. Clamp, J. R., Bhatti, T., and Chambers, R. E., The determination of carbohydrate in biological materials by gas-liquid chromatography, in *Methods of Biochemical Analysis*, Glick, D., Ed., Vol. 19, Interscience, New York, 1971, 229.
4. Clamp, J. R., Gas-liquid chromatographic approach to analysis of carbohydrates, *Biochem. Soc. Trans.*, 5, 1693, 1977.
5. Drawert, F. and Leupold, G., Quantitative gas-liquid chromatographic determination of neutral carbohydrates, *Chromatographia*, 9, 447, 1976.
6. Dutton, G. G. S., Applications of gas-liquid chromatography to carbohydrates, Part I, in *Advances in Carbohydrate Chemistry and Biochemistry*, Vol. 28, Tipson, R. S. and Horton, D., Eds., Academic Press, New York, 1973, 11.
7. Dutton, G. G. S., Applications of gas-liquid chromatography to carbohydrates, Part II, in *Advances in Carbohydrate Chemistry and Biochemistry*, Vol. 30, Tipson, R. S. and Horton, D., Eds., Academic Press, New York, 1974, 9.
8. Holligan, P. M., Routine analysis of gas-liquid chromatography of soluble carbohydrates in extracts of plant tissue. I. A review of techniques used for the separation, identification, and estimation of carbohydrates in gas-liquid chromatography, *New Phytol.*, 70, 239, 1971.

9. Jones, H. G., Gas-liquid chromatography of methylated sugars, in *Methods in Carbohydrate Chemistry*, Vol. 6, Whistler, R. L. and BeMiller, J. N., Eds., Academic Press, New York, 1972, 25.

10. Sweeley, C. C., Wells, W. W., and Bentley, R., Gas chromatography of carbohydrates, in *Methods in Enzymology*, Vol. 8, Neufeld, E. F. and Ginsburg, V., Eds., Academic Press, New York, 1966, 95.

Liquid Chromatography*

11. Churms, S. C., Gel chromatography of carbohydrates, in *Advances in Carbohydrate Chemistry and Biochemistry*, Vol. 25, Tipson, R. S. and Horton, D., Eds., Academic Press, New York, 1970, 13.

12. Conrad, E. C. and Palmer, J. K., Rapid analysis of carbohydrates by high-pressure liquid chromatography, *Food Technol. Chicago*, 30, 84, 1976.

13. Goldstein, I. J. and Hayes, C. E., The lectin: carbohydrate-binding proteins of plants and animals, in *Advances in Carbohydrate Chemistry and Biochemistry*, Vol. 35, Tipson, R. S. and Horton, D., Eds., Academic Press, New York, 1978, 127.

14. Granath, K. A., Gel filtration: fractionation of dextran, in *Methods in Carbohydrates Chemistry*, Vol. 5, Whistler, R. L., BeMiller, J. N., and Wolfrom, M. L., Eds., Academic Press, New York, 1965, 20.

15. Jandera, P. and Churáček, J., Ion-exchange chromatography of aldehydes, ketones, ethers, alcohols, polyols and saccharides, *J. Chromatogr.*, 98, 55, 1974.

16. Kristiansen, T., Group-specific separation of glycoproteins, in *Methods in Enzymology*, Vol. 34, Jakoby, W. B. and Wilchek, M., Eds., Academic Press, New York, 1974, 331.

17. Neukom, H. and Kuendig, W., Fractionation on diethylaminocellulose columns: fractionation of neutral and acidic polysaccharides by ion-exchange column chromatography on diethylaminoethyl (DEAE) — cellulose, in *Methods in Carbohydrate Chemistry*, Vol. 5, Whistler, R. L., BeMiller, J. N., and Wolfrom, M. L., Eds., Academic Press, New York, 1965, 14.

18. Samuelson, O., Partition chromatography on ion-exchange resins, in *Methods in Carbohydrate Chemistry*, Vol. 6, Whistler, R. L. and BeMiller, J. N., Eds., Academic Press, New York, 1972, 65.

Paper Chromatography

19. Hough, L., Analysis of mixtures of sugars by paper and cellulose column chromatography, in *Methods of Biochemical Analysis*, Glick, D., Ed., Vol. 1, Interscience, New York, 1954, 205.

20. Hough, L. and Jones, J. K. N., Chromatography on paper, in *Methods in Carbohydrate Chemistry*, Vol. 1, Whistler, R. L. and Wolfrom, M. L., Eds., Academic Press, New York, 1962, 21.

21. Kowkabany, G. N., Paper chromatography of carbohydrates and related compounds, in *Advances in Carbohydrate Chemistry*, Vol. 9, Wolfrom, M. L. and Tipson, R. S., Eds., Academic Press, New York, 1954, 303.

Thin-Layer Chromatography

22. Ghebregzabher, M., Rufini, S., Monaldi, B., and Lato, M., Thin-layer chromatography of carbohydrates, *J. Chromatogr.*, 127, 133, 1976.

23a. Wing, R. E. and BeMiller, J. N., Qualitative thin-layer chromatography, in *Methods in Carbohydrate Chemistry*, Vol. 6, Whistler, R. L. and BeMiller, J. N., Eds., Academic Press, New York, 1972, 42.

23b. Wing, R. E. and BeMiller, J. N., Quantitative thin-layer chromatography, in *Methods in Carbohydrate Chemistry*, Vol. 6, Whistler, R. L. and BeMiller, J. N., Eds., Academic Press, New York, 1972, 54.

23c. Wing, R. E. and BeMiller, J. N., Preparative thin-layer chromatography, in *Methods in Carbohydrate Chemistry*, Vol. 6, Whistler, R. L. and BeMiller, J. N., Eds., Academic Press, New York, 1972, 60.

Electrophoresis

24. Foster, A. B., Zone electrophoresis of carbohydrates, in *Advances in Carbohydrate Chemistry*, Vol. 12, Wolfrom, M. L. and Tipson, R. S., Eds., Academic Press, New York, 1957, 81.

25. Foster, A. B., Zone electrophoresis (ionophoresis), in *Methods in Carbohydrate Chemistry*, Vol. 1, Whistler, R. L. and Wolfrom, M. L., Eds., Academic Press, New York, 1962, 51.

26. Northcote, D. H., Zone electrophoresis: qualitative, quantitative, and preparative electrophoretic separations of neutral polysaccharides, in *Methods in Carbohydrate Chemistry*, Vol. 5, Whistler, R. L., BeMiller, J. N., and Wolfrom, M. L., Eds., Academic Press, New York, 1965, 49.

27. Weigel, H., Paper electrophoresis of carbohydrates, in *Advances in Carbohydrate Chemistry*, Vol. 18, Wolfrom, M. L. and Tipson, R. S., Eds., Academic Press, New York, 1963, 61.

* The term "liquid chromatography" includes HPLC, partition chromatography, ion-exchange, gel permeation, and affinity chromatography.

Important Papers on Specific Aspects
Gas Chromatography

1. Aspinall, G. O., Gas-liquid partition chromatography of methylated and partially methylated methyl glycosides, *J. Chem. Soc.*, 1676, 1963.
2. Bhatti, T., Chambers, R. E., and Clamp J. R., The gas chromatographic properties of biologically important *N*-acetylglucosamine derivatives, monosaccharides, disaccharides, trisaccharides, tetrasaccharides and pentasaccharides, *Biochim. Biophys. Acta*, 222, 339, 1970.
3. Bishop, C. T. and Cooper, F. P., Separation of carbohydrate derivatives by gas-liquid partition chromatography, *Can. J. Chem.*, 38, 388, 1960.
4. Björndal, H., Lindberg, B., and Svensson, S., Gas-liquid chromatography of partially methylated alditols as their acetates, *Acta Chem. Scand.*, 21, 1801, 1967.
5. Clamp, J. R., Dawson, G., and Hough, L., The simultaneous estimation of 6-deoxy-L-galactose (L-fucose), D-mannose, D-galactose, 2-acetamino-2-deoxy-D-glucose (*N*-acetyl-D-glucosamine) and N-acetylneuraminic acid (sialic acid) in glycopeptides and glycoproteins, *Biochim. Biophys. Acta*, 148, 342, 1967.
6. Dmitriev, B. A., Backinowsky, L. V., Chizhov, O. S., Zolotarev, B. M., and Kochetkov, N. K., Gas-liquid chromatography and mass spectrometry of aldononitrile acetates and partially methylated aldononitrile acetates, *Carbohydr. Res.*, 19, 432, 1971.
7. Freeman B. H., Stephen, A. M., and van der Bijl, P., Gas-liquid chromatography of some partially methylated alditols as their trimethylsilyl ethers, *J. Chromatogr.*, 73, 29, 1972.
8. Gerwig, G. J., Kamerling, J. P., and Vliegenthart, J. F. G., Determination of the D and L configuration of neutral monosaccharides by high-resolution capillary g.l.c., *Carbohydr. Res.*, 62, 349, 1978.
9. Haverkamp, J., Kamerling, J. P., and Vliegenthart, J. F. G., Gas-liquid chromatography of trimethylsilyl disaccharides, *J. Chromatogr.*, 59, 281, 1971.
10. Holligan, P. M. and Drew, E. A., Routine analysis by gas-liquid chromatography of soluble carbohydrates in extracts of plant tissues. II. Quantitative analysis of standard carbohydrates, and the separation and estimation of soluble sugars and polyols from a variety of plant tissues, *New Phytol.*, 70, 271, 1971.
11. Kärkkäinen, J., Analysis of disaccharides as permethylated disaccharide alditols by gas-liquid chromatography-mass spectrometry, *Carbohydr. Res.*, 14, 27, 1970.
12. Kärkkäinen, J., Structural analysis of trisaccharides as permethylated trisaccharide alditols by gas-liquid chromatography-mass spectrometry, *Carbohydr. Res.*, 17, 11, 1971.
13. Laine, R. A. and Sweeley, C. C., *O*-Methyl oximes of sugars. Analysis as *O*-trimethylsilyl derivatives by gas-liquid chromatography and mass spectrometry, *Carbohydr. Res.*, 27, 199, 1973.
14. Leontein, K., Lindberg, B., and Lönngren, J., Assignment of absolute configuration of sugars by g.l.c. of their acetylated glycosides formed from chiral alcohols, *Carbohydr. Res.*, 62, 359, 1978.
15. Lönngren, J. and Pilotti, A., Gas-liquid chromatography of partially methylated alditols as their acetates. II., *Acta Chem. Scand.*, 25, 1144, 1971.
16. Perry, M. B. and Hulyalkar, R. K., The analysis of hexuronic acids in biological materials by gas-liquid partition chromatography, *Can. J. Biochem.*, 43, 573, 1965.
17. Petersson, G., Gas-chromatographic analysis of sugars and related hydroxy acids as acylic oxime and ester trimethylsilyl derivatives, *Carbohydr. Res.*, 33, 47, 1974.
18. Sawardeker, J. S., Sloneker, J. H., and Jeanes, A., Quantitative determination of monosaccharides as their alditol acetates by gas-liquid chromatography, *Anal. Chem.*, 37, 1602, 1965.
19. Seymour, F. R., Plattner, R. D., and Slodki, M. E., Gas-liquid chromatography-mass spectrometry of methylated and deuteriomethylated per-*O*-acetyl-aldononitriles from D-mannose, *Carbohydr. Res.* 44, 181, 1975.
20. Seymour, F. R., Slodki, M. E., Plattner, R. D., and Jeanes, A., Six unusual dextrans: methylation structural analysis by combined g.l.c.-m.s. of per-O-acetyl-aldononitriles, *Carbohydr Res.*, 53, 153—166, 1977.
21. Sweeley, C. C., Bentley, R., Makita, M., and Wells, W. W., Gas-liquid chromatography of trimethylsilyl derivatives of sugars and related substances, *J. Am. Chem. Soc.*, 85, 2497, 1963.
22. Sweet, D. P., Albersheim, P., and Shapiro, R. H., Partially ethylated alditol acetates as derivatives for elucidation of the glycosyl linkage-composition of polysaccharides, *Carbohydr. Res.*, 40, 199, 1975.
23. Vilkas, M., Hiu, I-Jan, Boussac, G., and Bonnard, M. C., Vapour phase chromatography of sugars as trifluoroacetates, *Tetrahedron Lett.*, 1441, 1966.
24. Yoshida, K., Honda, N., Iino, N., and Kato, K., Gas-liquid chromatography of hexapyranosides and hexofuranosides, *Carbohydr. Res.*, 10, 333, 1969.
25. Zanetta, J. P., Breckenridge, W. C., and Vincendon, G., Analysis of monosaccharides by gas-liquid chromatography of the *O*-methyl glycosides as trifluoroacetate derivatives. Application to glycoproteins and glycolipids, *J. Chromatogr.*, 69, 291, 1972.

Liquid Chromatography

Affinity Chromatography

26. Aspberg, K. and Porath, J., Group-specific adsorption of glycoproteins, *Acta Chem. Scand.*, 24, 1839, 1970.
27. Kennedy, J. F. and Rosevear, A., An assessment of the fractionation of carbohydrates on Concanavalin A-Sepharose 4B by affinity chromatography, *J. Chem. Soc. Perkin Trans. 1*, 2041, 1973.
28. Lloyd, K. O., The preparation of two insoluble forms of the phytohaemagglutinin, concanavalin A, and their interactions with polysaccharides and glycoproteins, *Arch. Biochem. Biophys.*, 137, 460, 1970.

Gel-Permeation Chromatography

29. Anderson, D. M. W., Dea, I. C. M., Rahman, S., and Stoddart, J. F., The use of "Bio-Gel P" in the gel filtration of polysaccharides, *Chem. Commun.*, 145, 1965.
30. Anderson, D. M. W., and Stoddart, J. F., The use of molecular-sieve chromatography in studies on *Acacia senegal* gum, *Carbohydr. Res.*, 2, 104, 1966.
31. Arturson, G. and Granath, K., Dextrans as test molecules in studies of the functional ultrastructure of biological membranes. Molecular weight distribution analysis by gel chromatography, *Clin. Chim. Acta*, 37, 309, 1972.
32. Brown, W., The separation of cellodextrins by gel-permeation chromatography, *J. Chromatogr.*, 52, 273, 1970.
33. Brown, W., The influence of solvent and temperature in dextran gel chromatography, *J. Chromatogr.*, 59, 335, 1971.
34. Dellweg, H., John, M., and Trenel, G., Gel-permeation chromatography of maltooligosaccharides at different temperatures, *J. Chromatogr.*, 57, 89, 1971.
35. Granath, K. A. and Flodin, P., Fractionation of dextran by the gel filtration method, *Makromol. Chem.*, 48, 160, 1961.
36. Granath, K. and Kvist, B. E., Molecular-weight distribution analysis by gel chromatography on Sephadex, *J. Chromatogr.*, 28, 69, 1967.
37. Havlicek, J. and Samuelson, O., Chromatography of oligosaccharides from xylan by various techniques, *Carbohydr. Res.*, 22, 307, 1972.
38. Hopwood, J. J. and Robinson, H. C., The molecular-weight distribution of glycosaminoglycans, *Biochem. J.*, 135, 631, 1973.
39. John, M., Trenel, G., and Dellweg, H., Quantitative chromatography of homologous glucose oligomers using polyacrylamide gel, *J. Chromatogr.*, 42, 476, 1969.
40. Porath, J. and Flodin, P., Gel filtration: a method for desalting and group separation, *Nature (London)*, 183, 1657, 1959.
41. Sabbagh, N. K. and Fagerson, I. S., Gel-permeation chromatography of glucose oligomers using polyacrylamide gels, *J. Chromatogr.*, 86, 184, 1973.
42. Sabbagh, N. K. and Fagerson, I. S., Gel chromatography of glucose oligomers. A study of operational parameters, *J. Chromatogr.*, 120, 55, 1976.
43. Wasteson, Å., A method for the determination of the molecular weight and molecular-weight distribution of chondroitin sulfate, *J. Chromatogr.*, 59, 87, 1971.

High-Performance Liquid Chromatography

44. Goulding, R. W., Liquid chromatography of sugars and related polyhydric alcohols on cation exchangers. The effect of cation variation, *J. Chromatogr.*, 103, 229, 1975.
45. Ladisch, M. R., Huebner, A. L., and Tsao, G. T., High-speed liquid chromatography of cellodextrins and other saccharide mixtures using water as the eluent, *J. Chromatogr.*, 147, 185, 1978.
46. Linden, J. C. and Lawhead, C. L., Liquid chromatography of saccharides, *J. Chromatogr.*, 105, 125, 1975.
47. McGinnis, G. D. and Fang, P., The separation of substituted carbohydrates by high-performance liquid chromatography, *J. Chromatogr.*, 130, 181, 1977, Part II, *J. Chromatogr.*, 153, 107, 1978.
48. Nachtmann, F. and Budna, K. W., Sensitive determination of derivatized carbohydrates by high-performance liquid chromatography, *J. Chromatogr.*, 136, 279, 1977.
49. Palmer, J. K., A versatile system for sugar analysis via liquid chromatography, *Anal. Lett.*, 8, 215, 1975.
50. Rabel, F. M., Caputo, A. G., and Butts, E. T., Separation of carbohydrates on a new polar bonded-phase material, *J. Chromatogr.*, 126, 731, 1976.

Ion-Exchange Resins: Partition Chromatography

51. Havlicek, J. and Samuelson, O., Separation of oligomeric sugar alcohols by partition chromatography on ion-exchangers, *Chromatographia*, 7, 361, 1974.

52. Havlicek, J. and Samuelson, O., Separation of oligosaccharides by partition chromatography on ion-exchange resins, *Anal. Chem.*, 47, 1854, 1975.
53. Martinsson, E. and Samuelson, O., Partition chromatography of sugars on ion-exchange resins, *J. Chromatogr.*, 50, 429, 1970.

Ion-Exchange Chromatography

54. Derivitskaya, V. R., Arbatsky, N. P., and Kochetkov, N. K., The use of anion-exchange chromatography to separate complex mixtures of oligosaccharides, *Dokl. Akad. Nauk SSSR*, 223, 1137, 1975.
55. Donald, A. S. R., Separation of hexosamines, hexosaminitols, and hexosamine-containing di- and trisaccharides on an amino acid analyser, *J. Chromatogr.*, 134, 199, 1977.
56. Hough, L., Jones, J. V. S., and Wusteman, P., On the automated analysis of neutral monosaccharides in glycoproteins and polysaccharides, *Carbohydr. Res.*, 21, 9, 1972.
57. Hough, L., Ko, A. M. Y., and Wusteman, P., The simultaneous separation and analysis of aminodeoxyalditols and alditols by automated ion-exchange chromatography, *Carbohydr. Res.*, 44, 97, 1975.
58. Kennedy, J. F. and Fox, J. E., The fully automatic ion-exchange and gel-permeation chromatography of neutral monosaccharides and oligosaccharides with a Jeolco JLC-6AH analyser, *Carbohydr. Res.*, 54, 13, 1977.
59. Kesler, R. B., Rapid anion-exchange chromatography of carbohydrates, *Anal. Chem.*, 39, 1416, 1967.
60. Mopper, K., Improved chromatographic separations on anion-exchange resins. II. Separation of uronic acids in acetate medium and detection with a noncorrosive reagent, *Anal. Biochem.*, 86, 597, 1978.
61. Mopper, K., Improved chromatographic separations on anion-exchange resins. III. Sugars in borate medium, *Anal. Biochem.*, 87, 162, 1978.
62. Torii, M. and Sakakibara, K., Column chromatographic separation and quantitation of *a*-linked glucose oligosaccharides, *J. Chromatogr.*, 96, 255, 1974.

Paper Chromatography

63. Angus, H. J. F., Briggs, J., Sufi, N. A., and Weigel, H., Chromatography of polyhydroxy compounds on cellulose impregnated with tungstate: determination of pseudo-stability constants of complexes, *Carbohydr. Res.*, 66, 25, 1978.
64. Hough, L., Jones, J. K. N., and Wadman, W. H., Quantitative analysis of mixtures of sugars by the method of partition chromatography. V. Improved methods for the separation and detection of the sugars and their methyl derivatives on the paper chromatogram, *J. Chem.*, Soc., 1702, 1950.
65. Jermyn, M. A. and Isherwood, F. A., Improved separation of sugars on a paper partition chromatogram, *Biochem. J.*, 44, 402, 1949.
66. Partridge, S. M. and Westall, R. G. Filter-paper partition chromatography of sugars. I. General description and application to the qualitative analysis of sugars in apple juice, egg white and fetal blood of sheep, *Biochem. J.*, 42, 238, 1948.
67. Teichmann, B., Paper chromatographic behaviour of homologous linear polymers. The isomaltodextrin, maltodextrin and cellodextrin series and their 1-(*m*-substituted phenyl)-flavazoles, *J. Chromatogr.*, 70, 99, 1972.

Thin-Layer Chromatography

68. Covacevich, M. T. and Richards, G. N., Continuous quantitative thin-layer chromatography of oligosaccharides, *J. Chromatogr.*, 129, 420, 1976.
69. Damonte, A., Lombard, A., Tourn, M. L., and Cassone, M. C., A modified solvent system and multiple detection technique for the separation and identification of mono- and oligosaccharides on cellulose thin layers, *J. Chromatogr.*, 60, 203, 1971.
70. Ghebregzhaber, M., Rufini, S., Ciuffini, G., and Lato, M., A two-dimensional thin-layer chromatographic method for screening carbohydrate anomalies, *J. Chromatogr.*, 95, 51, 1974.
71. Hansen, S. A., Thin-layer chromatographic method for identification of oligosaccharides in starch hydrolyzates, *J. Chromatogr.*, 105, 388, 1975.
72. Hansen, S. A. Thin-layer chromatographic method for the identification of mono-, di- and trisaccharides, *J. Chromatogr.*, 107, 224, 1975.
73. Mezzetti, T., Lato, M., Rufini, S., and Ciuffini, G., Thin-layer chromatography of oligosaccharides with tungstic or molybdic acid as impregnant, *J. Chromatogr.*, 63, 329, 1971.
74. Ovodov, Yu. S., Evtushenko, E. V., Vaskovsky, V. E., Ovodova, R. G., and Solo'eva T. F., Thin-layer chromatography of carbohydrates, *J. Chromatogr.*, 26, 111, 1967.
75. Raadsveld, C. W. and Klomp, H., Thin-layer chromatographic analysis of sugar mixtures, *J. Chromatogr.*, 57, 99, 1971.

76. **Shannon, J. C. and Creech, R. G.**, Thin-layer chromatography of maltooligosaccharides, *J. Chromatogr.*, 44, 307, 1969.
77. **Stahl, E. and Kaltenbach, U.**, Thin-layer chromatography. VI. Trace analysis of sugar mixtures on Kieselguhr G layers, *J. Chromatogr.*, 5, 351, 1961.

Electrophoresis

78. **Bourne, E. J., Foster, A. B., and Grant, P. M.**, Ionophoresis of carbohydrates. IV. Separations of carbohydrates on fibre glass sheets, *J. Chem. Soc.*, 4311, 1956.
79. **Bourne, E. J., Hutson, D. H., and Weigel, H.**, Paper ionophoresis of sugars and other cyclic polyhydroxy compounds in molybdate solution, *J. Chem. Soc.*, 4252, 1960.
80. **Frahn, J. L. and Mills, J. A.**, Paper ionophoresis of carbohydrates, *Aust. J. Chem.*, 12, 65, 1959.

INDEX

A

Acetamidodeoxy sugars
  trimethylsilyl ethers of, 64—68
Acetates, 221
  alditol, see Alditol acetates
  aldononitrile, 172, 220
  of partially methylated sugars, 23
Acetolysis, 215
Acetylated aldononitriles
  from methylated sugars, 18—19
  from neutral sugars, 16
Acetyl derivatives of sugars, 146—148
Acidic carbohydrates
  isolation of, 101
  separation of, 101
Acidic oligosaccharides, separation of, 94
Acidic plant-gum polysaccharides, 104
Affinity chromatography, 127
  media for, 249—250
  references on, 258—260, 263
Agar components, 102
Agarose gel columns, 119
Aldaric acid
  trimethylsilyl esters of, 57—63
  trimethylsilyl ethers of, 57—63
Alditol acetates, 219—220
  from partially ethylated sugars, 11—13
  from partially methylated sugars, 6—10, 172
Alditols, 220
  butaneboronates of, 21
  capacity factors of, 75
  electrophoresis of, 157—158
  from partially methylated sugars, 52—53
  high-performance liquid chromatography of,
    82
  liquid chromatography methods for, 182—183
  paper chromatography of, 133—134
  partition chromatography of, 85
  peracetylated, 4
  permethylated, 14—15
  thin-layer chromatography of, 138—143
  trifluoracetylated, 173
  trifluoroacetates of, 34—35
  trimethylsilylated, 225
  trimethylsilyl ethers of, 52—53
Aldonic acids
  trimethylsilyl esters of, 57—63, 226
  trimethylsilyl ethers of, 57—63
  uronic acid reduction to, 226
Aldono-1,4-lactones, 226
Aldononitrile acetates, 172, 220
Aldoses, o-isopropylidene derivatives of, 22
Amino acid analyser, 100
p-Aminobenzoic acid, 188
Aminodeoxyalditols, 4
  separation of, 100
Aminodeoxyhexoses, 16
Aminodeoxy sugars
  liquid chromatography methods for, 181—182

separation of, 100
  trimethylsilyl ethers of, 64—68
o-Aminodiphenyl, 188
Aminoguanidine sulfate-chromic acid, 188
p-Aminohippuric acid, 188
o-Aminophenol, 188
Amino sugars, automated chromatography of,
  182
Ammonium molybdate, 188
Ammonium sulfate, 189
Aniline citrate reagent, 206
Aniline oxalate/citrate, 189
Aniline phosphate, 189
Aniline phthalate, 189
Anion-exchange chromatography, 89—96
  D-glucose oligosaccharides, 93
  glycosides, 91
  oligosaccaharides, 93, 96
  polyols, 92
  reduced oligosaccharides, 93, 96
  saccharides, 89
  sugars, 90
Anisaldehyde-sulfuric acid, 189
p-Anisidine hydrochloride, 190
p-Anisidine-phthalic acid, 190
Anthrone, 176, 190
Aqueous ethanol, 83
Aryl derivatives of sugars, 146—148
Automated chromatography
  amino sugars, 182
  glycosaminoglycans, 95
  ketoses, 180

B

Benzidine, 190
Benzoylation, 226—227
Blood group substance H, 96
Blood sugar, 213
Bonded-phase packings, 70—74, 236
Boronates, 172
Bromophenol blue, 190
Butaneboronates, 221
Butaneboronic esters, 21

C

Cadoxen, 124
Calibrated agarose gel columns, 119
Calibrated Sephadex gel columns, 118
Capsular polysaccharides, 165
Carbazole-sulfuric acid, 180, 190
Cellodextrins
  column chromatography of, 112, 113
  high-performance liquid chromatography of,
    73
Cellulose, 124